SO-AFF-624

Introduction
to
Electronics

Introduction to Electronics

THEODORE KORNEFF

Department of Physics
Temple University
Philadelphia, Pennsylvania

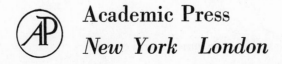

Academic Press

New York   London

We wish to thank the following organizations for permission to use the respective material: Westinghouse Electric Corporation for the schematic diagram Model H73 MPl as the basis for Fig. 8.26; Sylvania Electric Products, Inc., for material from their Remote Temperature Indicator as the basis for Fig. 11.33; The General Electric Company for material from their Electronic Tube Manual Receiving Types as the basis for Appendix IV, pages 518–534; and the Radio Corporation of America for material from their RCA Transistor Manual, TS/SC–10 as the basis for Appendix IV, pages 535–537.

ACADEMIC PRESS INC.
111 Fifth Avenue, New York, New York 10003

United Kingdom Edition published by
ACADEMIC PRESS INC. (LONDON) LTD.
Berkeley Square House, London W.1

LIBRARY OF CONGRESS CATALOG CARD NUMBER: 66–14470

First Printing, 1966
Second Printing, 1967

PRINTED IN THE UNITED STATES OF AMERICA

Preface

This text is designed for the undergraduate course generally taken by physics and other science majors. It presupposes only a general physics course and an introductory course in differential and integral calculus.

It is the intent of the author to instill such a basic, intuitive knowledge of electronics and electronic devices that the student can go on to more inclusive and advanced subjects with confidence and assurance. This means that the student should have a feel for the orders of magnitude involved. To this end detailed graphical approaches are used and the many examples are worked out in detail.

The basic approach is operational. Devices are first introduced by describing them in relation to their dc and ac characteristics. For small signal applications, the appropriate linear ac equivalent circuits are then derived and used in several detailed examples. In nonlinear or large signal operation, the approach is graphical.

The first three chapters are included to quickly bring the student to the point where he can use the complex plane in the solution of elementary ac networks. Chapter 3, on filters and tuned circuits, gives the student excellent practice on ac networks as well as the necessary basic concepts on frequency response and phase shift.

Transistors and vacuum tubes are dealt with separately. Although their functions are similar, the circuit considerations are not. For this reason, transistors and associated circuits are described in three comprehensive chapters after vacuum tube devices have been used to introduce the basic circuits.

With a compassion for the student's problems in mastering a new subject, the text leans heavily on diagrams, curves, and detailed examples, which many texts relegate to the problems section. The extreme detail used in working out examples allows the student to do more in his self-study, thereby allowing the instructor to concentrate on the material he considers to be in need of further exposition.

I wish to thank Professor T. C. Daniels, of Gettysburg College, for his detailed critique of the manuscript. His suggestions were most helpful.

I greatly appreciate the many hours of "spare time" spent by Mrs. Sarah Shaner and my wife Irene in the typing of the manuscript.

Burlington, New Jersey   T. K.

Contents

5. Power Supplies

6. Vacuum Tubes—Graphical (Triodes and Pentodes)

7. Voltage Amplifiers—Analytic

8. Power Amplifiers

9. Oscillators

10. Pulse and Wave Shaping

11. Transistors—Graphical

12. Transistors—Analytical

13. Multistage and Power-Amplifier Circuits

14. Oscilloscope and Its Uses—Electronic Regulated Power Supply

Introduction
to
Electronics

1.1 Resistance

The dc resistance of an electrical component can be defined operationally in the following manner. Place a difference of potential E across the component and measure the resultant current I. The dc resistance R is then defined as

$$R = \frac{E}{I} \qquad (1.1)$$

and is measured in ohms if E is in volts and I in amperes. The internal constitution of R need not be defined. It may be a single component or an aggregate of components forming an electric circuit. We can still speak of the resistance of R as E/I, where E is the voltage applied across R, and I is the resultant current through R.

GRAPHICAL APPROACH

To speak about R completely, we must vary E and see what happens to I. In this manner we obtain the dc E-I characteristics of R. Figure 1.1 shows the test circuit used and the resultant dc characteristics for a typical electrical component. As can be seen from the graph, the resultant current is directly proportional to the applied voltage. In a graph of this sort, the significance of the negative voltages and currents in the third quadrant of the graph is that the battery terminals have been reversed and that the current now flows in the opposite direction. If we form the ratio E/I for several points on the curve such as A, B, and C, we obtain the following results:

Point A: $\qquad R = \dfrac{E}{I} = \dfrac{-10 \text{ volts}}{-2.3 \times 10^{-3} \text{ amp}} \cong 4800 \text{ ohms}$

Point B: $\qquad R = \dfrac{E}{I} = \dfrac{20 \text{ volts}}{4.6 \times 10^{-3} \text{ amp}} \cong 4800 \text{ ohms}$

Point C: $\qquad R = \dfrac{E}{I} = \dfrac{30 \text{ volts}}{6.9 \times 10^{-3} \text{ amp}} \cong 4800 \text{ ohms}$

As has been demonstrated, and may be intuitively evident from the E–I curve, R has a constant value. It does not depend on E or I. An electrical component that behaves in this manner is called a *linear component*. As is well known, the resistance of a material is temperature dependent, and the temperature of an electrical component depends in turn upon the current flowing through that component. In this discussion we are assuming that the component does not heat up significantly as we increase the current; that is, the heat is dissipated to the surroundings quickly enough so as to keep the element essentially at a constant temperature.

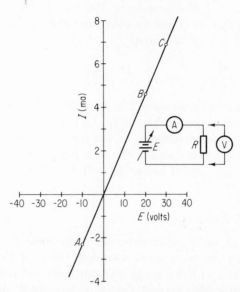

Fig. 1.1 *E–I* characteristics for a linear component.

A more general type of *E–I* characteristic is the one shown in Fig. 1.2(a). Here the resultant current is not directly proportional to the applied voltage. If we use the definition of resistance $R = E/I$, we obtain the following values for the points shown:

$$R_A = \frac{E_A}{I_A} = \frac{-15 \text{ volts}}{-2 \times 10^{-3} \text{ amp}} \cong 7500 \text{ ohms}$$

$$R_B = \frac{E_B}{I_B} = \frac{10 \text{ volts}}{6 \times 10^{-3} \text{ amp}} \cong 1670 \text{ ohms}$$

$$R_C = \frac{E_C}{I_C} = \frac{20 \text{ volts}}{22 \times 10^{-3} \text{ amp}} \cong 909 \text{ ohms}$$

This element has a dc resistance that is not constant, but which does depend upon the voltage and current in some complicated fashion. This is a nonlinear element.

Most components in nature are nonlinear. In order to deal with them analytically, however, we sometimes sacrifice some accuracy by approximating the curves by linear segments and deal, then, with linear elements. What we lose in accuracy we gain in predictability. Analytically we can usually predict the operation of the substitute linear circuit with the answer agreeing within a few percent of the actual answer, as obtained experimentally, for the original nonlinear circuit.

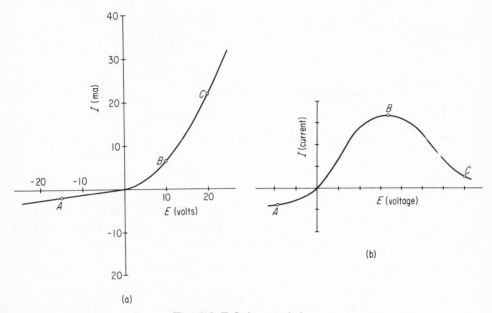

Fig. 1.2 *E–I* characteristics
for nonlinear components.

In the two cases that have been considered, it should be noted
that as the voltage was increased, the current increased; and when
the voltage was decreased, the current decreased. This is called a
positive resistance. It is by far the most common type of resistance
characteristic. But carbon, for example, is one of the few materials
that has a negative resistance region. Specifically, if the voltage is
increased and the current decreases, we have a negative resistance.
A hypothetical element having positive and negative resistance
regions is illustrated in Fig. 1.2(b). In the region between A and B
the current increases as the voltage increases and the element
exhibits positive resistance. From B to C, the resultant current
decreases even though the voltage is being increased. Here the
element exhibits negative resistance.

CIRCUITS

Two basic circuits occur very often in electrical work. They are
the series and parallel circuits. These are illustrated in Fig. 1.3.
A series circuit is shown in Fig. 1.3(a). In this case the current I
due to the battery E is the same in each succeeding element, R_1
and R_2. For this case, the total voltage E is equal to the sum of the
voltage drops across R_1 and R_2:

$$E = E_1 + E_2 \qquad (1.2)$$

The voltage drops E_1 and E_2 are given by

$$E_1 = IR_1$$

$$E_2 = IR_2$$

Substituting these values into Eq. (1.2),

$$E = I (R_1 + R_2)$$

Solving for the ratio E/I,

$$\frac{E}{I} = R_1 + R_2$$

From the definition for resistance given by Eq. (1.1), it is seen that the ratio E/I is the total resistance of the series combination. Calling this total resistance R_T,

$$R_T = R_1 + R_2$$

For two or more resistors in series, it is simple to show that the total resistance is equal to the sum of the individual resistances.

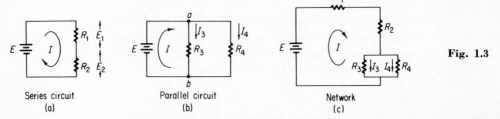

Series circuit
(a)

Parallel circuit
(b)

Network
(c)

Fig. 1.3

A simple parallel circuit is shown in Fig. 1.3(b). In this case, the current I splits up at junction a, with part going through R_3 and part going through R_4. For this case,

$$I = I_3 + I_4 \tag{1.3}$$

The current through a component is given by the voltage across the component divided by the resistance of that component. Applying this to the circuit of Fig. 1.3(b),

$$I_3 = \frac{E}{R_3}$$

$$I_4 = \frac{E}{R_4}$$

Substituting these values into Eq. (1.3),

$$I = E \left(\frac{1}{R_3} + \frac{1}{R_4} \right)$$

Solving for the ratio E/I, to obtain the total resistance,

$$\frac{E}{I} = \frac{R_3 R_4}{R_3 + R_4}$$

$$R_T = \frac{R_3 R_4}{R_3 + R_4}$$

Thus, for a parallel circuit consisting of two resistors, the total resistance is given by the product of the individual resistances, divided by the sum of the individual resistances.

Example 1.1 A 2000-ohm resistor is placed in parallel with a 500-ohm resistor. Find the total resistance.

1. Using the product divided by the sum rule, as derived above,

$$R_T = \frac{R_1 R_2}{R_1 + R_2}$$

$$= \frac{2000 \times 500}{2000 + 500} = 400 \text{ ohms}$$

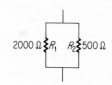

2. *Comments*
(a) As a useful check on your calculations, it should be noted that R_T will always be less than the smallest individual resistor. For example, $R_T < R_2$ in the problem just worked out. To show that this will be true in general, rewrite the equation for R_T as

$$R_T = R_2 \left(\frac{R_1}{R_1 + R_2} \right)$$

The ratio $R_2/(R_1 + R_2)$ is always less than unity; therefore, $R_T < R_2$.

(b) For two similar resistors in parallel, the total resistance is equal to one-half the resistance of either resistor. If $R_1 = R_2 = R$, then the equation for two resistors in parallel reduces as follows:

$$R_T = \frac{R_1 R_2}{R_1 + R_2} = \frac{R^2}{R + R} = \frac{R^2}{2R}$$

$$= \tfrac{1}{2} R$$

Example 1.2 Derive the general equation for N resistors in parallel and apply it to three resistors in parallel, $R_1 = 5000$ ohms, $R_2 = 1000$ ohms, and $R_3 = 500$ ohms.

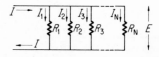

1. For the general case, the circuit is as shown. The total current I is split up into the components I_1, I_2, I_3, $\cdots$, I_N, so that

$$I = I_1 + I_2 + I_3 + \cdots + I_N \tag{1}$$

Now

$$I_1 = \frac{E}{R_1}$$

$$I_2 = \frac{E}{R_2} \tag{2}$$

$$\cdot$$
$$\cdot$$
$$\cdot$$

$$I_N = \frac{E}{R_N}$$

Rewriting (1) in terms of (2),

$$I = E\left(\frac{1}{R_1} + \frac{1}{R_2} + \frac{1}{R_3} + \cdots + \frac{1}{R_N}\right) \tag{3}$$

Solving for I/E rather than E/I puts Eq. (3) into simple form:

$$\frac{I}{E} = \frac{1}{R_1} + \frac{1}{R_2} + \frac{1}{R_3} + \cdots + \frac{1}{R_N}$$

I/E will be equal to the reciprocal of the total resistance,

$$\frac{1}{R_T} = \frac{1}{R_1} + \frac{1}{R_2} + \frac{1}{R_3} + \cdots + \frac{1}{R_N} \tag{4}$$

2. For the case of three resistors in parallel, the total resistance can be obtained from Eq. (4):

$$\frac{1}{R_T} = \frac{1}{R_1} + \frac{1}{R_2} + \frac{1}{R_3}$$

where

$$R_1 = 5000 \qquad R_2 = 1000 \qquad R_3 = 500$$

$$\frac{1}{R_T} = \frac{1}{5000} + \frac{1}{1000} + \frac{1}{500}$$

$$\frac{1}{R_T} = \frac{16}{5000}$$

$$R_T = \frac{5000}{16} \cong 312 \text{ ohms}$$

3. *Comments.* It should be noted that the equation for two resistors in parallel can be put into the general form show in (4):

$$R_T = \frac{R_1 R_2}{R_1 + R_2}$$

$$\frac{1}{R_T} = \frac{R_1 + R_2}{R_1 R_2}$$

$$= \frac{R_1}{R_1 R_2} + \frac{R_2}{R_1 R_2}$$

$$= \frac{1}{R_2} + \frac{1}{R_1}$$

The circuit shown in Fig. 1.3(c) is seen to be a composite circuit of series and parallel combinations. This is usually called a *network*. There are several ways of solving networks. Some of these methods will be discussed later in the chapter.

1.2 Some Useful Relations

Calculations can be speeded up with the use of two equations that will be derived in this section. The first equation deals with voltage dividers—that is, resistors in series—and the second equation deals with resistors in parallel.

VOLTAGE DIVIDER

Consider the voltage divider in Fig. 1.4(a). It is desired to obtain

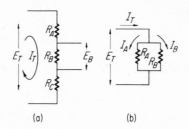

Fig. 1.4 (a) A voltage divider. (b) A two-resistor parallel circuit.

(a)   (b)

E_B, the voltage drop across R_B, in terms of E_T and the resistances R_A, R_B, and R_C.

$$E_T = I_T R_T = I_T (R_A + R_B + R_C) \tag{1.4a}$$

Also

$$E_B = I_T R_B \tag{1.4b}$$

Dividing Eq. (1.4a) by Eq. (1.4b) and solving for E_B,

$$\frac{E_T}{E_B} = \frac{I_T(R_A + R_B + R_C)}{I_T R_B} \tag{1.5}$$

$$E_B = E_T \left(\frac{R_B}{R_A + R_B + R_C}\right)$$

Thus, the voltage drop across R_B in a chain of resistors in series is given by the total voltage applied to the chain, multiplied by a unitless ratio of resistances; that is, the resistor across which one desires the voltage drop, divided by the total resistance in the series chain. Using this definition, Eq. (1.5), the voltage drop across R_C in Fig. 1.4(a) can be written by inspection as

$$E_C = E_T \left(\frac{R_C}{R_A + R_B + R_C}\right)$$

The method shown will be used extensively in later chapters and should be mastered.

PARALLEL CIRCUIT

An equation for two resistors in parallel can be derived, wherein the current through either resistor can be written down in terms of the total current and individual resistor. Consider the two-resistor parallel circuit shown in Fig. 1.4(b). The voltage drop across each branch is given by

$$E_A = I_A R_A \tag{1.6a}$$

$$E_B = I_B R_B \tag{1.6b}$$

and also

$$E_A = E_B = E_T = I_T R_T = I_T \left(\frac{R_A R_B}{R_A + R_B}\right) \tag{1.6c}$$

Dividing Eq. (1.6a) by Eq. (1.6c) and solving for I_A,

$$\frac{E_A}{E_A} = \frac{I_A R_A}{I_T \left[R_A R_B/(R_A + R_B)\right]} \tag{1.7}$$

$$I_A = I_T \left(\frac{R_B}{R_A + R_B}\right)$$

Similarly for I_B, divide Eq. (1.32) by Eq. (1.33) and solve for I_B:

$$\frac{E_B}{E_B} = \frac{I_B R_B}{I_T \left[R_A R_B / (R_A + R_B) \right]}$$

$$I_B = I_T \left(\frac{R_A}{R_A + R_B} \right)$$

A pattern can be seen. For the current through one of two resistors in parallel, multiply the total current by a unitless ratio of resistances; that is, the opposite resistor divided by the sum of the two resistors. This is also a useful trick and should be memorized. This latter result for currents holds true only for two resistors in parallel.

1.3 Galvanometer

The basic movement used in meters is called a *galvanometer*. It is essentially a coil of wire suspended in a permanent magnetic field and free to turn in that magnetic field, Fig. 1.5. When a current

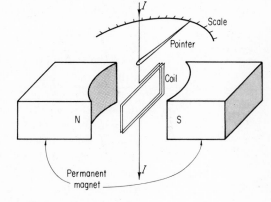

Fig. 1.5 A simplified galvanometer movement.

passes through the coil, the coil is surrounded by a magnetic field that is attracted or repelled by the field already present. This causes the coil to rotate on its suspension until the twisted suspension, or a spring inserted for this purpose, produces a torque that counteracts the torque due to the interaction of the magnetic fields. At this point the coil will be in equilibrium and remain at this new angle as long as the current I remains constant. A pointer attached to the coil is used to indicate the angle through which the coil has turned. The manufacturer, by properly shaping the coil and pole pieces of the permanent magnet, ensures (to a fraction of a percent) that the deflection is directly proportional to the current flowing through the coil.

For purposes of constructing other types of meters, the galvanometer is specified in terms of three interrelated quantities. These are the internal resistance R_g, the full scale current I_g, and the full scale voltage E_g.

The internal resistance of the galvanometer is due to the wire used in winding the coil. In order to gain sensitivity, the coil would have to contain many turns so that a given current would produce an appreciable magnetic field. But to do this while keeping the physical size of the coil down would require a small-diameter wire. This, in turn, raises the resistance of the coil. For a coil of moderate sensitivity, fewer turns would be used and the resistance would be correspondingly lower. In general, therefore, a high internal resistance (R_g of about 1000 ohms) implies a very sensitive galvanometer, one that would require a few microamperes for full-scale deflection. A galvanometer with moderate sensitivity, say, 1 milliampere (ma) for full-scale deflection, would have an internal resistance of about 10 to 50 ohms. The full-scale current I_g is defined as the current that must flow through the galvanometer coil in order to produce full-scale deflection. The full-scale voltage E_g is the voltage that must be applied to the terminals of the galvanometer in order to produce the full-scale current. The three quantities are related in the following simple manner:

$$E_g = I_g R_g$$

In the examples to follow, a voltmeter, ammeter, and ohmmeter will be discussed in turn. To make the discussion meaningful, assume that we have the following basic galvanometer at our disposal for the construction of the three instruments named: $R_g = 50$ ohms and $I_g = 1$ ma. This means that we have a galvanometer whose internal resistance is 50 ohms and whose sensitivity is such that if 1 ma flows through the galvanometer coil, the pointer attached to the coil will rotate until it points to the end of the scale—that is, full scale. The basic galvanometer will be represented by the symbol

where R_g represents the internal resistance and G represents the deflection system.

Using the values for our example galvanometer, it can be seen that the full-scale voltage is

$$E_g = I_g R_g = 0.001 \text{ amp} \times 50 \text{ ohms} = 0.05 \text{ volt}$$

or 50 mV (millivolts).

1.4 Voltmeter

Since only a small voltage is required to produce full-scale deflection for a galvanometer, in order to make the deflection proportional to a larger voltage, a series resistor R is placed in series with the galvanometer, Fig. 1.6(a). The resistor R in series with the galvanometer constitutes a voltmeter.

If R is very large, then a relatively large voltage will have to be impressed across the terminals AB to produce the small current needed to produce a deflection of the pointer in the galvanometer; see Fig. 1.6(b). The current I, of course, must be kept equal to or less than I_g, otherwise the galvanometer may be damaged. In order to calculate the voltage V that can be applied to the terminals AB to produce full-scale current, we solve the simple series circuit in Fig. 1.6(b):

$$E = I_g(R + R_g) \tag{1.8}$$

where E is the external voltage required for full-scale deflection. For a given galvanometer, I_g and R_g are known, so that the equation depends only upon V and R. It is usually the practice to choose V and calculate the necessary R. So, solving for R,

$$R = \frac{E}{I_g} - R_g \tag{1.9}$$

Example 1.3 For our particular galvanometer, let us make a 500-volt full-scale voltmeter. Since $R_g = 50$ ohms and $I_g = 1$ ma,

$$R = \frac{500}{10^{-3}} - 50 \cong 500 \text{ kilohms}$$

The 50 ohms was neglected here because it is so much smaller than the 500-kilohm value. The voltmeter then would have the circuit shown in Fig. 1.7(a). To use this voltmeter, we would have to calibrate the face of the meter in equal increments from 0 to 500 volts, as shown in Fig. 1.7(b). Now an unknown voltage can be placed across the terminals A and B and the deflection noted. For example, if we place the terminals of our voltmeter across a 300-volt battery, we shall have 0.6 ma flowing, and the pointer needle will deflect and come to rest at the 300-volt mark on the scale.

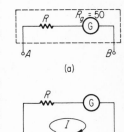

Fig. 1.6 (a) A basic dc voltmeter. (b) Measurement of an unknown voltage E.

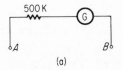

Fig. 1.7 (a) 0–500-volt voltmeter utilizing a 1-ma movement. (b) The scale for the 0–500-volt voltmeter.

Example 1.4 To make a voltmeter that can select several ranges, we simply insert a switching arrangement so that different values of R can be selected, and calibrate the face of the meter accordingly. The circuit for such a multirange meter is shown in Fig. 1.8(a) along with the face required on the meter, Fig. 1.8(b).

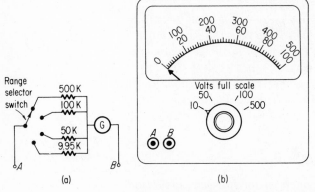

Fig. 1.8 (a) A multirange voltmeter. (b) The scale used in a multirange meter.

(a)

(b)

For this multirange meter, we have arbitrarily selected full-scale values of 10, 50, 100, and 500 volts. The values of R have been calculated using Eq. (1.9), for the four ranges. In the 0 to 10-volt range, it was decided to use 9.95 K, the actual value of R, rather than a rounded-off value of 10 K. The general practice is to use resistors accurate to 1 percent or $\frac{1}{2}$ percent for the series resistance R, which puts the probable error in the third significant figure. Any further accuracy in R will be lost because the galvanometer deflection proportionality is no better than this.

The scale on the meter should be explained also. As can be seen, a double scale of numbers is placed above the graduations. The reason for two number scales is due to the choice of two different full-scale values—that is, multiples of 10 and multiples of 50. For example, if the range switch is on 10, as indicated in Fig. 1.8(b), then we are measuring voltages in the range 0 to 10 volts, but the 0- to 100-volt scale would be read. The actual voltage would be the scale reading divided by 10. For the case where the range switch is switched to 100, the 0- to 100-volt scale would be read directly. Similarly, for the two range positions 50 and 500, the upper set of numbers would be used.

1.5 Loading

The voltmeter must have current passing through it in order to produce a pointer deflection. This current must be taken from the circuit being measured. To make this clear, take the example shown in Fig. 1.9(a). As can be seen, the voltage across R_2 is 50 volts.

However, if you try to determine this by putting the voltmeter across R_2, the voltmeter alters the circuit and changes the conditions. In actuality, the example voltmeter we have constructed would measure 24.8 volts if switched to the 50-volt range. The reason for this is that we are in effect shunting R_2 with a 50-K resistor, the internal resistance of the voltmeter. Figure 1.8(a) shows that this is the total resistance in series with the galvanometer. The resistance of R_2 and 50 K in parallel is equal to 33.3 K. Therefore, after the voltmeter is inserted, the circuit has been changed to a 100-K resistor, R_1, in series with an effective 33.3-K resistor, Fig. 1.9(b).

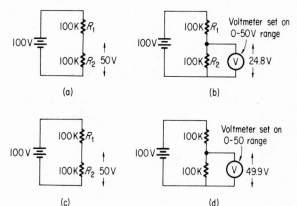

Fig. 1.9 The loading effect in a circuit due to a voltmeter.

The voltage division across these two resistors is such that 24.8 volts appears across the effective 33.3-K resistor. If the measurement is made with the range switch in the 100-volt position, the voltmeter will read 33.3 volts. This is an improvement, but still very far from the actual voltage before the voltmeter was inserted. Let us try measuring the voltage drop in the circuit of Fig. 1.9(c) before we come to any conclusions.

Since the total resistance of our voltmeter in the 50-volt range is 50 K, the total resistance of R_2 and 50 K in parallel is 99.8 ohms— very little change from the 100 ohms present before the voltmeter was inserted. Therefore, the new circuit does not differ from the original circuit appreciably, and the voltage drop across the combination of R_2 and the voltmeter in parallel is very close to the original 50 volts. From these two experiments we can conclude that the degree of loading by the voltmeter depends upon the resistance of the component with which the voltmeter is placed in parallel. In general, the greater the internal resistance of the voltmeter, the less the loading of the voltmeter in a particular measurement. The relative degree of loading for a given voltmeter can be ascertained from the ohms per volt rating. This figure can be obtained by dividing the total internal resistance for a given range by the full-scale reading for that range. Thus, for our example voltmeter, the ohms per volt rating is 1000 ohms per volt on all four ranges.

From the previous discussion a voltmeter having a higher ohms per volt rating will have less of an effect on a circuit. A more sensitive galvanometer can accomplish this. For example, if a galvanometer having an I_g of 50 microamperes (μa) is used, a simple calculation using Eq.(1.9) gives us values of 200 K, 1 meg (megohm), 2 megs, and 10 megs for the R of the 10-, 50-, and 100- and 500-volt ranges, respectively. This means a higher internal resistance and an ohms per volt rating of 20,000 ohms per volt for the four ranges. A voltmeter using this more sensitive galvanometer will produce less loading when used to measure voltage drops across high-resistance components.

1.6 Ammeter

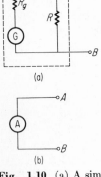

(a)

(b)

Fig. 1.10 (a) A simplified dc ammeter. (b) The symbol for an ammeter.

Normally, a small current will produce full-scale deflection in a galvanometer. In order to measure currents larger than full scale, the excess current must be shunted around the galvanometer. The most common method for doing this is to place a shunt resistor across the galvanometer, Fig. 1.10(a). The symbol for the ammeter is shown in Fig. 1.10(b). The resistor R is in parallel with the galvanometer. If R is equal to R_g, then it can be seen that the current entering terminal A can be twice the full-scale current of the galvanometer, since the current will split equally through R_g and R. For smaller values of R, more current can enter terminal A and not exceed the full-scale current in the path containing the galvanometer. If the scale of the galvanometer is calibrated in terms of the current entering terminal A, then we have an ammeter that will measure currents greater than the full-scale current of the galvanometer.

Example 1.5 Starting with a dc galvanometer whose internal resistance is 50 ohms and whose full-scale current is 1 ma, construct a 0 to 5-amp dc ammeter.

1. This means that when a 5-amp current enters terminal A, it must split up so that the required full-scale current flows through the galvanometer branch and the remainder flows through R. The problem is illustrated below:

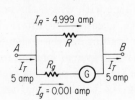

2. Since R and R_g are in parallel, the voltage drop across each must be the same; that is,

$$I_R R = I_g R_g$$

$$R = R_g \left(\frac{I_g}{I_R}\right) = 0.01 \text{ ohm}$$

This value of R in parallel with the galvanometer produces a 0- to 5-amp dc ammeter.

Example 1.6 Is the scale linear? If the current I_T is reduced to 2.5 amp, will the current through the galvanometer branch be reduced to one-half of the full scale?

To determine this, we shall call the current flowing through the galvanometer branch I, the current through resistor R, I_R, and the total current I_T. We know that

$$I_T = I + I_R$$

Also, because R and R_g are in parallel,

$$I R_g = I_R R$$

Eliminating I_R from these two equations and solving for I in terms of I_T, we obtain

$$I = I_T \left(\frac{R}{R_g - R}\right)$$

Since R and R_g are constants for a given-range ammeter, I is directly proportional to I_T. Therefore, the face of a galvanometer can be calibrated linearly in terms of the total current.

Example 1.7 Using the galvanometer of Example 1.3, construct a multirange ammeter to measure the following full-scale currents: 0 to 10 ma, 0 to 50 ma, 0 to 100 ma, 0 to 500 ma, and 0 to 1 amp.

First, it should be made clear how we must construct a circuit that will switch ranges for us. In much the same way as with the multirange voltmeter, we construct a switching circuit that switches different values of R across the galvanometer. A typical circuit is described in the sketch.

Using the equation of Example 1.5,

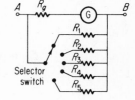

$$R = R_g \left(\frac{I_g}{I_R}\right) = R_g \left(\frac{I_g}{I_T - I_g}\right)$$

we calculate the required values of R: For R_1,

$$I_T = 10 \text{ ma}, \qquad I_g = 1 \text{ ma}, \qquad R_g = 50 \text{ ohms}$$

$$R_1 = 50 \left(\frac{1}{10 - 1}\right) = 5.56 \text{ ohms}$$

For R_2,

$$I_T = 50 \text{ ma}, \qquad I_g = 1 \text{ ma}, \qquad R_g = 50 \text{ ohms}$$

$$R_2 = 50\left(\frac{1}{50 - 1}\right) = 1.02 \text{ ohms}$$

For R_3,

$$I_T = 100 \text{ ma}, \qquad I_g = 1 \text{ ma}, \qquad R_g = 50 \text{ ohms}$$

$$R_3 = 50\left(\frac{1}{100 - 1}\right) = 0.505 \text{ ohm}$$

For R_4,

$$I_T = 500 \text{ ma}, \qquad I_g = 1 \text{ ma}, \qquad R_g = 50 \text{ ohms}$$

$$R_4 = 50\left(\frac{1}{500 - 1}\right) = 0.100 \text{ ohm}$$

For R_5,

$$I_T = 1000 \text{ ma}, \qquad I_g = 1 \text{ ma}, \qquad R_g = 50 \text{ ohms}$$

$$R_5 = 50\left(\frac{1}{1000 - 1}\right) = 0.050 \text{ ohm}$$

The constructed multirange ammeter circuit, with all circuit values, can now be shown as in the accompanying drawing below (left).

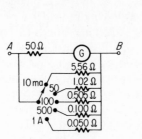
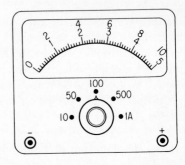

The scale on the face of the meter is made dual, as was done for the voltmeter. When using the 0- to 50-ma and 0- to 500-ma ranges, the scale markings of 0 to 5 are used with the appropriate multiplier. When using the 0- to 10-ma, 0- to 100-ma, and 0- to 1-amp ranges, the numbers 0- to 10 are used with the appropriate multiplier. In both cases, the same graduated scale is used as shown in the illustration above (right).

1.7 Internal Resistance

For an ammeter to have a negligible effect on a circuit, it should have a small internal resistance. This is due to the fact that the ammeter is physically connected into a circuit when measuring the current. As can be seen in Fig. 1.11, the ammeter is placed in series with R_1, and hence will measure the current flowing through R_1. This can be calculated as

$$I_{R_1} = \frac{E}{R_1 + R_A} \tag{1.10}$$

where R_A is the internal resistance of the ammeter. The current before the ammeter was introduced would have been E/R_1. To have a negligible effect on the circuit, $R_A \ll R_1$. Since an ammeter is constructed by shunting R_g with R, the internal resistance of the galvanometer is given by

$$R_A = \frac{R_g R}{R_g + R} \tag{1.11}$$

and this is always less than the smallest resistance in the parallel combination. Therefore, because of its construction, the ammeter is an inherently low-resistance device.

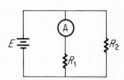

Fig. 1.11 The ammeter is inserted into the circuit to measure the current through R_1.

1.8 Ohmmeter

There are several basic ohmmeter circuits in use. The principle, however, is the same for all. An ohmmeter measures resistance by indicating how much the current is changed when an unknown resistance is placed into a previously known circuit. This known circuit for one type of ohmmeter consists of a dc voltage, galvanometer, and series resistance, arranged as shown in Fig. 1.12.

The unknown resistance, R_x, is inserted between A and B, forming a simple series circuit. The resistor R is put in so as to limit the current flow and help determine the range of the ohmmeter. The battery E is used as the source of emf for this circuit, while the galvanometer G measures the current flowing in this simple series circuit. Since R_g, R, and V are fixed for a given ohmmeter range, the maximum current condition occurs for the case $R_x = 0$, that is, when there is a short circuit from A to B. See Fig. 1.13(a). On the other hand, if there is an open circuit between A and B, that is, $R_x = \infty$ [Fig. 1.13(b)], then no current can flow in the circuit and the galvanometer will read zero. To act as an ohmmeter, the galvanometer face must be calibrated in terms of R_x, the resistance inserted between A and B.

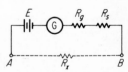

Fig. 1.12 A simplified ohmmeter.

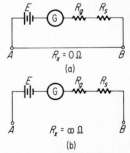

Fig. 1.13 (a) Maximum current occurs for $R_X = 0$. (b) For $R_X = \infty$, the current is zero.

It can be seen from the minimum and maximum conditions that the full scale on the galvanometer will correspond to zero ohms for R_x and that zero current corresponds to infinity for R_x. Evidently the scale must be nonlinear. To establish another point on the scale that will have some significance as far as the range of the ohmmeter is concerned, assume $R_x = R_s + R_g$. It can then be seen that the current through the circuit will be one-half of full scale. If the circuit is adjusted so that A is short-circuited to B, $R_x = 0$, and full-scale current flows, then, putting in $R_x = R_g + R_s$ will double the total resistance, which will reduce the total current to half-scale. This establishes a third point on our ohmmeter scale. When $R_x = R_g + R_s$, the galvanometer pointer will point to half-scale.

Figure 1.14 shows the three points determined so far. It is evident that readings will be easier to make on the right-hand side of the scale than on the left-hand side of the scale; for on the left-hand side of the scale lie all values of resistance from $R_x + R_g$ to ∞. Therefore, the use of a particular ohmmeter scale is limited to about three-quarters of the scale face, as shown in Fig. 1.14. The half-scale value is used to determine the range of resistance that can be easily read on the particular ohmmeter being used. Before any examples are worked, the scale calibration should be derived.

Region of "good" readability

$R_s + R_g$

Fig. 1.14 The region of good readability for an ohmmeter scale.

In the circuit of Fig. 1.13(a), the following equation holds true:

$$I = \frac{E}{R_x + R_g + R_s} \tag{1.12}$$

For the case of $R_x = 0$, the parameters are chosen so that full-scale current I_g flows. The equation becomes

$$I_g = \frac{E}{R_g + R_s} \tag{1.13}$$

The fractional scale deflection can be obtained by dividing Eq. (1.12) by Eq.(1.13), since I_g = full-scale deflection:

$$\frac{I}{I_g} = \frac{1}{1 + [R_x/(R_g + R_s)]} \tag{1.14}$$

Since $R_g + R_x$ is a constant for a given ohmmeter range, the only variable is R_x. The face of a galvanometer can now be calculated if $R_g + R_s$ is known.

Example 1.8 Using a galvanometer in which $R_g = 50$ ohms and $I_g = 1$ ma, find E for a half-scale value of 20,000 ohms, and construct the ohmmeter face to be used on the galvanometer.

For a particular ohmmeter, there are two variables, E and R_s. Choosing a value of R_s will determine E or, conversely, a particular value of E determines R_s. In this example, R_s has been chosen by specifying the half-scale value for the ohmmeter face.

1. Specifying half-scale means we can say

$$R_g + R_s = R_x \text{ (half-scale)}$$

Therefore

$$R_s = R_x \text{ (half-scale)} - R_g$$

$$= 19,950 \text{ ohms}$$

2. To find E, we know that when $R_x = 0$, we must have full-scale deflection for our galvanometer. For this condition, only R_g and R_s are in the circuit:

$$I_g = \frac{E}{R_g + R_s}$$

$$E = I_g(R_g + R_s)$$

Since $I_g = 1$ ma and $R_g + R_s = 20K$,

$$E = 20 \text{ volts}$$

3. The ohmmeter circuit becomes:

4. To calibrate the face, we use Eq. (1.14) and insert our values:

$$\frac{I}{I_g} = \frac{1}{1 + (R_x/20,000)}$$

Some representative values calculated by this formula are shown in Table 1.1.

TABLE 1.1 R_x Values (Example 1.8)

R_x, (K)	I/I_g*
0	1.00
1	0.95
2	0.91
3	0.87
4	0.83
5	0.80
10	0.67
20	0.50†
30	0.40
40	0.33
50	0.28
100	0.17

* Fraction of full-scale deflection.
† Half-scale.

5. The scale face can be constructed from Table 1.1 by super-imposing a linear scale to aid in the location of the tabulated values of R_x. See the accompanying drawing.

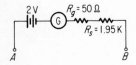

Linear scale for tabulation purposes

Ohmmeter scale

6. As can be seen from the face calibration, this ohmmeter would be most useful in measuring values of resistance from about 1 to 50 K. To measure resistances of about 1 K or smaller, a different E and R_s would be needed.

Example 1.9 Using the galvanometer of Example 1.8, construct a 2-K half-scale ohmmeter and draw the ohmmeter face.

1. Again the half-scale value of the ohmmeter has been specified, and we can write

$$R_s = R_x \text{ (half-scale)} - R_g$$

$$= 1950 \text{ ohms}$$

2. E can now be determined such that the full-scale current I flows when $R_x = 0$:

$$E = I_g(R_g + R_s)$$

$$= 2 \text{ volts}$$

3. The ohmmeter circuit thus becomes as shown in the sketch.

TABLE 1.2 R_x Values (Example 1.9)

R_x	I/I_g
100Ω	0.95
200Ω	0.91
300Ω	0.87
400Ω	0.83
500Ω	0.80
1K	0.67
2K	0.50*
3K	0.40
4K	0.33
5K	0.28
10K	0.17

* Half-scale.

4.   Using Eq.(1.14), a table of R_x values (Table 1.2) is constructed versus deflection so that an ohmmeter face can be calibrated.

5.   The scale face becomes as shown in the sketch. By comparing this scale to the one in Example 1.8, it can be seen that the two scales are exactly the same except that the numbers are smaller

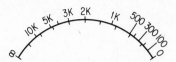

by a factor of 10 in Example 1.9. This was made possible by choosing the half-scale value as one-tenth of the former scale. Choices of this sort make the use of a single scale possible for different ranges. The ohmmeter constructed in this example can accurately determine resistance values in a range from about 5 K to 100 ohms.

To measure values less than 100 ohms, we should construct another ohmmeter with a half-scale value of one-tenth that of the one in Example 1.9. If this is done, the ohmmeter circuit is as shown in the drawing here.

The face would have values one-tenth of those shown in Example 1.9. In all three examples the quantities changed were E and R_s.

1.9   Multirange Ohmmeter

It is possible to construct a multirange ohmmeter by arranging the circuit so that the appropriate values of E and R_s are switched in for each range. A circuit that will accomplish this is shown in Fig. 1.15.

The selector switch is shown connecting the galvanometer to $R_s = 1.95$ K and $E = 2.0$ volts. The resistor R_s is formed by the series combination of the 1.85-K resistor and the 200-ohm rheostat labeled "Zero Adj." The voltage E is produced by the series combination of the 1.8-volt and 0.2-volt batteries. This is recognized as the circuit for the ohmmeter calculated in Example 1.9.

The Zero Adj. is inserted for practical purposes. The circuit as shown is correct only for the specified voltage. Since the voltage output of a battery declines with use, some way of compensating for this must be found. The simplest solution is to arrange the

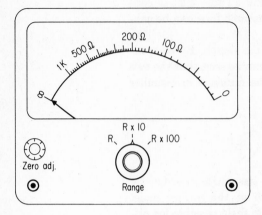

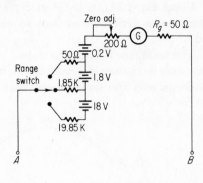

Fig. 1.15 A multirange ohmmeter and scale.

Zero Adj. rheostat so that it introduces in the middle of its range the correct amount of resistance in the circuit for the specified voltage E. As the output of the batteries decreases with use, the rheostat resistance must be decreased to allow the full-scale current to flow. This necessarily introduces an error in the ohmmeter accuracy. For this reason, the Zero Adj. rheostat is only a fraction of the total R_s needed. When the batteries are too low, reducing the Zero Adj. rheostat to zero ohms will not be enough to bring the current back to full scale, and then the old batteries must be replaced.

Before use, the ohmmeter must be "zeroed." This is accomplished by first short-circuiting the leads A and B. The galvanometer should now show full-scale deflection (zero ohms). If it does not, then the Zero Adj. rheostat should be adjusted until zero ohms is indicated. Now the ohmmeter is ready for use. Any unknown resistance placed between the leads A and B will now be indicated by the ohmmeter.

1.10 Low-Resistance Ohmmeter

By rearranging the components, E, R_s, and the galvanometer, an ohmmeter can be constructed to measure relatively small resistances. In essence the galvanometer is placed in parallel with the series combination E and R_s, and the unknown resistor R_x is placed across the galvanometer. The arrangement is shown in Fig. 1.16. The end points of the scale can be determined by inspection. If R_x is equal to zero (a short circuit), no current flows through the galvanometer branch. Therefore zero R_x is at the zero of the galvanometer scale. If $R_x = \infty$ (an open circuit between A and B), the circuit is a simple series circuit consisting of E, R_s, and

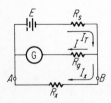

Fig. 1.16 The shunt-type ohmmeter.

R_g. The parameters should be such that full-scale deflection is produced. Therefore, infinite R_x is registered at the full-scale deflection point. This is just the opposite of the ohmmeter discussed in Secs. 1.8 and 1.9.

The working equation needed to calibrate the scale of the meter should be derived at this point so that other points can be filled in on the scale between zero ohms and infinity. In Fig. 1.16 the current through the galvanometer branch is given by

$$I = I_T \left(\frac{R_x}{R_g + R_x} \right) \tag{1.15}$$

I_T is given by

$$I_T = \frac{E}{R_s + [R_g R_x/(R_g + R_x)]} \tag{1.16}$$

The equation for the case $R_x = \infty$ is also needed:

$$I_g = \frac{E}{R_g + R_s} \tag{1.17}$$

eliminating I_T between Eqs.(1.15) and (1.16),

$$I = E \left(\frac{R_x}{R_s R_g + R_s R_x + R_g R_x} \right) \tag{1.18}$$

Taking the ratio Eq.(1.18) and Eq.(1.17) to obtain the fraction of full-scale deflection:

$$\frac{I}{I_g} = \frac{1}{1 + [R_s R_g/R_x(R_s + R_g)]} \tag{1.19}$$

and calling the combination

$$\frac{R_s R_g}{R_s + R_g} = R_{\parallel} \tag{1.20}$$

$$\frac{I}{I_g} = \frac{1}{1 + (R_{\parallel}/R_x)} \tag{1.21}$$

From this equation it can be seen, for example, that half-scale deflection is produced when $R_x = R_{\parallel}$.

Example 1.10 Using a galvanometer having $R_g = 50$ ohms and $I_g = 1$ ma, construct a shunt type of ohmmeter that utilizes a 1.5-volt battery for E.

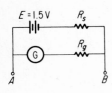

1. The circuit to be used is shown in the sketch.

2. R_s must be determined. In order to find R_s, it should be noted that when $R_x = \infty$, the current will be maximum. This maximum current is made to be the full-scale current I_g by picking the proper value of R_s. For this condition, a simple series circuit must be solved:

$$E = I(R_s + R_g)$$

$$R_s = \frac{E}{I} - R_g$$

Using the values given, R_s becomes

$$R_s = 1450 \text{ ohms}$$

3. The scale face must now be calibrated. Using Eq.(1.21) and the fact that $R_{||}$ is given by Eq.(1.20), the following working equation is obtained:

$$\frac{I}{I_g} = \frac{1}{1 + (48.3/R_x)}$$

Table 1.3 gives the calculated values from this equation and the constructed meter face appears in the sketch here.

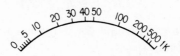

TABLE 1.3 Calculated Values of R_x

R_x	I/I_g
1	0.02
2	0.04
3	0.06
4	0.08
5	0.09
10	0.17
20	0.29
30	0.38
40	0.46
50	0.51
100	0.68
200	0.81
300	0.86
400	0.89
500	0.91
1 K	0.96

1.11 Positive and Negative Current

In dc circuit calculations, the current is treated as a mathematical construction. It has a direction and a magnitude. It may not seem evident at this point, but the direction chosen is not important. The magnitude, however, is important, for this is what a dc ammeter indicates when placed in a circuit.

Consider a simple circuit containing a battery and a resistor, Fig. 1.17. Current will flow in this circuit and produce a voltage drop across the resistor R. If the battery polarity is as shown, then the voltage drop across R will be as shown. Since the top of R is connected to the positive terminal of the battery, the top of R is positive with respect to the bottom of R. Knowing the direction of the current is not necessary to ascertain this. For this reason, the current direction can now be defined.

Fig. 1.17 The battery determines the polarity across R.

Current directions are defined in relation to the circuit external to a source of emf. A positive current is defined as moving through the external circuit, from the positive to the negative terminal of an emf. Applying this to the simple circuit of Fig. 1.18(a), the current will flow so that it enters the positive terminal of R and leaves the negative terminal. Conversely, if a current enters the positive terminal of a component and leaves the negative terminal, it is called a *positive current*. A negative current, on the other hand, is defined as a current that flows through the external circuit, from the negative to the positive of the battery. This is shown in Fig. 1.18(b). Conversely, a current that enters the negative terminal of a component and leaves by the positive terminal is called a *negative current*.

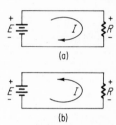

Fig. 1.18 (a) Positive current. (b) negative current.

In both cases, the polarity of the voltage drop across R is exactly the same. It can be seen that each circuit satisfies Kirchhoff's law; that is, the addition of the emf and the voltage drops about a closed loop is equal to zero. In Fig. 1.18, if the emf and voltage drops are added, starting with the battery and going counterclockwise, we obtain

$$-E + IR = 0 \quad \text{or} \quad E = IR$$

The signs are important. E had a negative sign in the preceeding equation because we moved down in potential (from the positive terminal to the negative terminal), and the voltage drop IR had a positive sign because we went up in potential in the counterclockwise (CCW) direction through R (from the negative to the positive terminal). The resultant equation is Ohm's law.

In more complicated circuits, the direction of the voltage drop may not be so evident as it is in the simple example used. In this case, the direction is obtained operationally by assuming a current direction through all components and solving the resultant equations. It does not matter whether positive or negative current is chosen, as long as only one of these is used throughout the problem. Also, more remarkable, the true direction of the current chosen need not be known. Any direction may be chosen. If the direction chosen is wrong, the answer will have the correct magnitude for I, but it will be preceded by a minus sign. If the direction chosen is correct, the sign for I will be positive. The use of current directions will be illustrated in Secs. 1.12 to 1.14.

1.12   Branch-Current Method

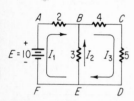

Fig. 1.19 A three-branch network.

A complex dc network can be broken down into a series of interconnected branches in which the branches are simple series circuits. Figure 1.19 illustrates this. There are three branches in this circuit. Branch 1 is given by the series circuit $EFAB$, in which the current is I_1. Branch 2 is given by EB, in which the current is I_2, and branch 3 is given by $BCDE$, in which the current is I_3. The currents I_1, I_2, and I_3 join at points B and E. A simple set of rules enables one to find the unknowns in circuits of this type.

1. Draw a good diagram.
2. Write all constants known on the diagram.
3. Label all unknowns clearly.
4. Choose either positive or negative current with which to work.
5. Use arrows to indicate current directions in all branches. Assume a direction when it is not given.
6. Mark the voltage drops for all resistors.
7. Use Kirchhoff's laws to set up as many independent equations as there are unknowns; that is,

$$\sum E_i = 0 \qquad \text{around a closed loop} \qquad (1.22)$$

$$\sum I_i = 0 \qquad \text{for any point in the circuit} \qquad (1.23)$$

Example 1.11   Given the circuit shown below, find the magnitude and direction of positive current through each resistor.

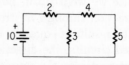

1.   Since the direction of the positive currents through each resistor are not given, assume directions in each branch. For this

example, a set of assumed directions is given in the accompanying sketch.

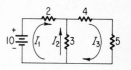

2.   Now the polarities of the voltage drops must be put into the diagram. Since this is positive current, the place where it enters a component must be the positive terminal, and the point where it leaves must be the negative terminal. This rule does not apply to emf sources. The polarity for the battery is determined by the way it is inserted in the circuit and not by the direction of current flow. The diagram becomes as shown in the drawing.

3.   Kirchhoff's laws can now be applied:

$$\sum E_i = 0$$

for all closed loops.

There are three closed loops in this circuit. They are $ABEFA$, $BCDEB$, and $ABCDEFA$. Going clockwise (cw) about each loop, the resultant equations become

$$-2I_1 + 3I_2 + 10 = 0 \qquad (1)$$

$$-4I_3 - 5I_3 - 3I_2 = 0 \qquad (2)$$

$$-2I_1 - 4I_3 - 5I_3 + 10 = 0 \qquad (3)$$

These are not independent equations. For example, Eq. (3) is really Eq. (1) plus Eq. (2). In all, there are only two independent equations present. We can choose any two; let these two be Eq. (1) and Eq. (2). Since there are three unknowns, we need three independent equations. The third equation is found by utilizing the other of Kirchhoff's circuital laws.

4. $\sum I_i = 0$ at any point in the circuit. Selecting a meaningful point such as B or E, we obtain

$$I_1 + I_2 - I_3 = 0$$

The equation $\sum I_i = 0$ is an algebraic equation, and signs for the I_i must be considered. A consistent choice, for example, would be one in which currents entering a point were plus and currents leaving a point were minus.

5.   The three independent equations chosen are

$$I_1 + I_2 - I_3 = 0$$

$$-2I_1 + 3I_2 + 10 = 0$$

$$-4I_3 - 5I_3 - 3I_2 = 0$$

Rearranging the equations for determinant use,

$$I_1 + I_2 - I_3 = 0$$

$$-2I_1 + 3I_2 \qquad = -10$$

$$- 3I_2 - 9I_3 = 0$$

and using Cramer's rule for solving simultaneous equations,

$$I_1 = \frac{\begin{vmatrix} 0 & 1 & -1 \\ -10 & 3 & 0 \\ 0 & -3 & -9 \end{vmatrix}}{\begin{vmatrix} 1 & 1 & -1 \\ -2 & 3 & 0 \\ 0 & -3 & -9 \end{vmatrix}} = \frac{120}{51} \text{ amp}$$

$$I_2 = \frac{\begin{vmatrix} 1 & 0 & -1 \\ -2 & -10 & 0 \\ 0 & 0 & -9 \end{vmatrix}}{51} = -\frac{90}{51} \text{ amp}$$

$$I_3 = \frac{\begin{vmatrix} 1 & 1 & 0 \\ -2 & 3 & -10 \\ 0 & -3 & 0 \end{vmatrix}}{51} = \frac{30}{51} \text{ amp}$$

6. The negative answer for I_2 means that I_2 actually is down through the 3-ohm resistor and not up, as was arbitrarily chosen at the beginning. Its magnitude is 90/51 amp. The correct picture for the positive current, then, is that shown in the accompanying illustration. As can be seen, I_1 splits up at junction B into I_2 and I_3, and then I_2 and I_3 recombine at junction E to form I_1 again.

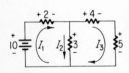

7. *Answers*
(a) Current through 2-ohm resistor = 120/51 amp
(b) Current through 3-ohm resistor = 90/51 amp
(c) Current through 4-ohm resistor = 30/51 amp
(d) Current through 5-ohm resistor = 30/51 amp

1.13 Loop-Current Method

A more abstract method for solving network problems is to make use of loop currents. A loop current is an artificial current made to flow in closed loops. In this case there will be some components with more than one current flowing through them. In all cases the total current is the algebraic sum of the currents flowing through the component. If the loop current is the only current flowing

through a component, then this is also the true current through
that component.

The use of loop currents usually reduces the complexity of a
problem. Also, the only equation needed for solution is

$$\sum E_i = 0 \qquad \text{for all closed loops} \qquad (1.24)$$

Example 1.12 Solve the circuit shown, using the loop current
method. Find the magnitude and direction of the current through
each resistor.

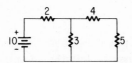

1. Drawing enough loop currents so that each component has
at least one current through it (both loop currents were arbitrarily assumed to be CW), the diagram becomes as shown.

2. Now assume that we are using positive current. The voltage
drops are shown in the next illustration.
Since both I_1 and I_2 pass through the 3-ohm resistor, two voltage
drops must be put in, one for each current.

3. Now the equation $\sum E_i = 0$ around a closed loop can be
used. The equation for the voltage drops clockwise around the
closed loop containing I_1 becomes

$$10 - 2I_1 - 3I_1 + 3I_2 = 0$$

and the equation for the closed loop containing I_2 becomes

$$3I_1 - 3I_2 - 4I_2 - 5I_2 = 0$$

Thus two currents pass through the 3-ohm resistor so two voltage
drops must be recorded in each equation for the 3-ohm resistor.

4. Rewriting the equation and solving,

$$-5I_1 + 3I_2 = -10$$

$$3I_1 - 12I_2 = 0$$

$$I_1 = \frac{\begin{vmatrix} -10 & 3 \\ 0 & -12 \end{vmatrix}}{\begin{vmatrix} -5 & 3 \\ 3 & -12 \end{vmatrix}} = \frac{120}{51} \text{ amp}$$

$$I_2 = \frac{\begin{vmatrix} -5 & -10 \\ 3 & 0 \end{vmatrix}}{\begin{vmatrix} -5 & 3 \\ 3 & -12 \end{vmatrix}} = \frac{30}{51} \text{ amp}$$

5. *Answers*

(a) For the 2-ohm resistor, I_1 is the only current flowing through it, so the current must be 120/51 amp in the direction of I_1.

(b) For the 4- and 5-ohm resistors, I_2 is the only current flowing through each of them, so the current must be 30/51 amp in the direction of I_2 for each resistor.

(c) For the 3-ohm resistor, I_1 and I_2 flow through; therefore the total current through the 3-ohm resistor is the resultant of I_1 and I_2. Since the currents are in opposite directions and $I_1 > I_2$, the final current is in the direction of I_1 and is given by

$$| I | = I_1 - I_2 = \tfrac{120}{51} - \tfrac{30}{51} = \tfrac{90}{51} \text{ amp}$$

6. As can be seen, the answers are identical to those obtained when using branch currents, but the complexity of the equations has been reduced.

1.14 Thévenin's Theorem

Thévenin's theorem will be a very useful mathematical tool when considering vacuum tube and transistor equivalent circuits. In effect, Thévenin's theorem takes a whole branch or component in a complicated network and replaces it with an equivalent, simple series circuit; all calculations are carried out in this equivalent circuit.

Consider the network in Fig. 1.20(a). The current through the 3-ohm resistor is desired. By Thévenin's theorem the remainder of the network can be converted into the simple resistor R_0 and voltage source E_0, as shown in Fig. 1.20(b). The current through the 3-ohm resistor is now calculated from the new circuit as

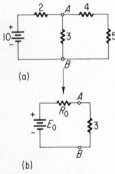

(a)

(b)

$$I_3 = \frac{E_0}{R_0 + 3} \tag{1.25}$$

The rules for choosing R_0 and E_0 are simple, and a few examples will suffice to show their application.

Fig. 1.20 The network in (a) can be converted by Thévenin's theorem into circuit (b).

APPLICATION OF THÉVENIN'S THEOREM

1. Network components must be linear, that is, must not depend upon voltage or current.
2. It will be applied here only to voltage sources and resistors.
3. To find R_0:
   (a) Starting with the old circuit, remove the load component or branch.
   (b) "Short" all emf.

(c) Find the resistance of this new circuit with respect to the points to which the load was connected. This is R_0, that is, $R_{AB} = R_0$.

4. To find E_0:
(a) Starting with the old circuit, "remove" the load component or branch.
(b) Find the voltage drop between the points to which the load was connected. This is E_0.

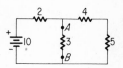

Example 1.13 In the circuit shown, find the current through the 3-ohm resistor.
To do this problem utilizing Thévenin's theorem, R_0 and E_0 must be found.

1. *To find R_0:*
(a) Remove the load (the 3-ohm resistor) and obtain the circuit shown here.

(b) Short all emf as shown here.
(c) Find $R_{AB} = R_0$. Redraw the circuit so that the resistance is equal to

$$R_{AB} = \tfrac{18}{11} \text{ ohms}$$

Then the circuit should be as shown in the next illustration.

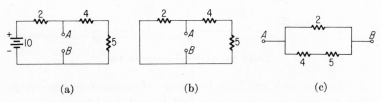

| (a) | (b) | (c) |

2. *To find E_0:*
(a) Remove the load and obtain the circuit shown.
(b) Find $E_{AB} = E_0$.

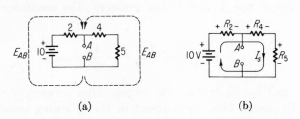

| (a) | (b) |

In this example, E_{AB} is either the voltage drop across the 4- and 5-ohm resistors, or the voltage drop across the 10-volt battery and the 2-ohm resistor. (E_{AB} means: Start at point B and sum up all voltage drops to point A.) To do this, the current flowing in this secondary circuit must be calculated:

$$I_s = \frac{E}{R_2 + R_4 + R_5} = \tfrac{10}{11} \text{ amp}$$

For E_{AB} (right-hand branch),

$$E_{AB} = +I_s R_5 + I_s R_4 = \tfrac{90}{11} \text{ volts}$$

For E_{AB} (left-hand branch),

$$E_{AB} = +E - I_s R_2 = \tfrac{90}{11} \text{ volts}$$

E_{AB} is independent of the path taken to go from B to A.

(c)   The new equivalent circuit is shown in the sketch.

The current through the 3-ohm resistor is now given by

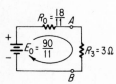

$$I_3 = \frac{E_0}{R_0 + 3} = \frac{90}{51} \text{ amp}$$

Example 1.14   Convert the following circuit into a voltage equivalent circuit by means of Thévenin's theorem.

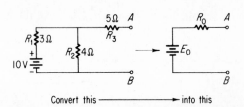

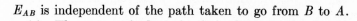

Convert this ⟶ into this

1.   In this case, the load is already removed from the circuit. All that needs to be done is to find R_0 and E_0.

2.   To find R_0: After "shorting" the emf E, the circuit becomes as shown in the illustration.

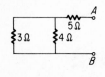

This is seen to be a 5-ohm resistor in series with the parallel combination of the 3- and 4-ohm resistors. The resistance R_{AB} becomes

$$R_{AB} = R_0 = 5 + \frac{3 \times 4}{7} = \frac{47}{7} \text{ ohms}$$

3.   To find E_0: $E_0 \equiv E_{AB}$, the voltage drop between points A and B. To obtain this, the current in the following secondary circuit must be solved for

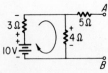

$$I = \frac{10}{3 + 4} = \frac{10}{7} \text{ amp}$$

E_{AB} can now be calculated. It will be instructive to obtain E_{AB} by two possible routes:

Route 1: $E_{AB} = +V - R_1 I$

$\qquad E_{AB} = +10 - \frac{30}{7} = \frac{40}{7}$ volts

There is no current through R_3, so there is no voltage drop across R_3.

Route 2: $E_{AB} = +IR_2$

$\qquad E_{AB} = +\frac{10}{7} \times 4 = \frac{40}{7}$ volts

The plus sign signifies that the top terminal of the voltage source is positive.

4. The equivalent circuit appears in the diagram at the right.

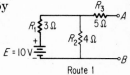

1.15 Power

The power dissipated in a resistor is given by the product of the voltage across the resistor and the current through the resistor. If E is in volts and I in amperes, the power will be in watts:

$$P = EI \qquad (1.26)$$

Due to Ohm's law, there is a relationship between E and I:

$$E = IR \qquad (1.27)$$

This allows one to put Eq. (1.26) into several forms:

$$P = I^2 R$$

$$P = \frac{E^2}{R} \qquad (1.28)$$

The physical significance of Eq. (1.26) or Eq. (1.28) is that this is the rate at which heat energy is developed in a resistor. To keep the resistor at a constant temperature, it must dissipate this heat, usually into the air around it. The amount of power to be dissipated by a resistor determines its physical size. Thus there are two "sizes" to consider in a resistor—its electrical "size," or resistance, and its physical size, or power-handling capacity.

MAXIMUM ENERGY TRANSFER

In some applications, it is desired that a resistance value be selected so that it dissipates maximum power in a given circuit.

Consider the circuit shown in Fig. 1.21. It is desired that maximum power be dissipated in R. The power that is dissipated in R is given by

Fig. 1.21

$$P = I^2 R \qquad (1.29)$$

where

$$I = \frac{E_0}{R_0 + R} \qquad (1.30)$$

Putting Eq. (1.30) in Eq. (1.29),

$$P = E_0{}^2 \frac{R}{(R_0 + R)^2} \qquad (1.31)$$

Equation (1.31) can now be maximized with respect to R if we solve

$$\frac{dP}{dR} = 0 \qquad (1.32)$$

Equation (1.32) should be recognized as the mathematical relation that defines the maxima and minima for Eq. (1.31) with respect to R:

$$\frac{d}{dR}\left[E_0{}^2 \frac{R}{(R_0 + R)^2} \right] = E_0{}^2 \left[\frac{1}{(R_0 + R)^2} - \frac{2R}{(R_0 + R)^3} \right] = 0$$

Getting a lowest common denominator,

$$E_0{}^2 \left[\frac{R_0 + R - 2R}{(R_0 + R)^3} \right] = 0 \qquad (1.33)$$

This equation will be true if the numerator is equal to zero. From this condition, Eq. (1.33) can be solved for R:

$$R = R_0 \qquad (1.34)$$

Thus, if the resistor R is made equal to the resistance R_0, maximum power will be dissipated in R.

Example 1.15 Given a 10-volt battery with an internal resistance of 2 ohms. What value of resistance must be connected across the battery in order to dissipate maximum power in the externally applied resistor?

Battery

1. Since the diagram shown conforms to Fig. 1.21, Eq. (1.34) is the solution to the problem. R must be equal to R_B for maximum power dissipation:

$$R = R_B$$

2. The power dissipated by R is given by

$$P = I^2 R = \left(\frac{E}{R_B + R}\right)^2 R$$

where

$$R = R_B = 2 \text{ ohms}$$

$$E = 10 \text{ volts}$$

$$P = (\tfrac{10}{4})^2 \times 2 = 12.5 \text{ watts}$$

3. The general result can be shown graphically by plotting Eq. (1.31) versus values of R. This is done in the graph here for the example. As can be seen, the curve has a maximum at $R = 2$ ohms, that is, $R = R_B$.

TABLE 1.4

R	P
0	0
1	11.1
2	12.5
3	12.0
4	11.1
5	10.2
6	9.4

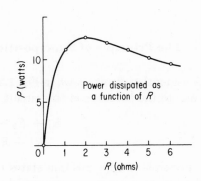

Power dissipated as a function of R

Example 1.16 Given the following circuit, determine R_4 so that it dissipates maximum power in the circuit.

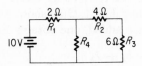

1. In order to solve this problem, it is necessary to make the circuit look like Fig. 1.21, with R_4 in the place of R. Then R_4 can be determined as equal to the resistance in series with it.

2. Thévenin's theorem must be used to accomplish this. Considering R_4 as the load, the circuit given below can be reduced to a voltage source E_0 in series with a resistor R_0.

(a) Determination of R_0:

$$R_0 = R_{AB} = \frac{2 \times 10}{12} = \frac{5}{3} \text{ ohms}$$

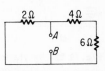

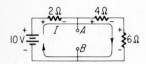

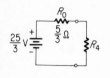

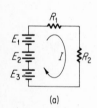

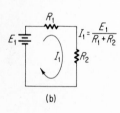

(a)

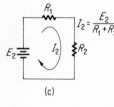

(b)

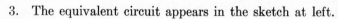

(c)

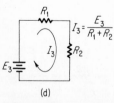

(d)

Fig. 1.22 By the reciprocity theorem, $I_1 + I_2 + I_3 = I$.

(b) Determination of E_0:

$$E_0 = E_{AB} = 10 - 2I$$

where

$$I = \frac{10}{2 + 4 + 6} = \frac{5}{6} \text{ amp}$$

$$E_0 = 10 - 2 \times \tfrac{5}{6} = \tfrac{25}{3} \text{ amp}$$

3. The equivalent circuit appears in the sketch at left.

Thus $R_4 = R_0 = 5/3$ ohms for maximum power dissipation in R_4.

4. It should be noted that for this problem it was not necessary to calculate E_0, but only R_0.

1.16 The Principle of Superposition

Consider the circuit shown in Fig. 1.22(a). The current I through R_2, due to the batteries in this circuit, is given by

$$I = \frac{E_1 + E_2 + E_3}{R_1 + R_2} \tag{1.35}$$

The principle of superposition states that this current can be obtained by summing up the currents due to each individual voltage source. The individual currents are found by shorting out all voltage sources except the one being investigated, and by calculating the current for this new circuit. Thus Fig. 1.22(b) shows the current through R_2 due to E_1; Fig. 1.22(c) shows the current through R_2 due to E_2; and Fig. 1.22(d) shows the current through R_2 due to E_3. If these are added, one obtains Eq. (1.35):

$$I = I_1 + I_2 + I_3$$

$$= \frac{E_1}{R_1 + R_2} + \frac{E_2}{R_1 + R_2} + \frac{E_3}{R_1 + R_2}$$

$$= \frac{E_1 + E_2 + E_3}{R_1 + R_2}$$

This principle applies only to linear circuits and linear equations. For example, the calculation of power involves squaring current or voltage. This means that superposition will not work for power calculations. In the case of the example cited above,

$$P = I^2 R_2 \neq (I_1^2 + I_2^2 + I_3^2) R_2$$

Example 1.17 For the circuit given, show that the superposition principle holds for the current through R_2.

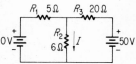

1. The superposition principle states that the currents I_1 and I_2, as shown below, will add up to I.

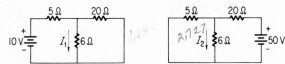

Note that in each case, the other voltage source was replaced by a short circuit.

2. Calculate I from the original diagram first, using branch currents.
The equations become

$$10 - 5I_3 - 6I = 0$$
$$50 - 20I_4 - 6I = 0$$
$$I - I_3 - I_4 = 0$$

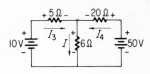

Ordering the equations:

$$6I + 5I_3 \qquad\quad = 10$$
$$6I \qquad\quad + 20I_4 = 50$$
$$I - I_3 - I_4 = 0$$

Solving for I:

$$I = \frac{\begin{vmatrix} 10 & 5 & 0 \\ 50 & 0 & 20 \\ 0 & -1 & -1 \end{vmatrix}}{\begin{vmatrix} 6 & 5 & 0 \\ 6 & 0 & 20 \\ 1 & -1 & -1 \end{vmatrix}} = \frac{9}{5} \text{ amp}$$

3. Solve the auxiliary circuit for I_1, shown in the accompanying sketch.
The equations become

$$-6I_1 - 5I_5 + 10 = 0$$
$$6I_1 \qquad\quad - 20I_6 = 0$$
$$I_1 - I_5 + I_6 = 0$$

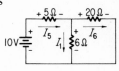

Ordering the equations:

$$6I_1 + 5I_5 \qquad\qquad = 10$$

$$6I_1 \qquad\qquad - 20I_6 = 0$$

$$I_1 - I_5 + I_6 = 0$$

Solving for I_1:

$$I_1 = \frac{\begin{vmatrix} 10 & 5 & 0 \\ 0 & 0 & -20 \\ 0 & -1 & 1 \end{vmatrix}}{\begin{vmatrix} 6 & 5 & 0 \\ 6 & 0 & -20 \\ 1 & -1 & 1 \end{vmatrix}} = \frac{4}{5}\ \text{amp} \ \checkmark$$

Solve the auxiliary circuit for R_2, as given in the drawing here. The equations become

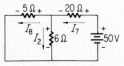

$$50 - 20I_7 - 5I_8 = 0$$

$$6I_2 \qquad\qquad - 5I_8 = 0$$

$$I_2 - I_7 + I_8 = 0$$

Ordering the equations:

$$20I_7 + 5I_8 = 50$$

$$6I_2 \qquad\qquad - 5I_8 = 0$$

$$I_2 - I_7 + I_8 = 0$$

Solving for I_2:

$$I_2 = \frac{\begin{vmatrix} 50 & 20 & 5 \\ 0 & 0 & -5 \\ 0 & -1 & 1 \end{vmatrix}}{\begin{vmatrix} 0 & 20 & 5 \\ 6 & 0 & -5 \\ 1 & -1 & 1 \end{vmatrix}} = 1\ \text{amp} \ \checkmark$$

4. The superposition principle can be checked as follows:

$$I = I_1 + I_2$$

$$\tfrac{9}{5} = \tfrac{4}{5} + 1$$

$$\tfrac{9}{5} = \tfrac{9}{5}$$

5. Although the superposition principle has been rather cumbersome when applied to such a simple circuit, in later chapters it will be used to good effect in calculating more complicated circuits involving dc and ac voltage sources.

Problems

1.1 In the circuit shown, find R_1, R_2 and R_3. Assume the voltmeters do not load the circuit and that the ammeter has negligible resistance.

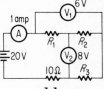
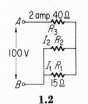

1.1

1.2 For the circuit shown:
(a) Find I_1 and I_2.
(b) Find R_2.
(c) What is the equivalent resistance from A to B across the 100-volt supply?

1.2

1.3 Find the current in the 3-, 5-, and 6-ohm resistors.

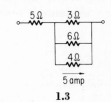

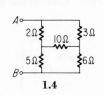

1.3

1.4

1.4 Find R_{AB}.

1.5 Find the equivalent resistance of the following circuits: Assume all resistors are 10 ohms each.

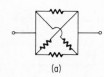

(a)

1.5

(b)

1.6 Find the currents I_1, I_2, and I_3 by:
(a) Branch current method
(b) Loop current method

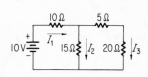

1.6

1.7 In the circuit shown:
(a) Find the magnitude of the current through each resistor.
(b) Find the direction of positive current through each resistor.

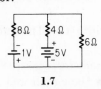

1.7

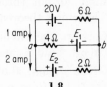

1.8

1.8 For the circuit shown:
(a) Find E_1 and E_2.
(b) Find the potential difference between a and b.

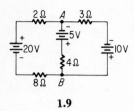

1.9

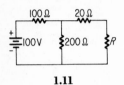

1.11

1.9 Find E_{AB} (left).

1.10 Find E_{AB} for the two cases shown below.

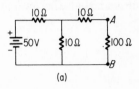

(a)

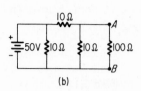

(b)

1.10

1.11 Find R for maximum power dissipation in R.

1.12 Find the magnitude and direction of the current through the ammeter. Assume the ammeter resistance is zero.

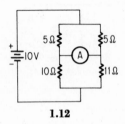

1.12

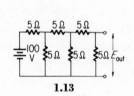

1.13

1.13 Find E_{out}.

1.14 Convert the circuit shown into a Thévenin's equivalent circuit.

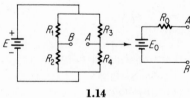

1.14

1.15 Using the circuit in Problem 1.8, find the current through the 4-ohm resistor by using Thévenin's theorem.

1.16 Find current through 20-ohm resistor in Problem 1.6 by Thévenin's theorem.

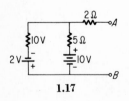

1.17

1.17 Convert the circuit shown into the Thévenin's equivalent circuit. Find E_0 and R_0.

1.18 Given a 1-ma, 50-ohm galvanometer, construct the following;
  (a) 0–1 amp ammeter. Draw the circuit and show the component values.
  (b) 0–100 volt voltmeter.
  (c) 1000-ohm half-scale ohmmeter. Draw the ohmmeter scale.

1.19 A 50-μa, 2000-ohm galvanometer is to be used as a

> 0–1   kV
>
> 0–500 volts
>
> 0–100 volts
>
> 0–50   volts

voltmeter. Draw the circuit and find the circuit values. What is its ohms per volt rating?

1.20 An ammeter in a simple series circuit reads 6 amp. A 4-ohm resistor is now inserted in the circuit and the ammeter reads 2 amp. Find the original resistance.

1.21 An ammeter reads 1 amp full scale when there is 0.01 volt across the ammeter. Construct a 0- to 10-amp ammeter with it.

1.22 A 20,000 ohm per volt voltmeter is on the 250-volt range. What is the internal resistance of the voltmeter?

1.23 A 20,000 ohm per volt voltmeter is being used. What is the reading on the 500-volt range?, On the 250-volt range?

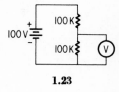

1.24 A 50-μa, 2000-ohm galvanometer is to be used along with a 1.5-volt battery to make an ohmmeter. Find the half-scale reading. Draw the circuit and show component values.

1.23

2.1 Sinusoidal Voltage and Current

2

AC Circuits

An important type of voltage encountered in electronics is the sinusoidal voltage. This is a voltage whose magnitude and polarity are cyclic functions of time. This type of voltage can be reproduced by means of special electronic circuitry or mechanically by means of a generator.

The output of an ac generator will be used to illustrate the parameters of a sinusoidal voltage. The diagram in Fig. 2.1(a) shows a simple ac generator. The potential induced between the two slip rings depends upon the geometry of the coil-magnetic-field system and the angular speed with which the coil rotates through the magnetic field. This potential is given by Faraday's induction law:

$$e = -N \frac{d\Phi}{dt} \qquad (2.1)$$

Where e is the potential in volts, N is the number of loops in the coil, and $d\Phi/dt$ is the time rate of change of the magnetic flux as intercepted by the loop. In terms of the geometry of this generator, and assuming the magnetic field to be constant over the area of the loop,

$$\Phi = BA_\perp$$

where $A_\perp$ is the area of the loop perpendicular to the magnetic field. From Fig. 2.1 (b), it can be seen that the area perpendicular to the B field is given by

$$A_\perp = ab \cos \theta$$

Since ab is the area of the loop,

$$A_\perp = A \cos \theta \qquad \text{and} \qquad \Phi = BA \cos \theta$$

Applying Eq. (2.1),

$$e = NBA \sin \theta \frac{d\theta}{dt}$$

The quantity $d\theta/dt$ is defined as the angular speed and is given the symbol ω:

$$\omega = \frac{d\theta}{dt}$$

The induction law, Eq. (2.1), reduces to the following equation for a loop rotating in a constant magnetic field:

$$e = N\omega BA \sin \theta \qquad (2.2)$$

where ω is the constant, angular speed of the loop, rotating in the magnetic induction, A is the area of the loop, and θ is the angle

42

through which the loop had rotated from its initial position, as indicated in Fig. 2.1.

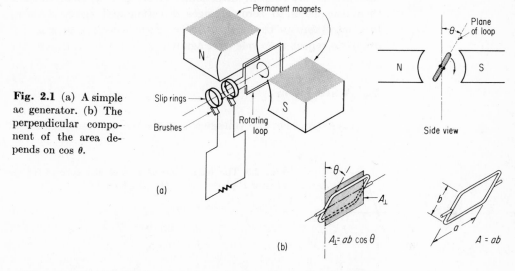

Fig. 2.1 (a) A simple ac generator. (b) The perpendicular component of the area depends on cos θ.

The quantity $N\omega BA$ is a constant, having the units of voltage. Let $N\omega BA = E_m$. The induced voltage can then be written as

$$e = E_m \sin \theta \qquad (2.3)$$

The plot of this equation is shown in Fig. 2.2. The quantity E_m is the maximum value that the function can take. The potential e

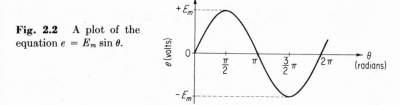

Fig. 2.2 A plot of the equation $e = E_m \sin \theta$.

is positive for the first half-cycle, then negative for the second half-cycle, and cyclically repeats this change in polarity. The significance of the positive and negative signs here is that for the first half-cycle, slip ring A is positive with respect to slip ring B, and during the second half-cycle, slip ring A is negative with respect to slip ring B. If we complete an external circuit between the two slip rings, current will flow in the external circuit. The magnitude of the current will depend upon the voltage e and the resistance in the circuit, R:

$$i = \frac{e}{R} \qquad (2.4)$$

Since the voltage applied to the external circuit changes polarity every half-cycle, the current will do likewise. For one half-cycle the current flows in one direction, increasing to a maximum and then decreasing to zero, changes direction and again increases to a maximum in the new direction, then decreases to zero and begins to repeat the cycle over again.

2.2 Wave Form

For a resistive load the voltage across the resistor and the current

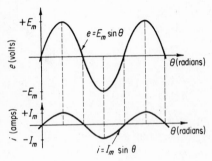

Fig. 2.3 The instantaneous voltage and current for the resistor R placed across the ac generator.

$$e = E_m \sin \theta$$

$$i = \frac{e}{R} + \frac{E_m}{R} = \sin \theta = I_m \sin \theta$$

$$I_m = \frac{E_m}{R}$$

$$i = I_m \sin \theta$$

through the resistor behave in the manner illustrated in Fig. 2.3, since

$$i = \frac{e}{R} \quad \text{and} \quad e = E_m \sin \theta \tag{2.5}$$

Combining, we obtain

$$i = \frac{E_m}{R} \sin \theta \tag{2.6}$$

Both E_m and R are constants, and the ratio E_m/R is in amperes. Equation (2.6) can be rewritten, using the definition $I_m = E_m/R$, as

$$i = I_m \sin \theta$$

So far, the voltage e has been written in terms of the angle θ through which the coil has turned. It is more useful to talk about e as a function of time t. Since ω is a constant, we can write

$$\theta = \omega t \tag{2.7}$$

Making this substitution, we obtain

$$e = E_m \sin \omega t \tag{2.8}$$

$$i = I_m \sin \omega t \tag{2.9}$$

where

$$I_m = \frac{E_m}{R} \tag{2.10}$$

In this case, the graphs for e and i appear identical to the previous ones except for the value of the abscissa, which is now the time t. This shown in Fig. 2.4.

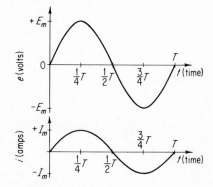

Fig. 2.4 The voltage and current for resistor R plotted as a function of time. T = time for one cycle in seconds.

In these graphs, the time is plotted in terms of T, the period. If ω is known, then numerical values for T can be put on the graph, since

$$T = \frac{2\pi}{\omega} \tag{2.11}$$

In many cases the frequency is known; in this case the period is given by

$$T = \frac{1}{f} \tag{2.12}$$

where f is the frequency in cycles per second.

2.3   Resistive Load

Ohm's law applies for a linear resistance. It is a relationship among applied voltage, resultant current, and resistance. If any two are known, the third quantity is determined. As was shown in Eqs. (2.8) and (2.9), this leads to the equations

$$e = E_m \sin \omega t \quad \text{and} \quad i = I_m \sin \omega t$$

for a resistor. The graphs of these functions, Fig. 2.4, show that both the current and voltage start at zero at the same time, reach maxima at the same time, and have the same sinusoidal shape. The voltage and current under these conditions are said to be in phase. This will always be true for a resistive load, since the relationship among voltage, current, and resistance is a direct proportionality:

$$e = iR$$

2.4 Instantaneous Power

Impressing a sinusoidal voltage across a resistor causes a sinusoidal current to flow in phase with the voltage. The resistor will heat up and power will be dissipated in the resistor. The instantaneous power is given by the product

$$p = ei \tag{2.13}$$

Applying this to the resistor, we obtain

$$p = E_m I_m \sin^2 \omega t \tag{2.14}$$

A graph of the power as a function of time, as given by Eq. (2.14), is shown in Fig. 2.5. As can be seen, the power varies from one instant of time to another.

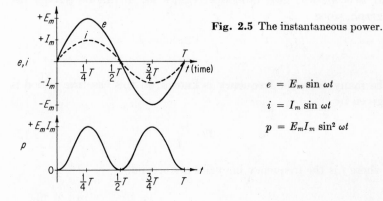

Fig. 2.5 The instantaneous power.

$$e = E_m \sin \omega t$$
$$i = I_m \sin \omega t$$
$$p = E_m I_m \sin^2 \omega t$$

The voltage and current for this case are also shown. The power curve is just the product of the e and i curves. The power curve shows that the resistor dissipates power in pulses at twice the applied frequency.

2.5 Average Power

In a case like this, where the waveform is cyclic, it is useful to talk about the *average power* dissipated in a cycle.

In general, the average value of a function $y = f(x)$ between the points X_1 and X_2 is given as the value of y, which when multiplied by the interval $X_2 - X_1$ will be equal to the area under the curve $f(x)$ in the same interval. This is shown in Fig. 2.6.

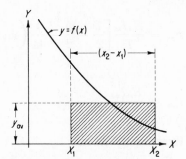

Fig. 2.6 y_{av} between X_1 and X_2.

Mathematically, the areas are equal for the rectangle thus formed and the area under the curve $f(x)$ in the interval $X_2 - X_1$. This can be written as

$$y_{av}(X_2 - X_1) = \int_{X_1}^{X_2} f(x) \, dx \tag{2.15}$$

Solving for y_{av}

$$y_{av} = \int_{X_1}^{X_2} f(x) \, dx / (X_2 - X_1) \tag{2.16}$$

This is the definition of average that will be used to calculate the average power. In terms of the quantities involved, this becomes

$$P_{av} = \int_0^T p \, dt / T \tag{2.17}$$

Example 2.1 Find the average power for the function $p = E_m I_m \sin^2 \omega t$ over 1 cycle. The interval is time, that is, one period, and therefore the integral is with respect to time. Expressing p in Eq. (2.17) as $E_m I_m \sin^2 \omega t$,

$$P_{av} = \frac{1}{T} \int_0^T E_m I_m \sin^2 \omega t \, dt$$

Using a trigonometric identity $\sin^2 \omega t = \frac{1}{2} - \frac{1}{2} \cos 2\omega t$,

$$P_{av} = \frac{E_m I_m}{T} \int_0^T \left(\frac{1}{2} - \frac{1}{2} \cos 2\omega t \right) dt$$

Integrating,

$$P_{av} = \frac{E_m I_m}{T} \left[\frac{t}{2} - \frac{1}{4\omega} \sin 2\omega t \right]_0^T$$

Now

$$\omega = 2\pi/T,$$

Eq. (2.11), and this is used to evaluate the upper and lower limits:

$$P_{\mathrm{av}} = \frac{E_m I_m}{T}\left[\frac{T}{2} - \frac{0}{2} - \frac{1}{4\omega}\left(\sin 2\times\frac{2\pi}{T}\times T - \sin 2\times\frac{2\pi}{T}\times 0\right)\right]$$

$$P_{\mathrm{av}} = \frac{E_m I_m}{2}$$

Effective Value

The 2 in the denominator is usually split up:

$$P_{\mathrm{av}} = \frac{E_m}{\sqrt{2}}\frac{I_m}{\sqrt{2}} = E_{\mathrm{eff}}I_{\mathrm{eff}} \tag{2.18}$$

and the resulting values are called *effective values*. The effective values of e and i for sinusoids are therefore equal to the maximum values divided by the square root of 2.

Mathematically, we can compare the heating effect of a sinusoidal function of power to a dc source; that is, the same heating effect could have been obtained if a battery of voltage E_{eff} were placed across the resistor. This is illustrated in Fig. 2.7.

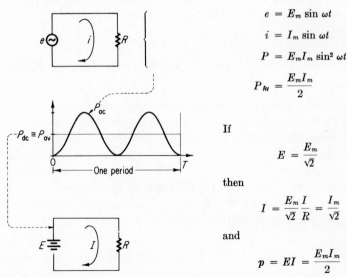

$$e = E_m \sin \omega t$$
$$i = I_m \sin \omega t$$
$$P = E_m I_m \sin^2 \omega t$$
$$P_{\mathrm{Av}} = \frac{E_m I_m}{2}$$

If

$$E = \frac{E_m}{\sqrt{2}}$$

then

$$I = \frac{E_m}{\sqrt{2}}\frac{I}{R} = \frac{I_m}{\sqrt{2}}$$

and

$$p = EI = \frac{E_m I_m}{2}$$

Fig. 2.7 The effective values of e and i related to their maximum values.

2.6   Differential Properties of R, L, and C

An ac network can be thought of as being made up of various combinations of three components: resistance, inductance, and capacitance. The following discussion will deal with the case in which a sinusoidal voltage is applied to these components. The problem, then, will be to find the resultant current. The complicating thing is that, in general, there will be a phase difference between applied voltage and resultant current. The saving factor is that the equations derived can be made to look like Ohm's law in the dc form.

RESISTANCE

For a resistor, the equation obeyed is Ohm's law:

$$e_R = i_R R \tag{2.19}$$

where e_R is in volts, i_R is in amperes, and R is in ohms. If i_R is sinusoidal, e_R must also be sinusoidal, since it is equal to i_R multiplied by a constant R. Therefore, if

$$i_R = I_m \sin \omega t \tag{2.20}$$

then, using Eq. (2.19),

$$e_R = I_m R \sin \omega t \tag{2.21}$$

Since e_R must be a sinusoid, it must have the form

$$e_R = E_m \sin (\omega t + \theta)$$

Comparing the two equations for e_R, it can be seen that

$$E_m = I_m R \quad \text{and} \quad \theta = 0 \tag{2.22}$$

Hence, the voltage across a resistor and the current through it are in phase, that is, $\theta = 0$, and the relationship between amplitudes is given by $E_m = I_m R$, an Ohm's law type of relationship, where E_m and I_m are maximum values.

This illustrates how sinusoids will be discussed, that is, in terms of their amplitudes and phase angles. The frequency is not changed by the differential properties. Therefore, if the amplitude A and phase angle θ are given for a sinusoid of angular frequency ω, the sinusoid can be reconstructed as

$$a = A_m \sin (\omega t + \theta) \tag{2.23}$$

In the examples to follow, A_m and θ are the coefficients to be determined.

INDUCTANCE

The voltage across an indicator depends upon the time rate of change of current in the inductor. As long as the current does not change, even though it is other than zero, it will produce no voltage drop across a pure inductance. Mathematically, this is written in the form

$$e_L = L \frac{di_L}{dt} \tag{2.24}$$

where e_L is in volts, i_L is in amperes, and L is a proportionality factor depending upon geometry and which is called the *inductance*, having the unit of henry.

If a sinusoidal generator is applied across an inductor, both the voltage and current will be sinusoids because the operator d/dt does not change the sinusoid; that is, the derivative of the sine is the cosine, which is sinusoidal, and the derivative of the cosine is a sine function. Therefore, in the equation, $e_L = L(di_L/dt)$, let

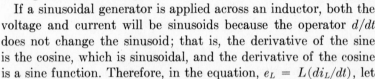

$$i_L = I_m \sin \omega t \tag{2.25}$$

Then, substituting Eq. (2.25) into Eq. (2.24) and differentiating,

$$e_L = \omega L I_m \cos \omega t \tag{2.26}$$

or, using a trigonometric substitution in Eq. (2.26) for $\cos \omega t$,

$$e_L = \omega L I_m \sin \left(\omega t + \frac{\pi}{2} \right) \tag{2.27}$$

Since e_L must be sinusoidal, in general it can be written as

$$e_L = E_m \sin (\omega t + \theta) \tag{2.28}$$

Comparing Eqs. (2.28) and (2.27),

$$E_m = I_m \omega L \quad \text{and} \quad \theta = \frac{\pi}{2} \tag{2.29}$$

The quantity ωL is defined as the *inductive reactance* X_L, and the equations can be written as

$$E_m = I_m X_L \quad \text{and} \quad \theta = \frac{\pi}{2} \tag{2.30}$$

The relationships again are in terms of the maximum values of the sinusoids, the phase angle between voltage and current, and a new quantity that is frequency dependent, namely, the inductive reactance. Solving for the current in Eq. (2.30),

$$I_m = \frac{E_m}{X_L} = \frac{E_m}{\omega L} = \frac{E_m}{2\pi f L} \tag{2.31}$$

The current, I_m, in ac circuits is dependent not only upon the applied voltage E_m, but also on the frequency of the applied voltage, f.

In summary, the relationships between voltage and current for an inductor are

$$i_L = I_m \sin \omega t$$

$$e_L = E_m \sin \left(\omega t + \frac{\pi}{2} \right)$$

This is shown graphically in Fig. 2.8 as a function of time.

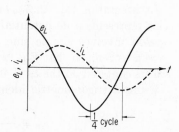

Fig. 2.8 The applied voltage leads the current by a quarter cycle in a pure inductor.

The fact that $\theta = +(\pi/2)$ for e_L, with respect to i_L, means that the voltage comes to a maximum before the current passes through a quarter-cycle. Thus the voltage leads the current in an inductor.

CAPACITANCE

For a capacitor, the voltage is given by

$$e_C = \frac{q}{C} \qquad (2.32)$$

where e_C is in volts, q is in coulombs, and C is a geometrical proportionality factor whose units are farads. Since the charge is continuously changing when a sinusoidal voltage is applied to a capacitor, this constitutes a current, and is given by

$$i_C = \frac{dq}{dt} \qquad (2.33)$$

The charge q can be found by integration of Eq. (2.33):

$$q = \int i_C \, dt \qquad (2.34)$$

Substituting Eq. (2.34) into the first equation, Eq. (2.32),

$$e_C = \frac{1}{C} \int i_C \, dt \qquad (2.35)$$

The defining equation for the capacitor is obtained. To obtain the phase and amplitude relations for a capacitor, let

$$i_C = I_m \sin \omega t \tag{2.36}$$

Then

$$e_C = \frac{1}{C} \int I_m \sin \omega t \, dt \tag{2.37}$$

$$e_C = -\frac{I_m}{\omega C} \cos \omega t \tag{2.38}$$

In integrating Eq. (2.37), one obtains a constant of integration. This represents a dc potential across the capacitor. Since the discussion involves simple sine waves, this dc value is arbitrarily chosen as zero, so that Eq. (2.38) will represent a cosine function that is centered on the $e_C = 0$ axis.

Using a trigonometric identity, Eq. (2.38) can be written as

$$e_C = \frac{I_m}{\omega C} \sin \left(\omega t - \frac{\pi}{2} \right) \tag{2.39}$$

Now we know that e_C must be sinusoidal, so that it must have the form

$$e_C = E_m \sin (\omega t + \theta) \tag{2.40}$$

Comparing Eqs. (3.39) and (2.40), we obtain

$$E_m = \frac{I_m}{\omega C} \quad \text{and} \quad \theta = -\frac{\pi}{2} \tag{2.41}$$

The quantity $1/\omega C$ is defined as X_C, the capacitive reactance. The equation between E_m and I_m now assumes a form like Ohm's law

$$E_m = I_m X_C \quad \text{and} \quad \theta = -\frac{\pi}{2} \tag{2.42}$$

Thus a sinusoidal voltage applied across a capacitor produces a current dependent not only upon the voltage but also upon the frequency of the applied voltage.

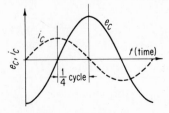

Fig. 2.9 The applied voltage lags the current by a quarter cycle in a pure capacitor.

$$I_m = \frac{E_m}{X_C} = E_m\omega C = 2\pi fC E_m \qquad (2.43)$$

Also, there is a phase difference between e_C and i_C for a capacitor. i_C leads e_C by a quarter-cycle, as shown in Fig. 2.9.

SERIES CIRCUIT

The most general case in a series circuit occurs when all three values (resistance, capacitance, and inductance) are present. The current i is the same for all components, since this is a series circuit. From Kirchhoff's laws

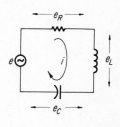

$$e = e_R + e_L + e_C \qquad (2.44)$$

In differential form,

$$e = iR + L\frac{di}{dt} + \frac{1}{C}\int i\,dt \qquad (2.45)$$

Let $i = I_m \sin \omega t$ and substitute into Eq. (2.45):

$$e = I_mR \sin \omega t + I_mX_L \cos \omega t - I_mX_C \cos \omega t \qquad (2.46)$$

Now e must be a sinusoid, so in general we can write

$$e = E_m \sin (\omega t + \theta) \qquad (2.47)$$

Expanding Eq. (2.47) by a trigonometric identity,

$$e = E_m \cos \theta \sin \omega t + E_m \sin \theta \cos \omega t$$

Comparing Eqs. (2.46) and (2.47) and matching coefficients of $\sin \omega t$ and $\cos \omega t$,

$$E_m \cos \theta = I_mR \qquad (2.48)$$

$$E_m \sin \theta = I_m(X_L - X_C) \qquad (2.49)$$

Equations (2.48) and (2.49) can be solved for E_m and θ by the following methods:

(a) Dividing Eq. (2.49) by Eq. (2.48),

$$\tan \theta = \frac{X_L - X_C}{R} \qquad (2.50)$$

This defines θ.

(b) Square Eqs. (2.49) and (2.48) and add the two resulting equations:

$$E_m^2 (\sin^2 \theta + \cos^2 \theta) = I_m^2[R^2 + (X_L - X_C)^2] \qquad (2.51)$$

$$E_m = I_m\sqrt{R^2 + (X_L - X_C)^2} \qquad (2.52)$$

This defines E_m. To put this into an Ohm's law type of formulation, the new quantity, $\sqrt{R^2 + (X_L - X_C)^2}$, is defined as the *impedance* of the circuit. An impedance has reactive terms (X_L, X_C) and resistive terms (R) in it.

$$Z = \sqrt{R^2 + (X_L - X_C)^2} \qquad (2.53)$$

so that

$$E_m = I_m Z \qquad (2.54)$$

For the general case, then, there is a phase angle between the applied voltage and resultant current, given by

$$\theta = \arctan\left(\frac{X_L - X_C}{R}\right) \qquad (2.55)$$

and a relationship between E_m and I_m, which also depends upon X_L, X_C, and R, given by Eq. (2.54).

The graphical solution can also be drawn qualitatively for the series circuit case. Since the current is common for all components, it is in phase with e_R, it leads e_C by a quarter-cycle, and it lags e_L by a quarter-cycle. This is shown in Fig. 2.10.

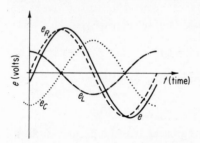

Fig. 2.10 The sum of two or more sinusoids, of the same frequency, is another sinusoid.

Also, $e = e_L + e_C + e_R$, the instantaneous applied voltage. This can be obtained graphically by adding e_L, e_C, and e_R, point by point, to obtain the curve labeled e. The result is a sinusoid. It can be seen that e and i are out of phase by the angle θ, given by $\theta = \arctan\left[(X_L - X_C)/R\right]$. The voltages e_C and e_L are seen to oppose each other, being a half-cycle out of phase with each other.

2.7 Phasor Concept

Using trigonometric identities to aid in adding sinusoidal quantities that are out of phase with each other, such as $e = e_L + e_C + e_R$ in the preceding section, is laborious. It turns out that a simpler method gives the same answer. To lead into this method, it will be instructive to describe the sinusoids in terms of *phasors*.

It is well known that the projection of a point moving on the circumference of a circle with constant speed will move with simple harmonic motion, that is, a sinusoidal motion. If this projection of motion is stretched out in time, it will describe a sinusoidal wave form. Consider a vector pivoting about its end with constant angular velocity ω. If the projection of the motion of the tip of this vector is swept out in time, it will produce a sinusoidal wave form. See Fig. 2.11. This vector, rotating with angular speed ω, is a phasor.

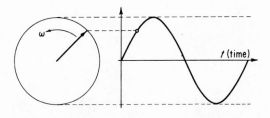

Fig. 2.11 The projection of the top of a phasor results in a sinusoid.

For a resistor, the voltage and current are in phase. This can be shown by having the phasors for e and i lie in the same line as they rotate with angular speed ω. The phasor and sinusoidal concepts for the resistor are shown in Fig. 2.12.

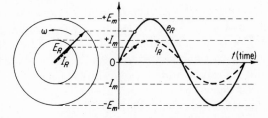

Fig. 2.12 The voltage and current phasors for a pure resistor.

For an inductor the voltage leads the current by a quarter-cycle, or 90 deg, in the phasor concept. The resulting phasor, with e_L ahead of i_L by a quarter-cycle, traces out the correct wave forms. See Fig. 2.13.

Fig. 2.13 The phasors for e and i for an inductor are shown tracing out the instantaneous values of e and i.

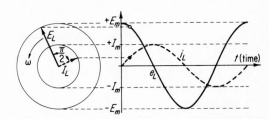

For a capacitor, the current leads the voltage by a quarter-cycle, or 90 deg, in the phasor concept. This is shown in Fig. 2.14.

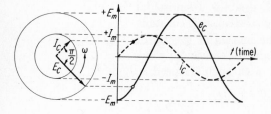

Fig. 2.14 The phasors for a capacitor are at right angles to each other with the current leading the voltage by one-quarter cycle. I_c leads E_c.

For the series circuit containing R, L, and C, the phasor diagram would look, qualitatively, like Fig. 2.15.

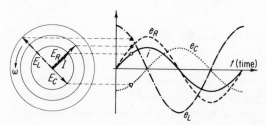

Fig. 2.15 The phasor diagram for a circuit containing pure L, C, and R components.

Once the current phasor I is drawn in, we can draw in the other phasors with the correct directions, since it is known from the previous exercises that E_R and I are in phase; therefore E_R is parallel to I. Also, E_L leads I by $\pi/2$ radians, so E_L is at right angles to I and leading. And E_C lags I by $\pi/2$ radians, so E_C is drawn at right angles to I and lagging. It is also clear that E_L and E_C can be added algebraically: they are collinear.

To show the equivalence of the phasor diagram to the trigonometric solutions obtained in Sec. 2.6, the phasor diagram in Fig. 2.15 will be solved for E, I, and θ.

For the series circuit of R, L, and C, the phasor diagram is shown in Fig. 2.16. For convenience, the phasor is "stopped" when I is horizontal so that we can look at it.

To obtain the applied voltage E, the total voltage (the phasors) will be added vectorially. Since the vectors are either collinear or at right angles, the summation will be done in terms of X and Y components. The components in the Y direction are $E_L - E_C$ The components in the X direction are E_R. Therefore

$$E = \sqrt{E_x{}^2 + E_y{}^2} \qquad (2.56)$$

and

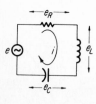

Fig. 2.16 A simple RCL series circuit and its phasor diagram.

$$\theta = \arctan \frac{E_y}{E_x} \qquad (2.57)$$

Equation (2.56) becomes

$$E = \sqrt{E_R{}^2 + (E_L - E_C)^2} \tag{2.58}$$

but

$$E_R = IR$$

$$E_L = IX_L \tag{2.59}$$

$$E_C = IX_C$$

Therefore

$$E = I\sqrt{R^2 + (X_L - X_C)^2} \tag{2.60}$$

or

$$E = IZ \tag{2.61}$$

and from Eq. (2.57), the angle θ becomes

$$\theta = \arctan \frac{E_L - E_C}{E_R} \tag{2.62}$$

Substituting Eq. (2.59) in Eq. (2.62),

$$\theta = \arctan \frac{X_L - X_C}{R} \tag{2.63}$$

These answers are exactly those obtained by trigonometric means in Sec. 2.6. The amazing conclusion is that adding vectors is the same, mathematically, as adding sinusoids of the same frequency.

From now on we shall discontinue the graphical and trigonometric solutions and use the phasor diagrams and vector algebra. The sinusoids can always be constructed, given the correct phasor diagram.

2.8 Average Power when Phase Angle Is Not Zero

For the case where the circuit has an impedance for a load, the applied voltage and resultant current in general are not in phase. The question then is, what is the average power in this case?

The voltage and current can be written for the general case, where the phase angle is not zero, as

$$e = E_m \sin \omega t$$

$$i = I_m \sin (\omega t + \theta) \tag{2.64}$$

The instantaneous power is given by

$$P = ei = E_m I_m \sin \omega t \sin (\omega t + \theta) \tag{2.65}$$

The average value for power over 1 cycle becomes

$$P_{av} = \frac{1}{T} \int_0^T E_m I_m \sin \omega t \sin (\omega t + \theta) \, dt \qquad (2.66)$$

Expanding,

$$P_{av} = \frac{E_m I_m}{T} \int_0^T \sin \omega t \, (\sin \omega t \cos \theta + \cos \omega t \sin \theta) \, dt \qquad (2.67)$$

$$P_{av} = \frac{E_m I_m \cos \theta}{T} \int_0^T \sin^2 \omega t \, dt + \frac{E_m I_m \sin \theta}{T} \int_0^T \sin \omega t \cos \omega t \, dt \qquad (2.68)$$

$$P_{av} = \frac{E_m I_m}{2} \cos \theta \qquad (2.69)$$

$$\frac{E_m}{\sqrt{2}} = E_{eff}$$

$$\frac{I_m}{\sqrt{2}} = I_{eff} \qquad (2.70)$$

where the subscript "eff" signifies effective values.

In terms of the effective values,

$$P_{av} = E_{eff} I_{eff} \cos \theta \qquad (2.71)$$

The quantity $\cos \theta$ is called the *power factor* and is always the cosine of the phase angle between a voltage and a current. In the phasor concept, θ is the angle between the voltage vector and the current vector.

Example 2.2   Given a series circuit having a 20-ohm resistor in series with a 0.5-henry inductance. The applied voltage is 100 volts at 60 cycles.

(a) Draw the phasor diagram for this circuit.

(b) Find the average power dissipated in this circuit.

Note: An understanding fortunately accepted by almost everyone is that in ac problems, *unless otherwise stated*, the quantities I and E are effective values. One of the reasons for this is that ac voltmeters and ammeters are calibrated in terms of effective values rather than of maximum values, and thus the equations written in terms of effective values can be compared directly with measurements. In the problem stated above, $E_{eff} = 100$ volts. For sinusoids, from Eq. (2.70),

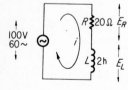

$$E_{eff} = \frac{E_m}{\sqrt{2}}$$

and the two quantities E_{eff} and E_m differ by a simple factor. Thus, if one is known, both are known. We shall conform to this standard. The relations, resembling Ohm's law, are deduced for the maximum values and hold true for the effective values also:

$$I_m = \frac{E_m}{Z}$$

Dividing both sides by $\sqrt{2}$,

$$\frac{I_m}{\sqrt{2}} = \frac{E_m}{\sqrt{2}} \frac{1}{Z}$$

$$I_{\text{eff}} = \frac{E_{\text{eff}}}{Z}$$

or

$$I = \frac{E}{Z}$$

Because these are effective values, now the average power is given simply by

$$P_{\text{av}} = IE \cos \theta$$

In contradistinction, *if maximum values had been used*, the average power would have been written as

$$P_{\text{av}} = \tfrac{1}{2} I_m E_m \cos \theta$$

1. To obtain the phasor diagram, we need E_R, E_L, I, and θ.

2. I can be obtained from the equation $I = E/Z$, where

$$Z = \sqrt{R^2 + X_L^2} \qquad (X_C = 0)$$

Since

$$R = 20 \text{ ohms}, \qquad X_L = 2\pi f L = 60\pi \text{ ohms}$$

then

$$Z = \sqrt{20^2 + 60^2 \pi^2} \cong 186 \text{ ohms}$$

and

$$I = \frac{E}{Z} = \frac{100}{186} = 0.547 \text{ amp}$$

Using Eqs. (2.22), (2.30), and (2.55),

$$E_R = IR = 0.547 \times 20 = 10.9 \text{ volts}$$

$$E_L = IX_L = 0.547 \times 60\pi = 97 \text{ volts}$$

and

$$\theta = \arctan \frac{X_L}{R} = \arctan 8.9$$

$$\theta \cong 83.6° \qquad \text{(the angle between } E \text{ and } I)$$

3. The phasor diagram becomes (arbitrarily setting I on the X axis) as shown in the drawing (left).

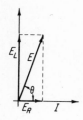

(a) E_R is in phase with I.

(b) E_L leads I by 90 deg.

(c) E leads I by 83.6 deg.

(d) Voltage and current are on separate scales.

4. The average power for the complete circuit is given by

$$P_{av} = EI \cos \theta$$

$$= 100 \times 0.547 \times \cos 83.6° \cong 6.0 \text{ watts}$$

5. Looking at the phasor diagram, it can be seen that

$$E \cos \theta = E_R$$

so that the average power can be written as

$$P_{av} = E_R I$$

Since $E_R = IR$, we obtain $P_{av} = I^2 R$, and all the power is dissipated by the resistor. None is dissipated by the inductor.

6. Comparing the value found in Par. 5 with the one found in Par. 4,

$$P_{av} = I^2 R = (0.547)^2 \times 20 \cong 6.0 \text{ watts}$$

which is the same as in Par. 4 above.

2.9   The Complex Plane

When the phasor diagram is solved, it resembles a vector diagram, and all methods applicable to the solution of vectors can be used. The complex plane method for solution of vector problems has been adopted as the standard method for ac network problems. In this method the vector is written in terms of its components along the real and imaginary axes. The component along the imaginary axis is multiplied by $j = \sqrt{-1}$. This is an excellent bookkeeping device to keep the components separated. In order to obtain the final answers, namely, magnitude and direction for the vectors, several operations must first be defined.

First of all, the vector is identified by stating its components along the real and imaginary axes. The vector I in Fig. 2.17 would thus be written as

$$I = 4 + j3 \tag{2.72}$$

where $j = \sqrt{-1}$. The symbol j is used here rather than i (as in mathematics texts) because the symbol i is reserved for current.

Fig. 2.17 The vector I can be thought of as the sum of two vectors at right angles to each other, I_1 and I_2.

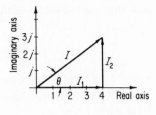

$$j = \sqrt{-1}$$
$$j^2 = -1$$
$$j^3 = -j$$
$$j^4 = 1$$

etc.

Thus the vector I is four units long in the positive real direction and three units long in the positive imaginary direction.

To obtain absolute values (in our work this is what has meaning, since an ac meter will read this value), the complex conjugate of a vector must be defined. To obtain the complex conjugate of a vector, change the signs of all the j terms. Thus, if

$$I = 4 + j3$$

then

$$I^* = 4 - j3$$

(where I^* is the complex conjugate of I).

With this definition, the absolute value of I is given as

$$|I| = \sqrt{II^*} \tag{2.73}$$

In evaluating this square root, the product under the square root is dealt with algebraically:

$$\sqrt{II^*} = \sqrt{(4 + j3)(4 - j3)} = \sqrt{16 + j12 - j12 - j^2 9} \tag{2.74}$$

The cross terms always cancel when a vector is multiplied by its complex conjugate, and since $j^2 = -1$,

$$|I| = \sqrt{II^*} = 5 \tag{2.75}$$

Also,

$$I^2 = II^* = \sqrt{II^*}\sqrt{II^*} = |I|\,|I| \tag{2.76}$$

The angle θ that the vector makes with the real axis is now given by

$$\tan \theta = \frac{\text{Im }(I)}{\text{Re }(I)} = \frac{3}{4} \tag{2.77}$$

where Im (I) is the imaginary component of I and Re (I) is the real component of I.

One of the important functions that can be handled by the complex plane notation is the division or product of phasors. Ohm's law will be used extensively in the solution of ac problems, and this necessitates taking products or dividing one phasor by another. For example,

$$E = IZ \quad \text{or} \quad I = E/Z$$

where I, E, and Z are now written in complex form.

In all calculations, the final form for a vector should be the rationalized form,

$$I = A + jB \tag{2.78}$$

Example 2.3   Given

$$I = a + jb \quad \text{and} \quad Z = c + jd$$

Find E.

Perform the operation indicated by Ohm's law and apply the rules of algebra:

$$E = IZ = (a + jb)(c + jd)$$
$$= (ac - bd) + j(ad + bc)$$

This is of the form $A + jB$, the rationalized form, since in this example

$$A = ac - bd$$
$$B = ad + bc$$

Example 2.4   Given

$$E = a + jb \quad \text{and} \quad Z = c + jd$$

Find I.

Again by Ohm's law,

$$I = \frac{E}{Z}$$

Therefore

$$I = \frac{a + jb}{c + jd}$$

To obtain the rationalized form, multiply top and bottom by the complex conjugate of the denominator. This makes the denominator a real number:

$$I = \frac{a + jb}{c + jd} \cdot \frac{c - jd}{c - jd} = \frac{ac + bd + jbc - jad}{c^2 + d^2}$$

$$= \frac{ac + bd}{c^2 + d^2} + j\,\frac{bc - ad}{c^2 + d^2}$$

This is of the form $A + jB$, the rationalized form, since in this example

$$A = \frac{ac + bd}{c^2 + d^2}$$

$$B = \frac{bc - ad}{c^2 + d^2}$$

Very little of the functions of a complex plane are needed to allow one to solve ac problems. The following rules have been established.

1. Any vector (phasor) can be written in terms of its components along the real and imaginary axes. This is the rationalized form:

$$I = A + jB$$

2. The complex conjugate of a vector is found by changing the signs of all the j terms: if

$$I = A + jB,$$

then

$$I^* = A - jB$$

3. The absolute value of a vector is defined as

$$|I| = \sqrt{II^*}$$

4.

$$I^2 = II^* \quad \text{or} \quad I^2 = |I|\,|I|$$

5. The angle between the vector and the real axis is given by

$$\tan \theta = \frac{I_m(I)}{R_e(I)}$$

6. By algebraic manipulation, any complex vector can be written in rationalized form:

$$I = A + jB$$

2.10 Z as a Complex Quantity

It was established previously that for a series circuit, the voltage across an inductor led the voltage across a resistor by 90 deg, while the voltage across a capacitor lags the voltage across a resistor by 90 deg. With this in mind, resistance is always placed on the real axis, inductive reactance is placed in the $+j$ direction, and the capacitive reactance is placed in the $-j$ direction.

Example 2.5 In the circuit given here, show that X_L and X_C will be at right angles to R.

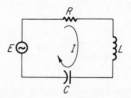

Since this is a series circuit, the individual voltage drops add up to the applied voltage, vectorially:

$$E = E_R + E_L + E_C$$

Now the voltage E_R is at right angles to E_L and E_C.

Since the current I is the same for each component (this is a series circuit), the orthogonality must be due to the vector directions for R, X_L, and X_C. To show that E_L leads E_R by 90 deg and that E_R leads E_C by 90 deg, we write the equation for E in the complex plane:

$$E = E_R + jE_L - jE_C$$

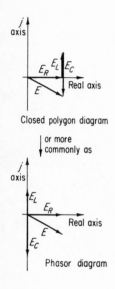

Closed polygon diagram

or more commonly as

Phasor diagram

This is equivalent to the accompanying phasor diagram. A voltage E_R on the positive real axis, a voltage E_L in the positive imaginary direction, and a voltage E_C in the negative imaginary direction are as shown.

Since I is the same for each component in this series circuit, E can be written as

$$E = IR + jIX_L - jIX_C$$
$$= I(R + jX_L - jX_C)$$
$$= IZ$$

where

$$Z = R + jX_L - jX_C$$

In all cases, Z can be factored out in rationalized form, and the real part will be the resistance of the circuit while the imaginary part will be the reactance. Whether the total reactance will be inductive or capacitive depends upon whether the j term is positive or negative.

Example 2.6 Given the circuit in the accompanying diagram, find I, E_L, and E_R, the phase angle θ between total current and voltage, and the power dissipated in the circuit, using calculations in the complex plane.

1. First, Z must be obtained. This can be written down for a series circuit as $Z = R + jX_L$

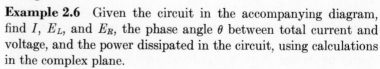

2.

$$I = \frac{E}{Z} = \frac{E}{R + jX_L} = \frac{E}{R + jX_L} \cdot \frac{R - jX_L}{R - jX_L}$$

$$= \frac{ER}{R^2 + X_L^2} - j\frac{EX_L}{R^2 + X_L^2}$$

and

$$|I| = \sqrt{II^*} = \frac{|E|}{\sqrt{R^2 + X_L^2}}$$

3. Since $\sqrt{R^2 + X_L^2} = |Z|$, Ohm's law can also be written in absolute value form:

$$|I| = \frac{|E|}{|Z|}$$

4. To find E_L,

$$E_L = I(jX_L) = \left(\frac{ER}{R^2 + X_L^2} - j\frac{EX_L}{R^2 + X_L^2}\right)(jX_L)$$

$$= \frac{EX_L^2}{R^2 + X_L^2} + j\frac{EX_L R}{R^2 + X_L^2}$$

and

$$|E_L| = \sqrt{E_L E_L^*} = \frac{|E||X_L|}{\sqrt{R^2 + X_L^2}}$$

or

$$|E_L| = |E|\frac{|X_L|}{|Z|}$$

5. E_R becomes

$$E_R = IR = \frac{ER^2}{R^2 + X_L^2} - j\frac{ERX_L}{R^2 + X_L^2}$$

and

$$|E_R| = \sqrt{E_R E_R^*} = \frac{|E||R|}{\sqrt{R^2 + X_L^2}}$$

or

$$|E_R| = |E|\frac{|R|}{|Z|}$$

6. At this point it should be noted that the voltages add vectorially. Therefore,

$$| E | \neq | E_R | + | E_L |$$

but $E = E_R + E_L$. Substituting for the vectors,

$$E = \underbrace{\frac{ER^2}{R^2 + X_L^2} - j \frac{ERX_L}{R^2 + X_L^2}}_{E_R} + \underbrace{\frac{EX_L^2}{R^2 + X_L^2} + j \frac{ERX_L}{R^2 + X_L^2}}_{E_L}$$

The j terms cancel and we obtain

$$E = \frac{ER^2 + EX_L^2}{R^2 + X_L^2} = E$$

an identity.

7. Since we wrote E as $E + j0$, we placed it on the real axis; therefore the angle between E and I is just θ, the angle between I and the real axis. Since

$$I = \frac{ER}{R^2 + X_L^2} - j \frac{EX_L}{R^2 + X_L^2}$$

$$\tan \theta = \frac{\text{Im } (I)}{\text{Re } (I)} = \frac{-X_L}{R}$$

Therefore

$$\cos \theta = \frac{R}{\sqrt{R^2 + X_L^2}}$$

$$\cos \theta = \frac{R}{\sqrt{R^2 + X_L^2}}$$

8. The power dissipated becomes

$$P_{\text{av}} = | I | \; | E | \cos \theta$$

$$= \frac{| E |}{\sqrt{R^2 + X_L^2}} \times | E | \times \frac{R}{\sqrt{R^2 + X_L^2}} = \frac{E^2 R}{R^2 + X_L^2}$$

This can be put into a familiar form by regrouping the quantities in the equation for P_{av}.

$$P_{\text{av}} = \underbrace{\left(\frac{| E |}{\sqrt{R^2 + X_L^2}} \right)}_{| I |} \underbrace{\left(| E | \times \frac{1}{\sqrt{R^2 + X_L^2}} \right)}_{| I |} \times \underbrace{R}_{R} = | I | \; | I | \, R$$

$$= I^2 R$$

Thus all the power is dissipated by the resistor.

2.11 Numerical Examples

Example 2.7 Given the circuit in the accompanying sketch, find the current and show the phasor diagram for this circuit.

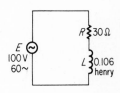

1. To find current, we apply Ohm's law:

$$I = \frac{E}{Z}$$

2. The impedance is found from

$$Z = R + jX_L$$

where $R = 30$ ohms and $X_L = 2\pi f L = 40$ ohms. Therefore

$$Z = (30 + j40) \text{ ohms}$$

3. The voltage E is chosen arbitrarily to lie along the real axis by writing it as

$$E = 100 + j0 \quad \text{or} \quad E = 100 \text{ volts}$$

4. The current I becomes

$$I = \frac{E}{Z} = \frac{100}{30 + j40}$$

and putting this in rationalized form:

$$I = \frac{100}{30 + j40} \times \frac{30 - j40}{30 - j40} = (1.2 - j1.6) \text{ amp}$$

$|I|$ is the value read by the ammeter; it becomes

$$|I| = \sqrt{II^*} = 2 \text{ amp}$$

5. To draw the phasor diagram, we need E_R and E_L.

$$E_R = IR = (1.2 - j1.6) \times 30 = (36 - j48) \quad \text{volts}$$
$$E_L = I(jX_L) = (1.2 - j1.6)j40 = (64 + j48) \quad \text{volts}$$

It can be shown also that

$$|E_R| = 60 \text{ volts}$$
$$|E_L| = 80 \text{ volts}$$

6. It should be noted at this point that the sums of voltmeter readings $|E_R|$ and $|E_L|$ do not add up to $|E|$, since $100 \neq 60 + 80$. But, vectorially, they add up

$$100 = \underbrace{36 - j48}_{E_R} + \underbrace{64 + j48}_{E_L}$$

7. Plotting the vectors E_R, E_L, and I on a common graph gives us the phasor diagram for this circuit.

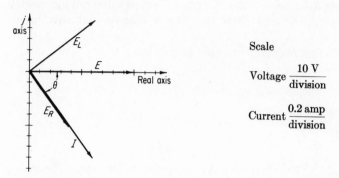

Scale

Voltage $\dfrac{10 \text{ V}}{\text{division}}$

Current $\dfrac{0.2 \text{ amp}}{\text{division}}$

The following facts can be seen from the phasor diagram: The current and the voltage for the resistor, E_R, and I are in phase; the voltage across the inductor leads the current through the inductor by 90 deg, the total voltage E is the vector sum of E_R and E_L; and the phase angle θ, between E and I, is the power-factor angle for the circuit.

The next example is introduced to point out the fact that the components of E can be arbitrarily chosen. As long as the absolute value of E is the same, the results will not change. The phasor diagram will be rotated, but the relative orientation of the vectors will be the same as in the preceding example.

Example 2.8   Given: The circuit shown in the sketch and that $|E| = 100$ volts, which is written as

$$E = 70.7 + j70.7 \quad \text{volts}$$

Find I and the phasor diagram for the circuit.

1. $Z = (30 + j40)$ ohms.

2. $$I = \frac{E}{Z} = \frac{70.7 + j70.7}{30 + j40}$$

$$= (1.98 - j0.28) \text{ amp}$$

and

$$|I| = 2 \text{ amp}$$

(Results are the same as in Example 2.7.)

3. $$E_R = IR = (1.98 - j0.28)30$$

$$= (59.4 - j8.4) \quad \text{volts}$$

$$|E_R| = 60 \quad \text{volts}$$

(Results are the same as in Example 2.7.)

4.
$$E_L = IjX_L = (1.98 - j0.28)j40$$

$$= (11.2 + j79.2) \quad \text{volts}$$

$$|E_L| = 80 \quad \text{volts}$$

(Results are the same as in Example 2.7.)

5. The phasor diagram is shown below.

$$E = 70.7 + j70.7$$
$$E_R = 59.4 - j8.4$$
$$E_L = 11.2 + j79.2$$
$$I = 1.98 - j0.28$$

Scale

Voltage $\dfrac{10 \text{ V}}{\text{division}}$

Current $\dfrac{0.2 \text{ amp}}{\text{division}}$

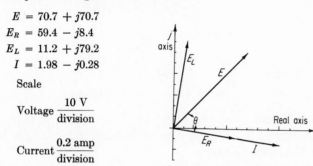

Comparing the absolute values of I, E_R, E_L, and E in Example 2.7 with those of Example 2.8, it can be seen that they are identical. Also, comparing phasor diagrams, it can be seen that these are identical. The only difference is that one diagram is rotated 45 deg with respect to the other. However, the relative angles between vectors and their lengths are exactly the same. Therefore the phasor diagrams are identical, since they will give the same answers.

It should be noted that the power-factor angle in the second example is not so easily obtained now, but it is still the angle between E and I.

2.12 Parallel Circuits

The numerical solution of ac network problems involving parallel circuits is performed in exactly the same manner, algebraically, as dc problems are solved, except that wherever an inductor is involved, the reactance is written as $+jX_L$, and wherever a capacitor is involved, the reactance is written as $-jX_C$.

Example 2.9 Given the parallel circuit shown at the right, find Z_T, the total impedance of the circuit.

R_1 is in series with a parallel combination of L and R_2. The impedance of the parallel combination is given by

$$Z_{11} = \frac{R_2(jX_L)}{R_2 + jX_L}$$

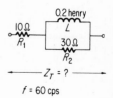

which is the product divided by the sum rule. This is then added to R_1 to obtain Z_T

$$Z_T = R_1 + \frac{jR_2X_L}{R_2 + jX_L}$$

and, numerically, $jX_L = 75.4$ ohms and

$$Z_T = 10 + \frac{j30 \times 75.4}{30 + j75.4}$$

Rationalizing the second term,

$$Z_T = 10 + 25.9 + j10.3$$

and combining,

$$Z_T = (35.9 + j10.3) \text{ ohms}$$

Example 2.10   Applying a 100-volt 60-cycle voltage to the circuit in Example 2.9, find the current through each component and draw the phasor diagram.

1. $Z_T = (35.9 + j10.3)$ ohms

2. $I_T = \dfrac{E_T}{Z_T} = \dfrac{100}{35.9 + j10.3} = (2.57 - j0.74)$ amp

and

$$|\,I_T\,| = 2.67 \text{ amp}$$

You should verify that $|\,I_T\,|$ could have been obtained from $|\,I_T\,| = |E_T|/|\,Z_T\,|$.

3.   Since the total current flows through R_1, this is the current through R_1.

4.   Current through R_2 and L. The rule used to evaluate I_{R_2} and I_L is that the dc circuit equations can be used if the inductive reactance is written as $+jX_L$ and the capacitive reactance is written as $-jX_C$. Applying the rule for two parallel branches, as derived in Chapter 1, that is,

$$I_{R_2} = I_T \left(\frac{jX_L}{R_2 + jX_L} \right)$$

$$I_L = I_T \left(\frac{R_2}{R_2 + jX_L} \right)$$

Putting in numerical values for I_{R_2}

$$I_{R_2} = (2.57 - j0.74) \left(\frac{j75.4}{30 + j75.4} \right)$$

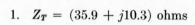

This is rationalized by multiplying numerator and denominator by the complex conjugate of the denominator:

$$I_{R_2} = (2.57 - j0.74) \left(\frac{j75.4}{30 + j75.4} \right) \left(\frac{30 - j75.4}{30 - j75.4} \right)$$

Carrying out the calculation,

$$I_{R_2} = (2.49 + j0.24) \text{ amp}$$

For I_L, the numerical values are

$$I_L = (2.57 - j0.74) \left(\frac{30}{30 + j75.4} \right)$$

Rationalizing,

$$I_L = (2.57 - j0.74) \left(\frac{30}{30 + j75.4} \right) \left(\frac{30 - j75.4}{30 - j75.4} \right)$$

Carrying out the calculations,

$$I_L = (0.097 - j0.98) \text{ amp}$$

5. The absolute values for I_{R_2} and I_L become

$$| I_{R_2} | = 2.49 \text{ amp}$$

$$| I_L | = 0.98 \text{ amp}$$

These are the readings one would obtain if ac ammeters were put in the inductive and resistive branches.

6. To complete the phasor diagram, it is necessary to find the voltage across each component.

(a)

$$E_{R_1} = I_{R_1}R_1 = (2.57 - j0.74)10$$

$$= (25.7 - j7.4) \quad \text{volts}$$

$$| E_{R_1} | = 26.7 \quad \text{volts}$$

(b)

$$E_{R_2} = I_{R_2}R_2 = (2.49 + j0.24)30$$

$$= (74.7 + j7.2) \quad \text{volts}$$

$$| E_{R_2} | = 75.1 \quad \text{volts}$$

(c)

$$E_L = I_L(jX_L) = (0.097 - j0.98)j75.4$$

$$= (74.8 + j7.3) \quad \text{volts}$$

$$| E_L | = 75.2 \quad \text{volts}$$

7. It should be noted here that, within slide-rule error, the

voltages across L and R_2 *are the same.* This must be so because the voltage is the same for all components in parallel. Also, the vector sum of I_{R_2} and I_L must equal I_T.

$$\underbrace{2.57 - j0.74}_{I_T} = \underbrace{2.49 + j0.24}_{I_{R_2}} + \underbrace{0.097 - j0.98}_{I_L}$$

8. The phasor diagram can now be drawn for this parallel circuit, using the vectors found for E_{R_1}, E_{R_2}, E_L, E_T, I_{R_1}, I_{R_2}, I_L, and I_T.

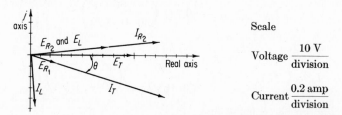

9. As can be seen, the rules for reactances and resistances are borne out. The voltage and current for R_1 are in phase, the voltage and current for R_2 are in phase, and the voltage leads the current by 90 deg for L. Vectorially,

$$E_T = E_{||} + E_{R_1}$$

and

$$I_T = I_L + I_{R_2}$$

The power factor for the complete circuit is the angle between the total current and the total voltage.

10. The power dissipated in this circuit is given by

$$P_{av} = |\,E_T\,|\;|\,I_T\,|\cos\theta$$

Numerically, this becomes

$$P_{av} = 100 \times 2.67 \times \frac{2.57}{2.67} = 257 \text{ watts}$$

Compare this to the I^2R losses in the two resistors. For R_1,

$$P_{av} = I_{R_1}^2 R_1 = (2.67)^2 \times 10 = 71.3 \text{ watts}$$

and for R_2,

$$P_{av} = I_{R_2}^2 R_2 = (2.49)^2 \times 30 = 186 \text{ watts}$$

Total power dissipated by the resistors $\cong 257$ watts. Thus the total power dissipated always can be found by summing the power dissipated in each resistor. The reactances dissipate no power.

The solution of ac networks will consist of reductions by Ohm's law to find the voltages and currents for various components and branches. In general, all the dc laws will apply for series and parallel components if the inductive reactance is written as $+jX_L$ and the capacitive reactance is written as $-jX_C$ whenever each appears in a circuit. The calculations are performed in the complex plane, and if left in vector form can be graphed as phasor diagrams.

To obtain the effective values that ammeters and voltmeters would read if placed in the circuit, the absolute values of the vectors are found. To find the power dissipated in the circuit, the cosine of the angle between total current and total voltage is found from the phasor diagram. Therefore, knowing the absolute values and phase angles involved, the solution is completed.

SUMMARY

Complex voltage or current	Mathematical trick used to solve ac circuit problems
Effective value of voltage or current	Value read on an ac voltmeter or ammeter Absolute magnitude of the complex voltage or current
Maximum value of voltage or current	Absolute magnitude of the complex voltage or current multiplied by $\sqrt{2}$
Actual instantaneous voltage or current	Can be constructed from the complex quantity:

$$e = \mid E \mid \sqrt{2}\ \sin\left[\omega t + \tan^{-1}\frac{\mathrm{Im}\,(E)}{\mathrm{Re}\,(E)}\right]$$

$$i = \mid I \mid \sqrt{2}\ \sin\left[\omega t + \tan^{-1}\frac{\mathrm{Im}\,(I)}{\mathrm{Re}\,(I)}\right]$$

Problems

2.1 What is the inductive reactance of a 100-mh coil at (a) 1,000 cps, (b) 1500 cps, (c) 2000 cps?

2.2 What is the capacitive reactance of a 0.1-μf capacitor at (a) 1000 cps, (b) 1500 cps, (c) 2000 cps?

2.3 Write down the complex impedance of each circuit. Assume $f = 100$ cps.

(a)

5   0.1 h

(b)

100   10 µf

(c)

0.1 h   5 µf

(d)

100   10 µf   0.3 h

(e)

100   5 µf   0.1 h   15 µf

2.4 (a) Find the sum of the following complex numbers:

$$A = 3 + j3$$
$$B = 5 - j10$$
$$C = -12 + j12$$

(b) Graph the individual numbers in the complex plane and find their sum graphically.

(c) Find the magnitude of each number and the sum.

2.5 Perform the indicated operations on the complex numbers shown:

$$A = 5 + j4$$
$$B = 3 + j6$$

(a) $A + B$ (e) AB (i) AA^*
(b) $A + B^*$ (f) AB^* (j) $-A/B$
(c) $A - B$ (g) A/B (k) $-A^*/B$
(d) $A - B^*$ (h) A/B^* (l) A^2

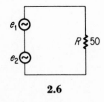

2.6

2.6 In the circuit shown, the voltages are given by

$$e_1 = 50 \sin (\omega t + \pi/6)$$
$$e_2 = 30 \sin \omega t$$

(a) Sketch $e_1 + e_2$.

(b) Find the average power dissipated in R.

2.7 Given the complex $I = 3 + j4$, determine the instantaneous current i.

2.8 In a given circuit, the complex current and voltage are given as $E = (5 + j10)$ volts and $I = 1$ amp. Rewrite these quantities with the voltage as reference, that is, $E = A + j0$, $I = B + jC$.

2.9 For a given reactive circuit, $e = 100 \sin (\omega t + \pi/6)$ volts and $i = (0.1 \sin \omega t)$ amp. Determine E and I in the complex form.

2.10 Using the complex form, calculate the total capacitance of three capacitors in parallel, C_1, C_2, and C_3. (*Hint:* Calculate the total impedance.)

2.11 In a given circuit, the complex current is given by $I = 3 + j6$ and the impedance is given by $Z = 8 + j20$.
(a) Find the complex voltage.
(b) Find the magnitude of the voltage.
(c) Find the power factor.

2.12 In a given circuit, the complex voltage is given by $E = 24 + j6$ and the impendance of the circuit is $Z = 3 + j4$.
(a) Find the complex current.
(b) Find the current that an ac ammeter will read.
(c) Find the power dissipated.

2.13 A 100-mh coil is placed in series with a 0.1-μf capacitor. What is the impedance of the series circuit at (a) 1000 cps, (b) 1500 cps, (c) 2000 cps?

2.14 A 100-mh coil is placed in parallel with a 0.1-μf capacitor. What is the impedance of this parallel circuit at (a) 1000 cps, (b) 1500 cps, (c) 2000 cps?

2.15 A voltage whose effective value is 110 volts, 60 cps, is applied to a 500-mh coil that has a dc resistance of 50 ohms.
(a) Draw the phasor diagram.
(b) Determine the power dissipated.
(c) Determine the current.

2.16 Two coils A and B are in parallel. A 100-volt, 60-cycle is impressed across this parallel combination. Coil A has an inductance of 100 mh and dc resistance of 20 ohms. Coil B has an inductance of 500 mh and a dc resistance of 40 ohms.
(a) Draw the phasor diagram.
(b) Find I_T, I_A, and I_B.
(c) Determine the power dissipated (two ways).

2.17 Determine I, E, Z, and P in complex form.

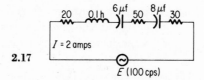

2.17

2.18 Coil A has a resistance of 50 ohms and inductance of 0.5 henry. Coil B has a resistance of 20 ohms and inductance of 0.2 henry. Find the impedance of the two coils, when in series and when in parallel, for a 100-cycle signal. Express the answers in complex form and as absolute values.

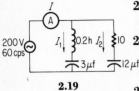
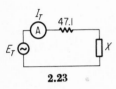

2.19

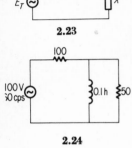

2.23

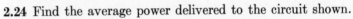

2.24

2.19 In the circuit shown, find the value of I that an ac ammeter would read.

2.20 A 110-volt, 60-cycle source is placed across a 200-ohm resistor in series with an unknown capacitor. An ac ammeter reads 0.1 amp. Find the capacitance.

2.21 Three pure components, R, L, and C, are in series, connected across a 110-volt, 60-cycle source. Voltmeter readings show 50 volts across R, 60 volts across L, and 80 volts across C. An ammeter reads 2 amp. Determine R, L, and C.

2.22 A 110-volt, 60-cycle source is placed across a 600-ohm resistor in series with a coil having inductance and resistance. An ac ammeter reads 0.1 amp. An ac voltmeter measures 60 volts across the resistor and 70 volts across the coil. Find L, the inductance, and R_L, the dc resistance, of the coil.

2.23 For the circuit shown on the left find X. It is capacitive or inductive; if capacitive, find C; if inductive, find L.

2.24 Find the average power delivered to the circuit shown.

2.25 A 50-ohm resistor dissipates 10 watts for a given dc current. Calculate the effective value of the ac current that will produce the same dissipation. What is the maximum value of the ac current?

2.26 Determine the effective value of a voltage given by

$$e = 20 \sin \omega t + 50 \sin (\omega t + \pi/4)$$

2.27 A rectangular voltage, as shown, is applied to a 200-ohm resistor. Plot the instantaneous power and determine the average power.

2.27

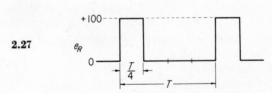

2.28 A triangular voltage, as shown, is applied to a 200-ohm resistor. Find the average power dissipated by the resistor. Plot the instantaneous power for a cycle of the voltage, e_R.

2.28

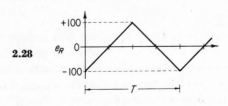

3.1 Introduction

There are only three basic components in electronics: resistance, inductance, and capacitance. Any circuit, no matter how simple or complicated, is made up of combinations of these three components. The fact that two of these components have characteristics that are frequency-dependent and introduce phase shifts gives rise to the variety of electronic circuits in use. This chapter introduces a certain family of circuits called *filters*. These circuits have the ability to pass or reject a band of frequencies. We shall deal with the electrical characteristics of some of the more common filter circuits.

Filters and Tuned Circuits

3.2 Frequency Response

A pure resistance, by definition, does not vary with frequency. A consequence of this is that the applied voltage and resultant current for a resistor are in phase. The two reactive components, inductance and capacitance, introduce phase shifts between applied voltage and resultant current. As a result, inductive reactance X_L and capacitive reactance X_C are functions of the applied frequency. This means that the ac current flowing through a reactive component depends not only upon the applied voltage, as is true for dc circuits, but also upon the applied frequency.

INDUCTANCE REACTANCE

The inductive reactance for a component is given by

$$X_L = 2\pi f L \tag{3.1}$$

where L is the inductance of the component and f is the frequency. The Ohm's law relationship for an inductance, relating the effective values of applied voltage and resultant current, is given by

$$E_{\text{eff}} = I_{\text{eff}} X_L \tag{3.2}$$

So, for an applied voltage E, the resultant current becomes

$$I_{\text{eff}} = \frac{E_{\text{eff}}}{X_L} = \frac{E_{\text{eff}}}{2\pi f L} \tag{3.3}$$

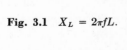

Fig. 3.1 $X_L = 2\pi f L$.

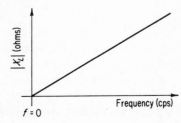

and the resultant current is therefore proportional to E/f; that is, the current depends not only upon the applied voltage, but on *frequency* too. For an inductor, the reactance increases linearly with frequency. Figure 3.1 shows the plot of X_L versus f. The fact that the inductive reactance increases with frequency means that, for a given applied voltage, an inductor will have less current flowing through it at higher frequencies than for lower frequencies.

Example 3.1 Find the current in the following circuit. Assume that the amplitude of the voltage does not change but that the frequency is varied from 60 to 240 cps.

1. Using the Ohm's law relation for the effective values,

$$I_{\text{eff}} = \frac{E_{\text{eff}}}{X_L} = \frac{E_{\text{eff}}}{2\pi f L}$$

2. Therefore $I_{\text{eff}} = 15.9/f$ amp.

3. The solution can be summarized in the form of a curve:

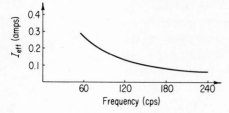

f	I
60	0.266
120	0.133
180	0.089
240	0.067

As can be seen, the current decreases from 0.266 amp at 60 cycles per second (cps) even though the applied voltage is held constant at 100 volts. An inductor, then, tends to retard high frequencies (offer a higher reactance to the current).

Until now, nothing has been said about phase shift. For a pure inductor, the current wave form is shifted with respect to the applied voltage so that the voltage leads the current by 90 deg in the phasor diagram.

CAPACITIVE REACTANCE

The capacitive reactance is given by

$$X_C = \frac{1}{2\pi f C} \tag{3.4}$$

where C is the capacitance in farads and f is the applied frequency in cycles per second. The curve in Fig. 3.2 shows how the capacitive reactance varies with frequency. As can be seen, the reactance is infinite at $f = 0$; that is, a capacitor does not pass dc current, but acts like an open circuit. For higher frequencies, however, the reactance goes down, approaching zero asymptotically.

Fig. 3.2 $X_C = \dfrac{1}{2\pi fC}.$

$|X_C|$ (ohms)

$f = 0$ Frequency (cps)

Example 3.2 Find the current through the capacitor for an applied voltage of 100 volts and frequencies of 60, 200, and 1000 cps.

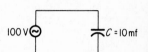

1. Using the Ohm's law relationship for effective values,

$$I_{\mathrm{eff}} = \frac{E_{\mathrm{eff}}}{X_C} = 2\pi EfC$$

2. Putting in numerical values,

$$I_{\mathrm{eff}} = 6.28 \times 10^{-3} f \quad \mathrm{amp}$$

3.

$$I_{\mathrm{eff}} = 0.377 \ \mathrm{amp \ at \ 60 \ cps}$$

$$= 1.26 \ \mathrm{amp \ at \ 200 \ cps}$$

$$= 6.38 \ \mathrm{amp \ at \ 1000 \ cps}$$

The results show that the current increases with increase in the frequency of the applied voltage, even though the amplitude of the voltage remains constant. This is due to the frequency dependence of X_C. The phase shift for a pure capacitance is such that the applied voltage lags the resultant current by 90 deg in the phasor diagram for the circuit.

Interesting effects come about when capacitors and inductors are present in the same circuit, since one component makes the voltage lead by 90 deg while the other makes the voltage lag by 90 deg. This will be the subject of the section on tuned circuits.

3.3 Low-Pass Filters

Filters are networks composed of resistive and reactive components. Their function is to pass or reject a band of frequencies. This characteristic of band rejection or admission is very important in electronic circuits. The low-pass filter is designed to pass all frequencies below a certain limit, called the *cutoff frequency*. Since the frequency characteristics of a filter do not change discontinuously, but vary smoothly from admission to rejection, the cutoff frequency has to be defined. Two simple low-pass filters will be considered, one containing an inductor and resistor, and the other containing a capacitor and resistor.

RL Filter

Because a filter has reactive components, the output voltage for such a device will depend upon two things, the input voltage, and the frequency of the input voltage. Also, a phase shift between input and output voltage will be introduced. The output voltage will always be directly proportional to the input voltage, so in measuring the characteristics of a filter network, the amplitude of the input voltage is kept constant and only the frequency is changed.

The two quantities that can be measured in an experiment of this type are (1) the output voltage as a function of frequency, and (2) the phase shift between input and output voltage. A schematic diagram illustrating the experimental setup used in measuring the characteristics of a filter is shown in Fig. 3.3.

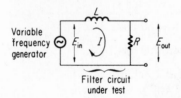

Variable frequency generator

Filter circuit under test

Fig. 3.3 Testing a simple *RL* filter.

The variable frequency generator is a sinusoidal oscillator whose voltage (E_{in}) is held constant as the frequency is varied. The output of the circuit under test (E_{out}) is measured with a high-input impedance voltmeter, to reduce loading.

This circuit is simple enough to lend itself to an analytical examination of its characteristics. The calculations are performed in the complex plane, and the answer is given in terms of the absolute value of E_{out} and the phase shift angle between input and output voltage. There are several ways in which these two quantities can be shown for a given filter. Two of the most common methods will be used in the examples. From the circuit of Fig. 3.3, consisting of L and R, the following equations result:

$$E_{in} = E_L + E_R \qquad (3.5)$$

$$E_{out} = E_R \qquad (3.6)$$

In terms of I, X_L, and R, the equations become

$$E_{in} = jIX_L + IR \qquad (3.7)$$

$$E_{out} = IR \qquad (3.8)$$

Dividing Eq. (3.8) by Eq. (3.7), to eliminate I, and solving for E_{out},

$$E_{\text{out}} = E_{\text{in}} \left(\frac{R}{R + jX_L} \right) \tag{3.9}$$

From this we can obtain the two quantities of interest: $|E_{\text{out}}|$ and the phase shift angle θ.

In this equation, E_{in} is written as

$$E_{\text{in}} = E_{\text{in}} + j0$$

Rationalizing,

$$E_{\text{out}} = E_{\text{in}} \left(\frac{R}{R + jX_L} \right) \left(\frac{R - jX_L}{R - jX_L} \right) \tag{3.10}$$

$$= \frac{E_{\text{in}}R}{R^2 + X_L{}^2} (R - jX_L) = A - jB \tag{3.11}$$

where

$$A = \frac{E_{\text{in}}R^2}{R^2 + X_L{}^2} \quad \text{and} \quad B = -\frac{E_{\text{in}}RX_L}{R^2 + X_L{}^2} \tag{3.12}$$

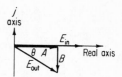

In the phasor representation, then, the angle θ measures the phase shift between E_{in} and E_{out} and shows that E_{out} lags E_{in}. The phase-shift angle is given by

$$\tan \theta = \frac{B}{A} = -\frac{X_L}{R} \tag{3.13}$$

$$\theta = \arctan \left(-\frac{X_L}{R} \right) = \arctan \left(-\frac{2\pi fL}{R} \right) \tag{3.14}$$

Also,

$$|E_{\text{out}}| = \frac{|E_{\text{in}}|}{\sqrt{1 + (X_L/R)^2}} = \frac{|E_{\text{in}}|}{\sqrt{1 + (2\pi fL/R)^2}} \tag{3.15}$$

As can be seen, both $|E_{\text{out}}|$ and θ are functions of the frequency f. Before going on to numerical examples, it will be instructive to examine the results qualitatively, to obtain some idea as to how the functions vary with frequency. This can be done by talking about extremes, that is, very low frequencies and very high frequencies. It can be seen by inspection that $|E_{\text{out}}|$ decreases with an increase in frequency, since in Eq. (3.15), the output voltage is of the form

$$|E_{\text{out}}| = \frac{A}{\sqrt{1 + Bf^2}} \tag{3.16}$$

so that as f increases, $|E_{out}|$ decreases. For $f = 0$, $|E_{out}| = |E_{in}|$, while as $f \to \infty$, $|E_{out}| \to 0$. In fact, the shape of the $|E_{out}|$ versus frequency curve has the form shown in Fig. 3.4.

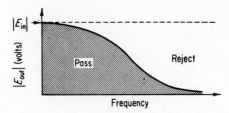

Fig. 3.4 E_{out} as a function of frequency for an RL, low-pass filter.

The phase shift between E_{out} and E_{in} can be examined qualitatively by looking at Eq. (3.14).

$$\theta = \arctan\left(-\frac{2\pi L}{R} f\right) \tag{3.17}$$

As $f \to 0$, $\theta \to 0$, and as $f \to \infty$, $\theta \to -90$ deg. Therefore, at low frequencies, which are essentially the frequencies passed, the phase shift is small, whereas for the highly attenuated frequencies, the high frequencies, the phase shift approaches -90 deg.

Cutoff Frequency

Since the attenuation changes smoothly from zero to maximum, there is no definite frequency beyond which one could say that all higher frequencies are rejected. All frequencies are passed; however, the higher frequencies are highly attenuated while the lower ones pass relatively unaffected. Yet a glance at Fig. 3.4 shows that for practical purposes, only a band of frequencies is allowed through the filter. For this reason, a cutoff frequency, f_{co}, is defined. The cutoff frequency is defined as the frequency at which the output voltage is reduced to such a value that the output power is reduced to one-half its maximum value.

For cutoff frequency,

$$P_{out} \text{ (cutoff)} = \tfrac{1}{2} P_{out} \text{ (max)} \tag{3.18}$$

The output power can be put in terms of the output voltage and load resistance R_L. In this case the load resistance R_L is the internal impedance of the voltmeter measuring the output voltage:

$$P_{out} = \frac{E_{out}^2}{R_L} \tag{3.19}$$

Using this result in Eq. (3.18), the definition becomes

$$\frac{E_{out}^2 \text{ (cutoff)}}{R_L} = \frac{1}{2} \frac{E_{out}^2 \text{ (max)}}{R_L} \tag{3.20}$$

Substituting the value of $| E_{\text{out}} |$, defined in Eq. (3.15), for the RL circuit illustrated in Fig. 3.3,

$$\frac{| E_{\text{in}} |^2}{R_L[1 + (2\pi f_{\text{co}}L/R)^2]} = \frac{1}{2}\frac{| E_{\text{in}} |^2}{R_L} \qquad (3.21)$$

Solving for f_{co},

$$f_{\text{co}} = \frac{R}{2\pi L} \qquad (3.22)$$

Using this definition for f_{co}, Eq. (3.15) can be rewritten as

$$| E_{\text{out}} | = \frac{| E_{\text{in}} |}{\sqrt{1 + (f/f_{\text{co}})^2}} \qquad (3.23)$$

and the phase angle, as given in Eq. (3.17), can be written as

$$\theta = \arctan\left(-\frac{f}{f_{\text{co}}}\right) \qquad (3.24)$$

PLOTTING RESULTS

The quantities $| E_{\text{out}} |$ and θ can be represented in either of two ways. In the first method, each is plotted as a function of frequency. This allows one to see the "spectrum" of frequencies admitted or rejected by a given filter. A typical graph is shown in Fig. 3.5. To accommodate the large frequency range usually needed to represent $| E_{\text{out}} |$ and θ, a semilog scale is used. In order to give as much information as possible, both $| E_{\text{out}} |$ and θ have been superimposed on the same frequency scale. One uses the ordinate scale on the left in Fig. 3.5 for $| E_{\text{out}} |$, and the ordinate scale on the right for values of θ.

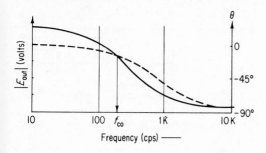

Fig. 3.5

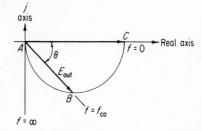

Fig. 3.6 The output voltage plotted in the complex plane for a low-pass filter.

In the second method, rather than plotting $| E_{\text{out}} |$ and θ, Eq. (3.11) is plotted in the complex plane. A qualitative representation of such a graph is shown in Fig. 3.6.

The curved line represents the locus of points given by the

magnitudes of the E_{out} vectors. The vector AC in Fig. 3.6, for example, represents E_{out} for $f = 0$. It shows that E_{out} is in phase with E_{in}, that is, $\theta = 0$. The magnitude of E_{out} is given by the length of the vector AC. Another vector AB is shown in Fig. 3.6. This is E_{out} at f_{co}, the cutoff frequency. For this case, $\theta = 45$ deg, and the length of the vector AB gives the magnitude of E_{out} at this frequency.

As can be seen, each method of graphing has its own merits. The graph versus frequency shows the band width better, while the method of summarizing Eq. (3.11), as shown in Fig. 3.6, gives information on magnitude and phase shift at a glance.

Example 3.3 Find E_{out} for an RL low-pass filter where $L = 0.1$ henry and $R = 10$ K.

1. The cutoff frequency is determined first.

$$f_{co} = \frac{R}{2\pi L} = 15.9 \text{ kc}$$

2. The equations for $| E_{out} |$ and θ become

$$| E_{out} | = \frac{| E_{out} |}{\sqrt{1 + (f/15.9 \times 10^3)^2}}$$

$$\theta = \arctan \left(-\frac{f}{15.9 \times 10^3} \right)$$

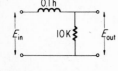

where f is in cycles per second (cps).

3. These results can be plotted on semilog paper as shown here.

4. The answer can be plotted in the complex plane also. This is shown below. Some of the frequencies are also shown.

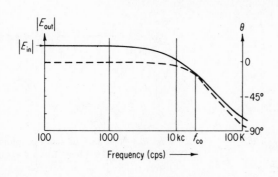

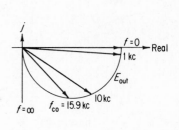

RC Filter

A low-pass filter can be constructed utilizing a capacitor and resistor. Figure 3.7 shows the circuit used.

Writing E_{in} as

$$E_{in} = E_{in} + j0 \qquad (3.25)$$

places it on the real axis of the complex plane, in terms of which the voltage E_{out} becomes

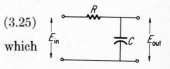

Fig. 3.7 An RC low-pass filter.

$$E_{out} = E_{in}\left(\frac{-jX_C}{R - jX_C}\right) \qquad (3.26)$$

Dividing top and bottom by $-jX_C$,

$$E_{out} = E_{in}\left[\frac{1}{1 + j(R/X_C)}\right] \qquad (3.27)$$

Rationalizing,

$$E_{out} = E_{in}\left[\frac{1}{1 + j(R/X_C)}\right]\left[\frac{1 - j(R/X_C)}{1 - j(R/X_C)}\right] \qquad (3.28)$$

$$= E_{in}\left[\frac{1 - j(R/X_C)}{1 + (R/X_C)^2}\right]$$

From this, $\mid E_{out} \mid$ and θ can be obtained:

$$\mid E_{out} \mid = \sqrt{E_{in}^2\left[\frac{1 - j(R/X_C)}{1 + (R/X_C)^2}\right]\left[\frac{1 + j(R/X_C)}{1 + (R/X_C)^2}\right]} \qquad (3.29)$$

$$= \frac{\mid E_{in} \mid}{\sqrt{1 + (R/X_C)^2}} = \frac{\mid E_{in} \mid}{\sqrt{1 + (2\pi fRC)^2}}$$

$$\theta = \arctan\left[\frac{-(R/X_C)}{1}\right] = \arctan\left(-\frac{R}{X_C}\right) \qquad (3.30)$$

$$= \arctan(-2\pi fRC)$$

The half-power point can also be calculated in order to find f_{co}:

$$P_{out}\text{ (cutoff)} = \tfrac{1}{2}P_{out}\text{ (max)} \qquad (3.31)$$

$$\frac{\mid E_{in} \mid^2}{R_L[1 + (2\pi f_{co}RC)^2]} = \frac{1}{2}\frac{\mid E_{in} \mid^2}{R_L} \qquad (3.32)$$

$$f_{co} = \frac{1}{2\pi RC} \qquad (3.33)$$

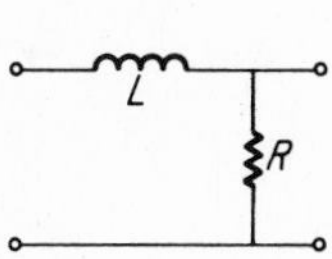

$$f_{co} = \frac{R}{2\pi L}, \qquad \frac{\mid E_{out} \mid}{\mid E_{in} \mid} = \frac{1}{\sqrt{1 + (f/f_{co})^2}}$$

$$f_{co} = \frac{1}{2\pi RC}, \qquad \frac{\mid E_{out} \mid}{\mid E_{in} \mid} = \frac{1}{\sqrt{1 + (f/f_{co})^2}}$$

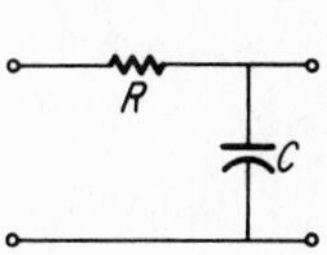

Fig. 3.8 Low-pass filters.

3.4 High-Pass Filters

By rearranging the components used in the low-pass filters, the characteristics can be changed so that the low frequencies will be rejected and only the high frequencies will pass through, with little attenuation. Two high-pass filters with typical characteristics are shown in Fig. 3.9. The analytical expression for $| E_{out} |/| E_{in} |$ and θ, the phase angle between E_{in} and E_{out} are obtained as in Sec. 3.3.

RL High-Pass Filter

For the RL high-pass filter shown in Fig. 3.9, the output voltage can be written as

$$E_{out} = E_{in} \left(\frac{jX_L}{R + jX_L} \right) \tag{3.34}$$

Rationalizing,

$$E_{out} = \frac{E_{in}}{1 + (R/X_L)^2} \left(1 + j \frac{R}{X_L} \right) \tag{3.35}$$

From this equation, θ and $| E_{out} |/| E_{in} |$ can be obtained in the form of Eqs. (3.36) and (3.37)

$$\frac{| E_{out} |}{| E_{in} |} = \frac{1}{\sqrt{1 + (R/X_L)^2}} \tag{3.36}$$

$$\theta = \arctan \frac{R}{X_L} \tag{3.37}$$

The two equations, Eqs. (3.36) and (3.37), can be put into a simpler form if the cutoff frequency is defined.

For cutoff, the ratio of $| E_{out} |^2$ at cutoff to that for $| E_{out} |^2$ at maximum must be equal to $\frac{1}{2}$:

$$\frac{| E_{out} |^2 \text{ (cutoff)}}{| E_{out} |^2 \text{ (max)}} = \frac{1}{2} \tag{3.38}$$

Therefore, from Eq. (3.36),

$$\frac{1}{1 + (R/X_L)^2} \text{ (cutoff)} = \frac{1}{2} \tag{3.39}$$

In terms of frequency,

$$\frac{1}{1 + (R/2\pi f_{co}L)^2} = \frac{1}{2} \tag{3.40}$$

$$\frac{R}{2\pi f_{co}L} = 1 \tag{3.41}$$

$$f_{co} = \frac{R}{2\pi L} \tag{3.42}$$

This is the same value as that found for the low-pass filter, using the same components, where the positions of L and R were interchanged.

Equations (3.36) and (3.37) can now be written as

$$\frac{|E_{out}|}{|E_{in}|} = \frac{1}{\sqrt{1 + (f_{co}/f)^2}} \tag{3.43}$$

$$\theta = \arctan\left(\frac{f_{co}}{f}\right) \tag{3.44}$$

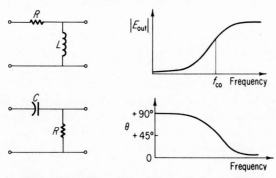

Fig. 3.9 High-pass filters and their characteristics.

RC High-Pass Filter

From Fig. 3.9 the output voltage for the RC high-pass filter can be written as

$$E_{out} = E_{in}\left(\frac{R}{R - jX_C}\right) \tag{3.45}$$

In rationalized form, this becomes

$$E_{out} = \frac{E_{in}}{1 + (X_C/R)^2}\left(1 + j\frac{X_C}{R}\right) \tag{3.46}$$

From this equation, the values for $|E_{out}|/|E_{in}|$ and θ become

$$\frac{|E_{out}|}{|E_{in}|} = \frac{1}{\sqrt{1 + (X_C/R)^2}} \qquad (3.47)$$

$$\theta = \arctan\left(\frac{X_C}{R}\right) \qquad (3.48)$$

Using the definition in Eq. (3.38), the cutoff frequency is given by

$$f_{co} = \frac{1}{2\pi RC} \qquad (3.49)$$

This is also the same value as that obtained for an RC low-pass filter, where R and C were interchanged. Equations (3.47) and (3.48) can now be written as

$$\frac{|E_{out}|}{|E_{in}|} = \frac{1}{\sqrt{1 + (f_{co}/f)^2}} \qquad (3.50)$$

$$\theta = \arctan\left(\frac{f_{co}}{f}\right) \qquad (3.51)$$

The two equations, Eqs. (3.43) and (3.50), can be graphed in a generalized form if the ordinate is given in terms of the ratio $|E_{out}|/|E_{in}|$ and the frequency is plotted in multiples of the cutoff frequency f_{co}.

The generalized plots for the frequency response and the phase angle are shown in Fig. 3.9. As can be seen, the low frequencies are highly attenuated, whereas the high frequencies are relatively unattenuated and shifted very little in phase in going through the filter.

Mechanistically, we can explain the behavior of these simple circuits by observing that for the RL filter, the reactance of L increases with frequency, and is zero at $f = 0$. Therefore, at low frequencies, $X_L \ll R$ and most of the input voltage is dropped across R. As the input frequency is increased, X_L increases and more and more voltage is dropped across X_L, relative to R, until at very high frequencies, $X_L \gg R$ and the output voltage is approximately equal to the input voltage.

The reasoning is carried out in the same manner for the RC filter. At low frequencies, $X_C \gg R$, and most of the input voltage is dropped across X_C. At the higher frequencies, $X_C \ll R$ and offers very little impedance to the ac current. Therefore, at high frequencies, the output voltage, which is the voltage across R, approaches E_{in} in amplitude.

GENERAL RESULTS FOR HIGH- AND LOW-PASS FILTERS

Because the frequency characteristics can be put in terms of the cutoff frequency f_{co}, general curves can be drawn which will be good for any RL or RC filter. These curves will be graphed in terms of f_{co}. The results are shown in Fig. 3.10. Both $|E_{out}|/|E_{in}|$ and θ are given for high- and low-pass filters. It should be noted

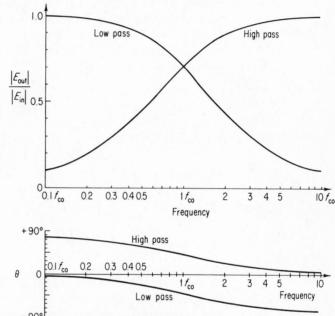

Fig. 3.10 Phase and output characteristics for single high- and low-pass filters.

that a logarithmic scale has been chosen for the horizontal axis so that a large frequency range may be compressed into a convenient dimension in order to show the form of $|E_{out}|/|E_{in}|$, not only at low frequencies, but also at high frequencies.

3.5 Tuned Circuits

Circuits containing both L and C give rise to frequency characteristics that differ from low- and high-pass filters. In this case a relatively narrow band of frequencies is admitted or rejected, depending upon the circuit used. An inductive reactance makes the applied voltage lead the current, and capacitive reactance makes the applied voltage lag the current. Coupled with this are the facts that both X_L and X_C are frequency-dependent and that X_L increases as frequency increases while X_C decreases as frequency increases. At a critical frequency this will give rise to a condition termed *resonance*, which is characterized by the fact that the applied voltage and current are in phase, even though the circuit contains reactive components.

Series Tuned Circuit

Consider a circuit containing R, L, and C in series. If a sinusoidal voltage generator is connected across this circuit, the resultant current will be a function of the frequency of the applied voltage. To test this frequency-dependent characteristic, assume that the amplitude of the applied voltage is held constant while the frequency is being varied. In Fig. 3.11 the effective value of the current is given by

$$I = \frac{E}{Z} \tag{3.52}$$

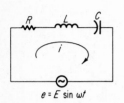

$e = E \sin \omega t$

Fig. 3.11 A simple *RCL* series circuit.

where Z is given by

$$Z = R + j(X_L - X_C) \tag{3.53}$$

Therefore

$$I = \frac{E}{R + j(X_L - X_C)} \tag{3.54}$$

Obtaining the absolute value

$$|I| = \frac{|E|}{\sqrt{R^2 + (X_L - X_C)^2}} \tag{3.55}$$

and putting this in terms of frequency,

$$|I| = \frac{|E|}{\sqrt{R^2 + [2\pi f L - (1/2\pi f C)]^2}} \tag{3.56}$$

The absolute value of I can be plotted as a function of frequency. It would be of use to look at Eq. (3.56) and deduce the shape of the curve obtained. As f approaches zero, the value of $|I|$ approaches zero also. Looking at the other end of the scale, as $f \rightarrow \infty$, $|I| \rightarrow 0$ again. Obviously, there must be at least one maximum if $|I|$ is to have any value at all. This maximum can be obtained if the quantity in the parentheses is set equal to zero:

$$2\pi f L - \frac{1}{2\pi f C} = 0 \tag{3.57}$$

In this case the current $|I|$ will be equal to

$$|I| = \frac{|E|}{R} \tag{3.58}$$

Any other value of f will produce a current $| \ I \ | \ < \ | \ E \ |/R$. The condition in Eq. (3.57), then, gives the frequency at which maximum current will flow in the circuit. This frequency will be labeled f_R. From Eq. (3.57), f_R can be found to be

$$f_R = \frac{1}{2\pi\sqrt{LC}} \tag{3.59}$$

Qualitatively, the frequency response of the series circuit will look like the curve in Fig. 3.12. The maximum current condition occurs for $f = f_R$, and the current is then given by Eq. (3.58). On either side of f_R, the current falls to zero as the frequency is either increased or decreased. Equation (3.54) is used to calculate the phase angle between applied voltage and the resultant current.

Rationalizing Eq. (3.54),

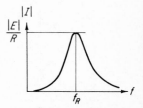

Fig. 3.12 Current as a function of frequency for the RCL series circuit.

$$I = \frac{E}{R^2 + (X_L - X_C)^2} \ [R - j(X_L - X_C)] \tag{3.60}$$

from which the phase angle is obtained:

$$\theta = \arctan\left(-\frac{X_L - X_C}{R}\right) \tag{3.61}$$

Since $f_R = 1/2\pi\sqrt{LC}$, Eq. (3.59), inserting this into the equation for θ gives

$$X_L = X_C \quad \text{at} \quad f = f_R$$

and

$$\theta = \arctan 0 \quad \text{or} \quad \theta = 0°$$

that is, the current and voltage are in phase at $f = f_R$. This singular condition, namely, that applied voltage and resultant current be in phase for a given frequency f_R, is called *resonance*.

For the series circuit shown in Fig. 3.11, then, the following can be summarized at resonance: $X_L = X_C$ and the net impedance is purely resistive. Thus the applied voltage and resultant current are in phase. Also, the current is a maximum at resonance; that is, the impedance $Z = \sqrt{R^2 + (X_L - X_C)^2}$ is a minimum.

Further manipulation of Eq. (3.56) can be performed to put it into an expression that can be used to give meaning to the band width for such a series circuit. Starting with Eq. (3.56),

$$| \ I \ | = \frac{| \ E \ |}{\sqrt{R^2 + [2\pi fL - (1/2\pi fC)]^2}} \tag{3.62}$$

and the condition that at resonance,

$$X_L = 2\pi f_R L = 2\pi\left(\frac{1}{2\pi\sqrt{LC}}\right)L = \sqrt{\frac{L}{C}} \equiv X_R \tag{3.63}$$

$$X_C = \frac{1}{2\pi f_R C} = \frac{2\pi \sqrt{LC}}{2\pi C} = \sqrt{\frac{L}{C}} \equiv X_R \qquad (3.64)$$

Then the following can be established:

$$2\pi L = 2\pi \sqrt{LC}\, \sqrt{\frac{L}{C}} = \frac{X_R}{f_R}$$

$$\frac{1}{2\pi C} = \frac{1}{2\pi \sqrt{LC}}\, \sqrt{\frac{L}{C}} = f_R X_R \qquad (3.65)$$

Putting the results of Eq. (3.65) into Eq. (3.62) and factoring out R,

$$|I| = \frac{|E|}{R}\, \frac{1}{\sqrt{1 + (X_R/R)^2[(f/f_R) - (f_R/f)]^2}} \qquad (3.66)$$

The quantity X_R/R, which is given the symbol Q, determines the shape of the curve:

$$|I| = \frac{|E|}{R}\, \frac{1}{\sqrt{1 + Q^2[(f/f_R) - (f_R/f)]^2}} \qquad (3.67)$$

Let $|E|/R = |I|_R$, the current at resonance. The equation can be put in normalized form:

$$\frac{|I|}{|I|_R} = \frac{1}{\sqrt{1 + Q^2[(f/f_R) - (f_R/f)]^2}} \qquad (3.68)$$

A plot can be made of this equation, where the variable f is plotted in multiples of f_R. The result is a general curve that is applicable

$$\frac{|I|}{|I|_R} = \frac{1}{\sqrt{1 + Q^2[(f/f_R) - (f_R/f)]^2}}$$

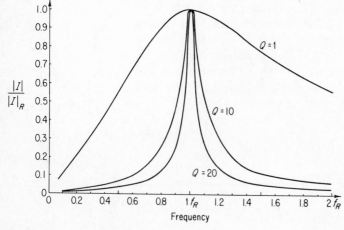

Fig. 3.13 The width of the resonance curve as a function of Q.

to any series circuit for which Q and f_R are known. This is illustrated in Fig. 3.13. To show the effect of the factor Q, a family of curves is plotted for different values of Q. Figure 3.13 shows that as Q is made larger, the band of frequencies for which appreciable current flows gets smaller. Also, the resultant curve becomes more symmetrical about the resonant frequency f_R.

3.6 Band Width

The band width for a curve such as this is defined as the difference between the frequencies that define the half-power points. On the $|I|/|I|_R$ versus frequency curve, this would occur for the condition

$$\frac{|I|}{|I|_R} = \frac{1}{\sqrt{2}} = 0.707 \tag{3.69}$$

because the condition for half-power is that

$$\frac{|I|^2}{R} = \frac{1}{2}\frac{|I|_{max}^2}{R} \tag{3.70}$$

Solving for $|I|/|I|_{max}$ in Eq. (3.70),

$$\frac{|I|^2}{|I|_{max}^2} = \frac{1}{2} \quad \text{or} \quad \frac{|I|}{|I|_{max}} = \frac{1}{\sqrt{2}} \tag{3.71}$$

Figure 3.14 illustrates this graphically.

The band width (BW) is defined as

$$BW = f_2 - f_1 \tag{3.72}$$

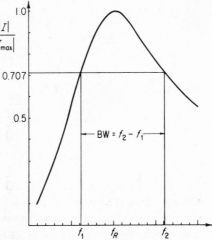

Fig. 3.14 The band width of a resonance curve.

CASE OF HIGH Q

For the special case where $Q \gg 1$, the curve is quite symmetrical about f_R and the following approximate approach can be used. For half-power points,

$$\frac{|I|}{|I|_{max}} = \frac{1}{\sqrt{2}} = \frac{1}{\sqrt{1 + Q^2[(f'/f_R) - (f_R/f')^2]}} \quad (3.73)$$

where f' denotes the half-power frequency.

$$\left(\frac{f'}{f_R} - \frac{f_R}{f'}\right)^2 = \frac{1}{Q^2} \quad (3.74)$$

$$\frac{f'}{f_R} - \frac{f_R}{f'} = \pm\frac{1}{Q} \quad (3.75)$$

$$f'^2 \pm \frac{f_R}{Q}f' - f_R^2 = 0 \quad (3.76)$$

This is a quadratic in f', and the solution can be written as

$$f' = \pm\frac{f_R}{2Q} \pm \frac{1}{2}\sqrt{\frac{f_R^2}{Q^2} + 4f_R^2} \quad (3.77)$$

$$f' = \pm\frac{f_R}{2Q}[1 \pm \sqrt{1 + 4Q^2}] \quad (3.78)$$

At this point, the condition that $Q \gg 1$ is put into the equation. Thus $4Q^2$ can be taken out of the radical, and we get the following four solutions:

$$f_1' = \frac{f_R}{2Q} + f_R$$

$$f_2' = \frac{f_R}{2Q} - f_R$$

$$f_3' = -\frac{f_R}{2Q} - f_R \quad (3.79)$$

$$f_4' = -\frac{f_R}{2Q} + f_R$$

The only solutions having physical meaning here are the values that are labeled f_1' and f_4'. These give positive values for f. The other two solutions are true mathematically, but have no physical significance because they are negative, and a negative frequency is not meaningful. Two solutions are accepted and the other two are rejected on the basis of the physics of the situation. Thus

$$f_1 = f_R - \frac{f_R}{2Q}$$

$$f_2 = f_R + \frac{f_R}{2Q}$$

(3.80)

Substituting these values for f_1 and f_2 into the band-width equation, $\text{BW} = f_2 - f_1$,

$$\left(\frac{f_R}{2Q} + f_R\right) - \left(-\frac{f_R}{2Q} + f_R\right) = \text{BW}$$

(3.81)

$$\text{BW} = \frac{f_R}{Q}$$

(3.82)

This, although good only for $Q \gg 1$, is an important result in electronics because the circuits commonly used have Q values of the order of 100 or higher, ensuring the validity of Eq. (3.82).

3.7 Parallel Tuned Circuits

The condition called *resonance* can also occur in parallel circuits containing R and C. As in series tuned circuits, the impedance and phase shift will vary greatly about the resonant frequency. Consider the following circuit, an inductor in parallel with a capacitor, with a sinusoidal voltage generator across this combination. The inductor is made up of turns of wire and so has an internal resistance associated with it. This is labeled R_L in Fig. 3.15. In most applications the capacitance can be considered pure. The current I is given by $I = E/Z$, where Z is the parallel combination of the two branches. The impedance of the inductive branch is $Z_L = R_L + jX_L$, and the impedance of the capacitive branch is $Z_C = -jX_C$. Using the "product divided by the sum" rule,

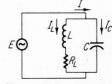

Fig. 3.15 A simple parallel RCL circuit.

$$Z = \frac{Z_L Z_C}{Z_L + Z_C} = \frac{(R_L + jX_L)(-jX_C)}{R_L + jX_L - jX_C}$$

(3.83)

Rationalizing,

$$Z = \frac{R_L X_C^2}{R_L^2 + (X_L - X_C)^2} - j\frac{R_L^2 X_C + X_L X_C(X_L - X_C)}{R_L^2 + (X_L - X_C)^2}$$

(3.84)

RESONANCE

The condition for a resonance is that E and I be in phase. For this to be true, Z must be a real number; that is, the j term must be zero. Setting up this condition for Eq. (3.84),

$$-\frac{R_L{}^2 X_C + X_L X_C (X_L - X_C)}{R_L{}^2 + (X_L - X_C)^2} = 0 \qquad (3.85)$$

This can be true only if the numerator is equal to zero:

$$R_L{}^2 X_C + X_L X_C (X_L - X_C) = 0 \qquad (3.86)$$

$$R_L{}^2 + X_L{}^2 - X_L X_C = 0 \qquad (3.87)$$

Substituting,

$$2\pi f_R L = X_L$$
$$\frac{1}{2\pi f_R C} = X_C \qquad (3.88)$$

$$R_L{}^2 + 4\pi^2 f_R{}^2 L^2 - \frac{L}{C} = 0 \qquad (3.89)$$

and solving for f_R (retaining only the positive value for f_R):

$$f_R = \frac{1}{2\pi} \sqrt{\frac{1}{LC} - \frac{R_L{}^2}{L^2}} \qquad (3.90)$$

For this parallel circuit, the condition for resonance is different from that for a series circuit. However, it should be noted that as $R_L \to 0$, f_R (parallel) $\to f_R$ (series), for if $R_L = 0$ in Eq. (3.90),

$$f_R = \frac{1}{2\pi \sqrt{LC}} \qquad (3.91)$$

just as for the series tuned case. So, if $R_L{}^2/L^2 \ll 1/LC$, the simpler equation for f_R can be used.

At resonance, Z is given by

$$Z_R = \frac{R_L X_C{}^2}{R_L{}^2 + (X_L - X_C)^2} \qquad (3.92)$$

and is a pure resistance, since the j term is equal to zero. In terms of the exact definition for f_R as expressed by Eq. (3.90), Z_R reduces to

$$Z_R = \frac{L}{R_L C} \quad \text{ohms} \qquad (3.93)$$

at resonance.

To show how the absolute value of the current behaves as a function of frequency, the equation $|I| = |E|/|Z|$ is used.

If $|E|$ is kept constant and the behavior of $|Z|$ is known, then the dependence of $|I|$ on f can be seen. From Eq. (3.63), the absolute value of Z is found to be

$$|Z| = \sqrt{\frac{(L^2/C^2) + (R_L^2/4\pi^2f^2C^2)}{R_L^2 + [2\pi fL - (1/2\pi fC)]^2}} \qquad (3.94)$$

The dependence of $|Z|$ on f can be explored by looking at extremes for f. For $f \to 0$, $|Z| \to R_L$. This makes sense, since for low frequencies the inductive reactance approaches zero and the capacitive reactance approaches infinity. Thus a very large impedance is placed in parallel with R_L and the total impedance will be R_L. This is illustrated in Fig. 3.16. For $f \to \infty$, $|Z| \to 0$; thus the impedance $|Z|$ is small at both ends of the spectrum. If the maximum points for $|Z|$ are desired, the following equation must be used:

$$\frac{d|Z|}{df} = 0 \qquad (3.95)$$

Evaluating this will show that maximum $|Z|$ occurs for $f = f_R$.

Thus, for a parallel tuned circuit, minimum total current will occur at resonance, the impedance is purely resistive, and E and I are in phase. Therefore a parallel tuned circuit offers maximum impedance at the resonant frequency.

A simpler parallel tuned circuit is shown in Fig. 3.17. It is an important circuit in electronics because many of the parallel tuned circuits used can be approximated by the circuit shown. In many cases the resistance of the coil comprising the inductance L is negligible, and both the capacitor and inductor can be thought of as perfect. The only resistance, then, must come from some other components. This resistance is due to resistors or tubes in parallel with the tuned circuit. Thus R in Fig. 3.17 stands for some resistance other than that due to L and C, which is in parallel with the tuned circuit. The quantities needed to describe the characteristics of this circuit are Z, the total impedance; f_R, the resonant frequency; and $I(f)$, the frequency dependence of the total current.

To find Z, the rule for components in parallel is used:

$$\frac{1}{Z} = \frac{1}{R} + \frac{1}{jX_L} + \frac{1}{-jX_c} \qquad (3.96)$$

$$Z = \frac{RX_LX_c}{X_L^2X_c^2 + R^2(X_L - X_c)^2}[X_LX_c - jR(X_L - X_c)] \qquad (3.97)$$

For resonance, the j term must equal zero. Therefore

$$\frac{-R^2X_LX_c(X_L - X_c)}{X_L^2X_c^2 + R^2(X_L - X_c)^2} = 0 \qquad (3.98)$$

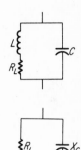

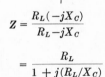

$$Z = \frac{R_L(-jX_c)}{R_L - jX_c}$$

$$= \frac{R_L}{1 + j(R_L/X_c)}$$

for $X_c \gg R_L$

$$Z \cong R_L$$

Fig. 3.16 At low frequencies, Z approaches R_L.

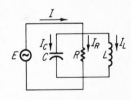

Fig. 3.17 A simple band rejection filter.

For this to be true, the numerator must equal zero:

$$-R^2 X_L X_C (X_L - X_C) = 0 \qquad (3.99)$$

$$X_L = X_C \qquad (3.100)$$

Solving for f_R,

$$2\pi f_R L = \frac{1}{2\pi f_R C} \qquad (3.101)$$

Keeping the positive root,

$$f_R = \frac{1}{2\pi \sqrt{LC}} \qquad (3.102)$$

This is the same as that obtained for a series tuned circuit. Substituting this value of f into Eq. (3.97), the value at Z of resonance, Z_R, can be calculated. Since the j term will go out, the impedance at resonance is a pure resistance given by

$$Z_R = \left[\frac{R X_L^2 X_C^2}{X_L^2 X_C^2 + R^2 (X_L - X_C)^2} \right]_{f=f_R} = R \qquad (3.103)$$

So, by Eq. (3.102), X_L and X_C effectively cancel each other, and the total impedance is just equal to the parallel resistance R. To obtain the frequency dependence of I, the equation $|I| = |E|/|Z|$ is used, so that $|Z|$ must be obtained. Using Eq. (3.97),

$$|Z| = \sqrt{ZZ^*} = \frac{R X_L X_C}{\sqrt{X_L^2 X_C^2 + R^2 (X_L - X_C)^2}} \qquad (3.104)$$

Dividing numerator and denominator by $R X_L X_C$, $|Z|$ can be put into the form

$$|Z| = \frac{1}{\sqrt{(1/R)^2 + [(1/X_C) - (1/X_L)]^2}} \qquad (3.105)$$

It should be noted that this form for $|Z|$ could have been obtained from Eq. (3.96) directly. The frequency characteristics for $|I|$ can now be obtained by substituting Eq. (3.58) into Eq. (3.105):

$$|I| = |E| \sqrt{\left(\frac{1}{R}\right)^2 + \left(\frac{1}{X_C} - \frac{1}{X_L}\right)^2} \qquad (3.106)$$

Factoring out $1/R^2$,

$$|I| = \frac{|E|}{R} \sqrt{1 + R^2 \left(\frac{1}{X_C} - \frac{1}{X_L}\right)^2} \qquad (3.107)$$

From Eq. (3.103), X_C and X_L can be calculated for $f = f_R$. Since, from Eq. (3.100) they are equal, they are given by

$$X_R = \sqrt{\frac{L}{C}} \qquad (3.108)$$

Using this result and Eq. (3.102), Eq. (3.107) can be put into the form

$$|I| = \frac{|E|}{R} \sqrt{1 + \left(\frac{R}{X_R}\right)^2 \left(\frac{f}{f_R} - \frac{f_R}{f}\right)^2} \qquad (3.109)$$

The quantity $|E|/R$ is equal to the current at resonance. Also, the quantity $(R/X_R)^2$ is the form factor for the curve of $|I|$ versus f. If R/X_R is labeled Q', then Eq. (3.91) resembles the equation found for the series tuned circuit, except that Q' is the inverse of Q. Both Q and Q' perform the same function in the equations, however, and each determines the band width in its respective equation. The graph for Eq. (3.109) is shown in Fig. 3.18 for several values of Q'. As can be seen, the greater the value of Q', the narrower the band width. This is in agreement with the definition for Q in Eq. (3.68).

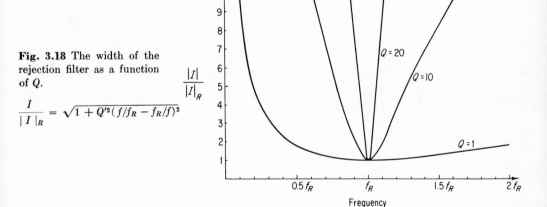

Fig. 3.18 The width of the rejection filter as a function of Q.

$$\frac{I}{|I|_R} = \sqrt{1 + Q'^2 (f/f_R - f_R/f)^2}$$

Since the current is minimum at f_R (see Fig. 3.18), the impedance must be maximum at f_R.

The band width should also be discussed at this point. It can be seen in Fig. 3.18 that the half-power point definition cannot be used directly because, at $f = 0$ and $f = \infty$, the power dissipated is infinite. Therefore, to obtain a band width for this circuit that is a counterpart of the series tuned circuit, the band width is defined as bounded by the frequencies where the power dissipated is twice that dissipated at resonance. There is still a relationship of $\frac{1}{2}$ between the resonant-frequency power dissipation and the power dissipated by the upper and lower frequency limits to the band width. For this reason, the upper and lower frequency limits are still called *half-power points*.

Applying the definition stated for f_1 and f_2 for this case,

$$\underset{\text{(half-power)}}{I^2 R} = \underset{\text{(min)}}{2 I_R^2 R} \tag{3.110}$$

$$\frac{I^2}{I_R^2} = 2 \tag{3.111}$$

Solving for $|I|/|I|_R$ and equating this to Eq. (3.100), and labeling the frequency as f' to remind one that the equation is being solved for a particular condition (that is, the band width),

$$\frac{|I|}{|I|_R} = \sqrt{2} = \sqrt{1 + Q'^2 \left(\frac{f'}{f_R} - \frac{f_R}{f'} \right)^2} \tag{3.112}$$

Solving for f',

$$Q' \left(\frac{f'}{f_R} - \frac{f_R}{f'} \right) = \pm 1 \tag{3.113}$$

$$f'^2 \pm \frac{f_R}{Q'} f' - f_R^2 = 0 \tag{3.114}$$

Four values of f' are obtained, as in Eq. (3.76) in Sec. 3.6. Again, the negative values for f' are discarded, leaving two values for f'. They are the upper and lower values for the band width, f_2 and f_1, respectively.

Lower frequency:

$$f_1 = -\frac{f_R}{2Q'} + \sqrt{1 + 4Q'^2} \tag{3.115}$$

Upper frequency:

$$f_2 = \frac{f_R}{2Q'} + \sqrt{1 + 4Q'^2} \tag{3.116}$$

For the case where $Q' \gg 1$, as is usually the case in circuits of this sort,

$$f_1 = 2Q' - \frac{f_R}{2Q'} \tag{3.117}$$

$$f_2 = 2Q' + \frac{f_R}{2Q'} \tag{3.118}$$

and band width becomes

$$BW = f_2 - f_1 = \left(2Q' + \frac{f_R}{2Q'}\right) - \left(2Q' - \frac{f_R}{2Q'}\right) \quad (3.119)$$

$$BW = \frac{f_R}{Q'} \quad (3.120)$$

where $Q' = R/X_R$.

Summing up for the parallel circuit in Fig. 3.17, the impedance is maximum at resonance and purely resistive, given by R. The resonant frequency is given by

$$f_R = \frac{1}{2\pi\sqrt{LC}} \quad (3.121)$$

and the band width is given by

$$BW = \frac{f_R}{Q'} \quad (3.122)$$

3.8 Bandpass Circuits

Tuned circuits are used to pass or reject a band of frequencies. This differs from low- or high-pass filters in that only a narrow band of frequencies is passed or rejected. These characteristics stem directly from the characteristics of Z for a tuned circuit. Two tuned circuits that admit a band of frequencies are shown in Fig. 3.19. One is a parallel tuned circuit, the other a series tuned circuit. Each is arranged so that the output voltage is a maximum at resonance. Since the characteristics of Z, the impedance of the tuned circuits, have been examined in Sec. 3.7, the discussion of the bandpass characteristics for circuits utilizing these tuned circuits is straightforward.

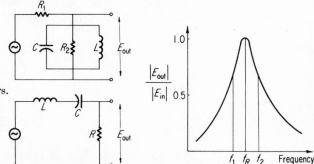

Fig. 3.19 Two simple bandpass filters.

PARALLEL TUNED CIRCUIT

In the parallel tuned circuit of Fig. 3.19, the output voltage is given by

$$E_{out} = E_{in}\left(\frac{Z_p}{R_1 + Z_p}\right) = E_{in}\left[\frac{1}{1 + (R_1/Z_p)}\right] \quad (3.123)$$

where Z_p is the impedance of the parallel tuned circuit, as a function of frequency. In this circuit R_1 and R_2 are arranged so that $R_2 \gg R_1$. Thus, at resonance, Eq. (3.123) reduces to

$$E_{out} = E_{in}\left[\frac{1}{1 + (R_1/R_2)}\right] \cong E_{in} \quad (3.124)$$

while for all other frequencies, Z_p drops off in value rapidly, for high-Q values, thus making E_{out} drop off rapidly on either side of resonance. This action can be described in a qualitative manner by remembering that a parallel tuned circuit has maximum impedance at resonance, equal to R_2, and the impedance drops to zero on both sides of resonance. The parallel tuned circuit and R_1 form a series circuit across the generator, and the output voltage is equal to the voltage drop across Z_p. Thus, at low frequencies Z_p is close to zero and there is a small voltage drop across Z_p, that is, most of E_{in} is dropped across R_1. As resonance is approached, Z_p increases and more of E_{in} is dropped across Z_p, with respect to R_1, since E_{R_1} and E_{Z_p} must always equal E_{in}. At resonance, Z_p is maximum and the maximum amount of E_{in} is dropped across Z_p. At the resonant frequency, $Z_p = R_2$. Since R_2 is made much larger than R_1, most of the input voltage is found across R_2. As the frequency is increased beyond resonance, Z_p decreases to zero again, and the output voltage also decreases to zero.

QUANTITATIVE RESULTS

The phase shift between E_{in} and E_{out} must now be considered. In order to obtain a quantitative value, the vector quantity E_{out}/E_{in} must be put in rationalized form:

$$\frac{E_{out}}{E_{in}} = \frac{Z_p}{R_1 + Z_p} = \frac{1}{1 + (R_1/Z_p)} \quad (3.125)$$

Now

$$\frac{1}{Z_p} = \frac{1}{R_2} + j\left(\frac{1}{X_C} - \frac{1}{X_L}\right) \quad (3.126)$$

Putting this into Eq. (3.125) and factoring R_1 out of the denominator,

$$\frac{E_{out}}{E_{in}} = \frac{1/R_1}{[(1/R_1) + (1/R_2)] + j[(1/X_C) - (1/X_L)]} \quad (3.127)$$

To rationalize this equation, multiply numerator and denominator by the complex conjugate of the denominator:

$$\frac{E_{out}}{E_{in}} = \frac{1/R_1}{[(1/R_1) + (1/R_2)]^2 + [(1/X_C) - (1/X_L)]^2}$$

$$\times \left[\left(\frac{1}{R_1} + \frac{1}{R_2}\right) - j\left(\frac{1}{X_C} - \frac{1}{X_L}\right)\right] \quad (3.128)$$

The phase angle can now be read off as

$$\theta = \arctan\left[-\frac{(1/X_C) - (1/X_L)}{(1/R_1) + (1/R_2)}\right] \quad (3.129)$$

Calling $(1/R_1) + (1/R_2) = 1/R_T$ and remembering that $f_R = 1/2\pi\sqrt{LC}$, the angle can be put in the form

$$\theta = \arctan\left[-Q'\left(\frac{f}{f_R} - \frac{f_R}{f}\right)\right] \quad (3.130)$$

where

$$Q' = \frac{R_T}{X_R}$$

To ascertain the phase shift qualitatively, Eq. (3.130) is examined for several limiting cases. At $f \to 0$, $X_C \gg X_L$ and θ approaches +90 deg. For $f \to \infty$, $X_L \gg X_C$ and θ approaches −90 deg. And at resonance, $X_L = X_C$ and θ equals 0 deg.

Equation (3.127) can also be used to obtain the equation that will be used to plot $|E_{out}|/|E_{in}|$.

Again using the definition that

$$\frac{1}{R_T} = \frac{1}{R_1} + \frac{1}{R_2} \quad (3.131)$$

$$\frac{E_{out}}{E_{in}} = \frac{R_T/R_1}{1 + jR_T[(1/X_C) - (1/X_L)]} \quad (3.132)$$

The quantity $R_T[(1/X_C) - (1/X_L)]$ can be put in the form

$$Q'\left(\frac{f}{f_R} - \frac{f_R}{f}\right) \quad (3.133)$$

where $Q' = R_T/X_R$.

The quantity

$$\frac{R_T}{R_1} = \frac{R_2}{R_1 + R_2} = \frac{1}{1 + (R_1/R_2)} \tag{3.134}$$

and since $R_2 \gg R_1$,

$$\frac{R_T}{R_1} \cong 1 \tag{3.135}$$

Therefore

$$\frac{E_{\text{out}}}{E_{\text{in}}} = \frac{1}{1 + jQ'[(f/f_R) - (f_R/f)]} \tag{3.136}$$

and

$$\frac{|E_{\text{out}}|}{|E_{\text{in}}|} = \frac{1}{\sqrt{1 + Q'^2[(f/f_R) - (f_R/f)]^2}} \tag{3.137}$$

Equation (3.137) is of the form already plotted in Sec. 3.7.

Series Tuned Circuit

The series tuned circuit in Fig. 3.19 can also be put into this form. starting with the equation for output voltage,

$$E_{\text{out}} = E_{\text{in}} \left[\frac{R}{R + j(X_L - X_C)} \right] \tag{3.138}$$

Then two results follow:

$$\frac{|E_{\text{out}}|}{|E_{\text{in}}|} = \frac{1}{\sqrt{1 + Q^2[(f/f_R) - (f_R/f)]^2}} \tag{3.139}$$

where $Q = X_R/R$ and

$$\theta = \arctan \left[-Q \left(\frac{f}{f_R} - \frac{f_R}{f} \right) \right] \tag{3.140}$$

These results, Eqs. (3.139) and (3.140), are easily obtained from Eq. (3.138). They are of exactly the same form as Eqs. (3.137) and (3.130), except for the definitions for Q. Therefore, one set of curves will summarize both circuits.

These curves are shown in Fig. 3.20 for a Q value of 10. The phase shift is practically a linear function of frequency between the half-power points. Also, these two circuits are called bandpass circuits, since only a band of frequencies, $f_2 - f_1$, is allowed through the filter circuit without appreciable attenuation.

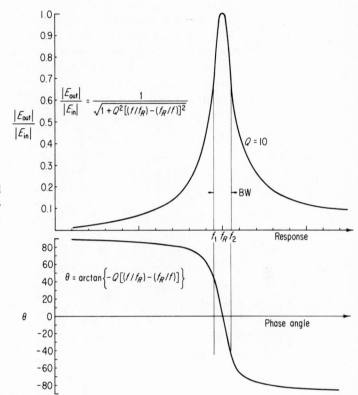

Fig. 3.20 Typical phase and output characteristics for a bandpass filter at $Q = 10$.

$$\frac{|E_{\text{out}}|}{|E_{\text{in}}|} = \frac{1}{\sqrt{1 + Q^2\,[(f/f_R) - (f_R/f)]^2}}$$

$Q = 10$

BW

$f_1\ f_R\ f_2$

Response

$$\theta = \arctan\left\{-Q\,[(f/f_R) - (f_R/f)]\right\}$$

θ

Phase angle

3.9   Band Rejection Filters or Wave Traps

By reversing components, the bandpass filters can be converted into wave traps; that is, they will admit all frequencies except for a narrow band that is rejected. The two circuits that will accomplish this are shown in Fig. 3.21. The calculations proceed in exactly the same manner as in Secs. 3.7 and 3.8. Write down E_{out} first, and then from this equation obtain $|E_{\text{out}}|/|E_{\text{in}}|$ and $\tan\theta$. The procedure is straightforward and the results are summarized below. The equations for $|E_{\text{out}}|/|E_{\text{in}}|$ and θ are plotted in Fig. 3.22 for a Q value of 10. The band width is obtained in exactly the same manner as shown in Sec. 3.7. This type of filter behaves as a wave trap. All frequencies are admitted with little or no attenuation except for a narrow band of frequencies centering on the resonant frequency of the tuned circuit. This narrow band of frequencies is rejected.

Filters such as these are used to get rid of objectionable or unwanted signals without affecting the remaining frequencies. The phase characteristics show that except for the rejected frequencies, all other frequencies pass through with little or no phase shift. About the resonant frequency there is a large change in phase, which appears discontinuous here because perfect, resistanceless components of L and C were assumed along with the condition $R_2 \gg R_1$, which allowed ratios such as R_1/R_2 to be dropped in the derivations.

For the real case, the phase angle would vary from -90 to $+90$ deg in a continuous fashion, if the resistances due to the

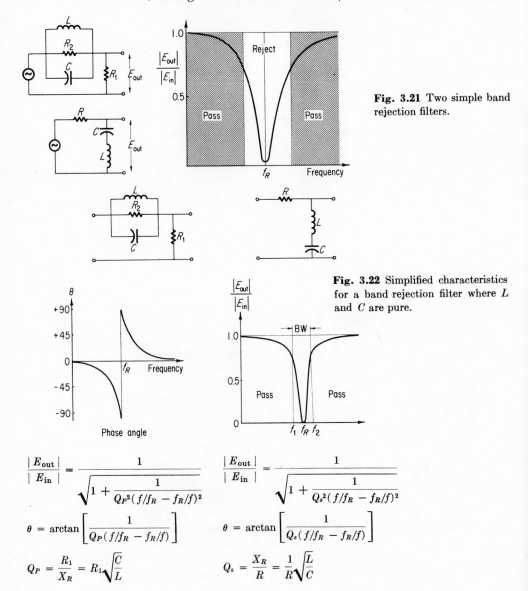

Fig. 3.21 Two simple band rejection filters.

Fig. 3.22 Simplified characteristics for a band rejection filter where L and C are pure.

$$\frac{|E_{\text{out}}|}{|E_{\text{in}}|} = \frac{1}{\sqrt{1 + \dfrac{1}{Q_P{}^2(f/f_R - f_R/f)^2}}}$$

$$\theta = \arctan\left[\frac{1}{Q_P(f/f_R - f_R/f)}\right]$$

$$Q_P = \frac{R_1}{X_R} = R_1\sqrt{\frac{C}{L}}$$

$$\frac{|E_{\text{out}}|}{|E_{\text{in}}|} = \frac{1}{\sqrt{1 + \dfrac{1}{Q_s{}^2(f/f_R - f_R/f)^2}}}$$

$$\theta = \arctan\left[\frac{1}{Q_s(f/f_R - f_R/f)}\right]$$

$$Q_s = \frac{X_R}{R} = \frac{1}{R}\sqrt{\frac{L}{C}}$$

reactances were taken into account. This is done at the expense of laborious algebraic manipulations, and the equations summarized in Fig. 3.22 suffice as excellent approximations. The results for the exact case are shown in Fig. 3.23. The differences are that the band rejection curve does not go to zero at f_R and that there is no discontinuity for the phase-angle relationship in the region of resonance. In this example, the Q value was chosen as 10 and R_1/R_2 was chosen as 0.01. The phase-angle curve is seen to peak at about 78.7 deg for values of f given by 0.995 f_R and 1.005 f_R. This is very close to the resonant frequency. Within this small frequency range the phase angle changes from -78.7 to $+78.7$ deg, going through zero at f_R. The curve for $|E_{out}|/|E_{in}|$ versus frequency does not change too much, since $|E_{out}|/|E_{in}| \cong$ 0.01 at $f = f_R$. So, even in this more exact case, the amplitude versus frequency curve does not change appreciably. This example suffices to show that, for most cases, the less exact equations diagrammed in Fig. 3.22 are close enough to the results shown in Fig. 3.23. As Q is made larger, the discrepancy becomes less between the two cases. Therefore, the equations shown in Fig. 3.22 are the ones usually used in calculations for these circuits.

Fig. 3.23 The band rejection characteristics when L and C are not pure.

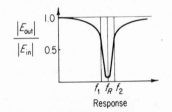

Response

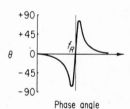

Phase angle

Finally, it is of interest to see how the data for the simple case would look when plotted in the complex plane for the two cases of bandpass and band rejection. These results are shown in Fig. 3.24. Surprisingly, the two graphs have the same form. However, the resonant frequency, f_R is seen to be in different positions. The points $f = 0$ and $f = \infty$ are also shown in each case.

Fig. 3.24 Band pass and band rejection characteristics plotted in the complex plane.

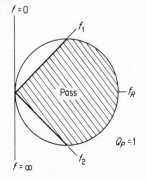

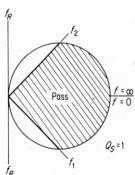

Problems

3.1 Find the cutoff frequency for the following filter. Is it a high-pass or low-pass filter?

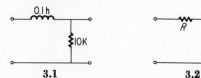

3.1   3.2

3.2 Find the frequency response of the above circuit.

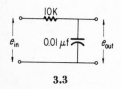

3.3

3.3 A complex voltage wave form given by $e_{in} = 100 \sin 2\pi$ $(1000t) + 50 \sin 2\pi(2000\ t)$. This is applied to the following filter.

Determine e_{out} and graph e_{in} and e_{out} on the same scale. (Remember that a phase shift must also be determined).

3.4 Determine the voltage across C as a function of frequency, keeping the amplitude of E_{in} constant.

$$L = 100 \ \mu h \qquad C = 250 \text{ pf}$$

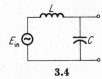

3.4   3.5

3.5 For the above circuit, the band width is 10 kc. Find R.

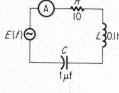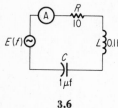

3.6 In the circuit shown, plot the current as a function of frequency at $f = 100, 200, 300, 400, 500, 600, 800,$ and 1000 cps. Use $R = 10$ ohms, $L = 0.1$ henry, $C = 1 \ \mu f$, and the amplitude of E is kept constant at 1 volt.

3.6

3.7 Repeat Problem 3.6 for $R = 100$ ohms and $R = 0$ ohm.

3.8 Calculate the voltage across L at resonance.

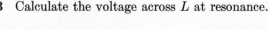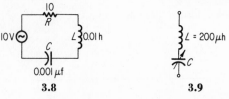

3.8   3.9

3.9 In the righthand circuit, C is variable from 100 pf to 300 pf. Plot the resonant frequency as a function of C.

3.10 Prove that

$$\frac{|E_{\text{out}}|}{|E_{\text{in}}|} = \frac{1}{\sqrt{1 + (f_{\text{co}}/f)^2}}$$

is a circle when plotted as a function of θ in the complex plane where

$$\theta = \tan^{-1}(f_{\text{co}}/f)$$

3.11 Find the resonant frequency of each of the following circuits:

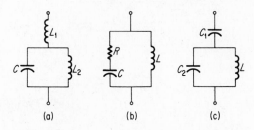

(a)   (b)   (c)

3.12 Find the impedance at resonance for the following circuits:

(a)    (d)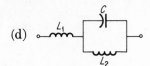

(b) 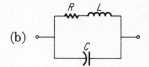   (e)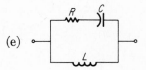

(c) 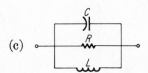   (f)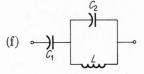

3.13 The Q' of the following circuit is $Q' = 20$. Find R and the upper and lower half-power points.

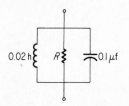

3.14 Find the equivalent series combination for the following parallel circuit:

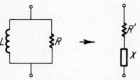

3.15 Find the frequency response of the following circuit:

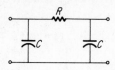

4.1 Vacuum Diodes

A vacuum diode consists of a *cathode* from which electrons leave; a *plate* to which the electrons are attracted; and an intervening *space* that is highly evacuated. The entire assembly is housed in a glass tube. Leads connected to the cathode and plate are brought out through the glass so that the diode can become part of an electric circuit. The charge is transported from the cathode to the plate by electrons. To keep electrons from being stopped or scattered by gas molecules, the intervening space must be highly evacuated. In essence, the electrons are forced to travel in a conductor, leave the conductor at the cathode, travel through space to the plate, strike the plate, be absorbed, and continue their travel through a metallic conductor, Fig. 4.1(a). The reason for going to all this trouble is the fact that a diode is a nonlinear device; its voltage-current characteristics do not obey Ohm's law. This makes for interesting and useful properties.

The schematic symbols for diodes are shown in Fig. 4.1(b), (c), and (d). In Fig. 4.1(a), a single diode is shown. It consists of a plate, a cathode, and a filament. The filament is used to heat the cathode so that the cathode will emit electrons. Figure 4.1(c) shows another type of commonly used diode. This tube has two separate plates, a common cathode, and a filament. Figure 4.1(d) shows the symbol for a dual diode. It is essentially two separate diodes in one envelope. The dual diode is constructed so that the diodes do not interact with one another. There are currently three popular tube types, the octal base and two miniature types. These bases and tube outlines are described in Chapter 6.

4.2 Carrier Production

In order to have a useful diode, there must be electron flow between cathode and plate. This means that the electrons must be induced to leave the conductor so that they can be accelerated in the vacuum between cathode and plate. The electrons can be produced in one of four ways: thermionic emission, field emission, photoelectric emission, and secondary emission. Of the four methods, thermionic emission is the most commonly used in vacuum diodes; secondary and photoelectric emission are used in phototubes; and field emission is the rarest of the four methods, being used in some high-voltage x-ray tubes and in the field-emission electron microscope.

4.3 Thermionic Emission

If the temperature of a solid is raised to about 1000 °K or higher, electrons are emitted by the solid. For a vacuum tube,

4

Diodes

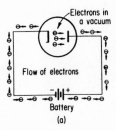

(a)

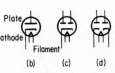

(b)   (c)   (d)

Fig. 4.1 (a) Electron flow in a diode circuit. (b) Single diode. (c) Duo-diode with common cathode. (d) Two separate diodes in a single envelope.

111

this emission should be high (a large number of electrons) and occur at as low a temperature as possible (to keep the operating temperature and power requirements down). At elevated temperatures, enough electrons leave the solid to produce an appreciable electric field between electrons and solid. This is due to the fact that as the number of positive ions remaining in the solid increases, a resultant retarding field is produced, which keeps the electrons in the vicinity of the solid, Fig. 4.2(a). This cloud of electrons around a heated solid is an example of a *space charge*. At a given temperature the number of electrons in the space charge is constant. If an external accelerating field is superimposed on the space-charge system, then the space-charge thickness is reduced. Some of the electrons will be swept out of the region by the accelerating field. This will constitute a current, Fig. 4.1.

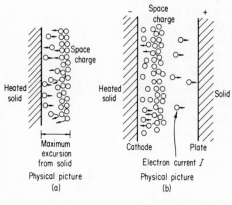

Fig. 4.2 (a) Space charge electrons near a heated solid. (b) Electron flow due to external electric field.

The heated solid is the cathode, and the electrode attracting and absorbing the electrons is the plate. In the overall picture at elevated temperatures, the cathode is surrounded by an invisible cloud of electrons, the space charge. If an electric field is established between the plate and cathode so that the plate is positive with respect to the cathode, a current is produced as electrons leave the space charge and are accelerated to the plate. The number of electrons leaving this space charge and contributing to the cathode-plate current is a function of the temperature of the cathode and the magnitude of the applied field. Higher temperatures place more electrons in the space charge and higher fields attract more electrons. If the field is increased enough, the current will become saturated, Fig. 4.3(b). This happens because as many electrons are leaving the space charge as are leaving the cathode. This saturation current has been investigated experimentally and theoretically by Richardson and Dushman.*

* Dushman, Saul., "Electron Emission," *Electrical Engineering*, **53**, No. 7, July 1934.

In the cylindrical geometries most often used for vacuum-tube work, the plate is concentric to the cathode, as shown in Fig. 4.3(a). This type of geometry allows one to put in guard rings for precise measurements. The resultant saturation current is found to obey the following equation for a large number of materials:

$$J = AT^2 e^{-B/T} \tag{4.1}$$

where

J = current density in amp/m²
A = a constant dependent upon the material used and of the order of 10^6 amp/m²/°K²
B = a constant for a given material of the order of 50,000 °K

Table 4.1 shows the value of A and B for several materials.

TABLE 4.1   Richardson-Dushman Constants for Several Materials

Material	A (amp/m²/°K²)	B (°K)
Carbon	0.60×10^6	46×10^3
Cesium	1.62×10^6	21×10^3
Nickel	0.26×10^6	32×10^3
Platinum	0.60×10^6	59×10^3
Tungsten	0.60×10^6	52×10^3

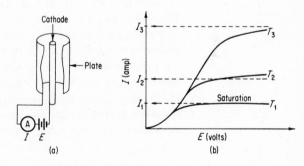

Fig. 4.3 (a) Diode construction. (b) Diode current as a function of cathode temperature.

To obtain the current in amperes, Eq. (4.1) is multiplied by the area of the emitting surface:

$$I_s = JS \tag{4.2}$$

where

I_s = saturation current in amp
J = current density given by Eq. (4.1)
S = area of emitter in m²

Figure 4.3(b) shows that $I_3 > I_2 > I_1$. As the temperature of the cathode is increased from T_1 to T_3, the saturation current is increased from I_1 to I_3.

It is found in practice that the saturation current does not level off, but is dependent upon the applied field to some extent. This behavior is shown in Fig. 4.4. The applied external field

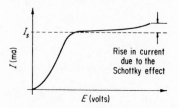

Fig. 4.4 Change in saturation current due to Schottky effect.

modifies the retarding field near the surface of the solid in such a way that more electrons can be emitted, and this produces the rise in saturation current. This behavior is called the *Schottky effect.*

PRACTICAL CATHODES

Tungsten was one of the original practical materials because (1) its melting point is high (3643 °K), which allows operation at high temperatures and therefore produces high emission; (2) it withstands contamination by residual gases; and (3) it does not evaporate easily and therefore lasts a long time. Thoriated tungsten filaments, in which a monatomic layer of thorium is diffused on a tungsten base, operate at low temperatures and higher emission currents and are used in many rectifier tubes. Highly efficient cathodes are used in most of the receiving types of tubes. These are constructed by coating some base metal with an alkali metal oxide such as barium or strontium oxide. They are efficient because they operate at temperatures much lower than that of tungsten and still give the same emission. Representative values are shown in Table 4.2.

TABLE 4.2   Cathode Properties

Material	Operating Temperature, °K	Current Density, amp/m²
Tungsten	2500	0.6×10^4
Thoriated Tungsten	1900	3.9×10^4
Oxide coated	1000	0.6×10^4

4.4   Field Emission

In field emission, the cathode is not heated. Instead, the electric field from plate to cathode is made so large that it alters the retarding field at the surface of a solid sufficiently; therefore electrons having surface-directed velocities now have a probability of penetrating the surface. The probability will be greater for the higher-speed electrons. These electrons will immediately be accelerated away from the solid by the external electric field.

FIELD-EMISSION MICROSCOPE

This effect is enhanced if the cathode is very sharply pointed, making the field very strong in the immediate vicinity of the cathode, Fig. 4.5. Electrons thus emitted move off in straight lines perpendicular to the cathode surface. Since the surface of the point is made up of a lattice of ions, the electrons that are

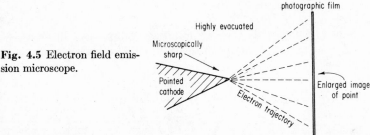

Fig. 4.5 Electron field emission microscope.

emitted will come from sites where they will most probably be found, that is, near the ions. Since the electron trajectories are straight lines, this pattern of ion sites is carried in the electron beam to the fluorescent screen, where the pattern is made visible. In this way, one can "see" the surface atoms move about on the point of the cathode.

4.5   Photoelectric Emission

Electrons in a solid can be made to escape by absorbing energy from incident radiation rather than from heat. Electron emission depends upon the intensity and the frequency of the incident radiation. The intensity governs the number of electrons that will be emitted, and the frequency determines whether the electrons will absorb enough energy to be emitted.

There is a cutoff frequency for all materials. Below the cutoff frequency an insignificant number of electrons will be emitted. Above the cutoff frequency, the probability of emission depends upon the frequency, and one can obtain a spectrum that relates the number of electrons emitted versus frequency, Fig. 4.6. The example shown happens to be most sensitive in the visible range. The curve was obtained by keeping the intensity of incident radiation constant. The significance of the cutoff frequency is that, below this frequency, the incident radiation can have very high intensities and still produce no significant emission of electrons. These properties can be used to detect incident radiation.

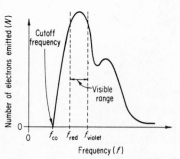

Fig. 4.6 Electron emission as a function of incident frequency for a photocell.

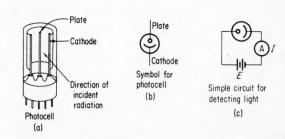

Fig. 4.7 (a) Photocell construction. (b) Schematic symbol. (c) Simple circuit for detecting light.

Fig. 4.8 Typical E–I characteristics for a vacuum photocell.

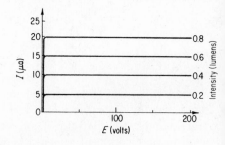

Figure 4.7(a) shows the construction of a photocell used to detect electromagnetic radiation. The cathode is made of a material that is chosen for the range of radiation to be detected. The cathodes can be selected so that they are most sensitive at infrared, visible, or ultraviolet radiation. The cathode shown is formed into half a cylinder, leaving an opening for incident radiation. The plate is a small-diameter wire placed on the axis of the cathode. The plate is positive with respect to the cathode, so that emitted electrons will be drawn to the plate. The circuit is shown completed externally through an ammeter and battery, or power supply, Fig. 4.7(c). The characteristics for such a device are obtained by varying the voltage E and measuring I for a

given intensity of incident radiation. Then the intensity is changed a known amount and E is varied again. The result for a vacuum photocell is shown in Fig. 4.8. The intensity is measured in lumens. The things to be noted are (1) the current is small, measured in microamperes (μa); (2) the device is extremely linear, that is, the constant-intensity curves are equally spaced; and (3) the device is a relatively high-voltage device, up to 200 volts on the plate.

4.6 Secondary Emission

Electrons are ejected from a solid when ions or other electrons strike the solid. The electron impacts are the more efficient because energy transfer is enhanced when the masses in a collision are nearly equal. Also, for incident electrons, the curve for the number of ejected electrons versus incident energy will peak at a characteristic energy for a given material.

The curve in Fig. 4.9 shows that, for the given material, the maximum number of secondary electrons is about 2, and occurs at an incident electron energy of about 400 electron volts (eV). The curves of N versus U vary from element to element and from compound to compound. The most efficient emitter of secondary electrons, cesium oxide, liberates about 10 electrons for every incident 500-eV electron. Usually, electron tubes having accelerating voltages greater than about 50 volts will have appreciable secondary electron emission. In most tube designs, this is an undesirable result and must be compensated for. However, some tubes are designed to take advantage of the effect.

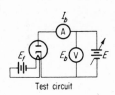

Fig. 4.9 Number of secondary electrons as a function of incident electron energy.

4.7 DC Characteristics of the Thermionic Vacuum Diode

The dc characteristics of a diode are obtained in the same manner as for any element, that is, apply a known voltage and measure the resultant current. The circuit and results are shown in Fig. 4.10. In effect, there are three parameters: E_F, the filament voltage; E_b, the plate to cathode voltage; and I_b, the plate current. In practice, E_F is held constant.

The set of curves in Fig. 4.10 is called a *family* of curves, since it shows how the E–I characteristics depend upon E_F. To obtain this set of curves, E_F is set for a low value, E_b varied, and the resulting values of I_b measured. This curve is labeled E_{F_1}. Then E_F is increased and another E–I curve is plotted. The increasing curves represent increasing values of E_F, that is, $E_{F_4} > E_{F_3} >$

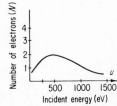

Test circuit

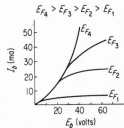

Fig. 4.10 A diode test circuit and characteristics obtained.

$E_{F_2} > E_{F_1}$. The desired characteristics are those for E_{F_4}. The other curves were shown to point out the effect of cathode temperature on saturation current. Ordinarily, thermionic tubes are operated at filament voltages such that the saturation current is far above the range of operation of the tube.

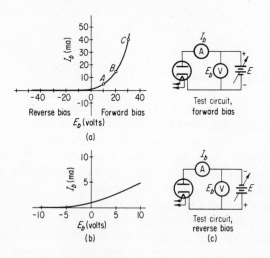

Fig. 4.11 (a) $E-I$ characteristics for a vacuum diode. (b) An enlargement of the $E_b = 0$ point. (c) Test circuits for forward and reverse bias.

The important set of characteristics is shown in Fig. 4.11 for a typical thermionic diode. The negative values of E_b correspond to the case where the plate is negative with respect to the cathode (reverse bias). For this case, there is no flow of electrons from cathode to plate. On closer inspection, Fig. 4.11(b), it can be seen that at $E_b = 0$ there is a small nonzero current flowing in the circuit. This is due to the kinetic energy spread of the thermionically emitted electrons in the space charge. The faster electrons move farther away from the cathode than the slower electrons. It is these fast electrons that reach the plate, are absorbed at the plate, and return to the cathode through the external circuit. At a reverse bias of about -5 volts, even the fastest electrons are stopped before they reach the plate, and the plate current falls to zero and remains there for more negative values of E_b.

Thus the diode characteristics show that there is a preferential direction of current flow and that the diode is a nonlinear device. The resistance of the diode is dependent upon the voltage impressed across it. For example, at points A, B, and C in Fig. 4.11(a), the resistance values can be calculated as

$$R_A = \frac{10 \text{ volts}}{5 \times 10^{-3} \text{ amp}} = 2000 \text{ ohms}$$

$$R_B = \frac{20 \text{ volts}}{15 \times 10^{-3} \text{ amp}} = 1330 \text{ ohms}$$

$$R_C = \frac{30 \text{ volts}}{40 \times 10^{-3} \text{ amp}} = 750 \text{ ohms}$$

In the reverse bias connection, the diode can be represented by an open circuit, that is, its resistance is infinite.

4.8 Dynamic Transfer Curve

The operation of a diode in a circuit can be predicted, once the dc characteristics are known. For example, the operation of the circuit shown in Fig. 4.12 can be predicted from a knowledge of the dc characteristics for tube V. Given e_{in} as a function of time, can e_{out} be predicted as a function of time? To answer this question, a dynamic transfer curve must be constructed for the circuit shown. The transfer curve shows the variation of I_b as a function of e_{in}. Once I_b is known, e_{out} can be calculated, since $e_{out} = I_b R_L$. To lead up to the solution, consider a static case first.

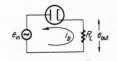

Fig. 4.12 Diode circuit with load resistor R_L.

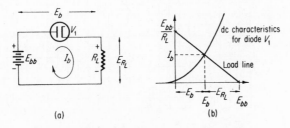

(a) (b)

Fig. 4.13 Graphical solution of the dc circuit.

In Fig. 4.13, the circuit consists of a battery E_{bb}, resistor R, and diode V. The voltage drops across the diode and resistor, as well as the current, can be determined. Summing up the voltage drops around the circuit in Fig. 4.13 and solving for I_b,

$$I_b = \frac{E_{bb} - E_b}{R_L} \tag{4.3}$$

Equation (4.3) is called the *load-line* equation for the diode circuit. There are two variables, I_b and E_b, where I_b is the current through V and E_b is the voltage drop across V. To solve for the variables

I_b and E_b, two equations are needed. The second equation relating I_b and E_b is the dc characteristic curve for the diode. The solution can be obtained graphically by superimposing the two curves on a single graph.

The solution is given by the intersection of the two curves. The load-line equation, Eq. (4.3) is linear. It is of the form $I_b = a + mE_b$, where $a = E_{bb}/R_L$ and $m = -1/R_L$. Two points are necessary to determine the load-line curve. Two easily obtained points are found for the conditions $I_b = 0$ and $E_b = 0$. At $E_b = 0$, $I_b = E_{bb}/R_L$, and at $I_b = 0$, $E_b = E_{bb}$. The resulting load line is shown in Fig. 4.13(b). The intersection of the two curves gives the value of I_b and E_b for the circuit shown. The voltage drop across R_L must then be $E_{R_L} = E_{bb} - E_b$. This is indicated on the diagram.

Example 4.1 Given the circuit shown, determine I_b, E_b, and E_{R_L} for the tube whose characteristics are given below.

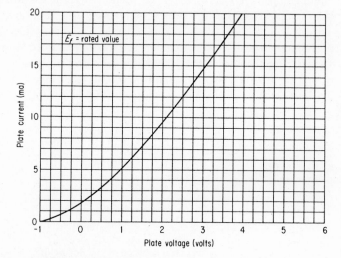

1. The load-line equation is

$$I_b = \frac{E_{bb} - E_b}{R_L}$$

In terms of the circuit, this becomes

$$I_b = \frac{30 - E_b}{2} \quad \text{ma}$$

2. (a) The intercept along $E_b = 0$ is found by

$$I_b = \frac{30 - 0}{2} = 15 \text{ ma}$$

and is labeled point A on the graph below. Its coordinates are (15, 0).

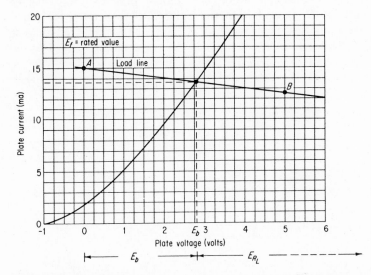

(b) The intercept along $I_b = 0$ cannot be plotted here; so, arbitrarily choosing another point that will fall on the graph, try $E_b = 5$ volts;

$$I_b = \frac{30 - 5}{2} = 12.5 \text{ ma}$$

This point is labeled B. Its coordinates are (12.5, 5). Drawing a straight line through these two points (A and B) produces the desired load line.

3. Thus $I_b = 13.5$ ma, $E_b = 2.8$ volts, and $E_{R_L} = 30 - E_b = 27.2$ volts.

In order to obtain the transfer curve, the variation of I_b with input voltage must be determined. Figure 4.14 shows how the constructions are obtained for various input voltages and a given circuit. In the circuit shown, R_L is kept constant at 1000 ohms and E_{in} is given several values. A load line is drawn on the tube characteristics for each circuit. Since R_L does not change, the slope of each load line is the same. Only the horizontal and vertical intercepts change. The load lines intersect the tube characteristics at points A, B, C, and D. To obtain the transfer characteristics, the plate current is plotted as a function of input voltage. This is shown in Fig. 4.14(c).

Once this construction is understood, a simpler method can be shown to obtain the transfer curve. In the curve shown in Fig. 4.14(b), the vertical components of the points A, B, C, and D are the values of I_b for the respective input voltages. Also, the intercepts of the load-line curves with the horizontal axis are the values of E_{in}. Thus the transfer curve can be drawn in on the characteristic curve of the original tube by projecting the points A, B, C, and D to the right until they intercept the respective vertical projections of E_{in} for each point. Relabeling the horizontal axis as E_{in} completes the construction. This is shown in Fig. 4.15. The transfer curve points are labeled A', B', C',

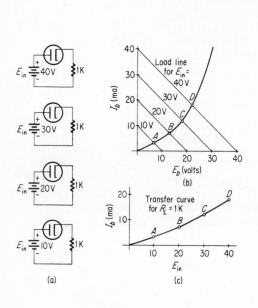

Fig. 4.14 Graphical determination of a dynamic transfer curve.

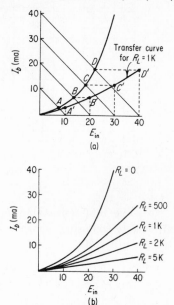

Fig. 4.15 (a) Simplified method curve. (b) Transfer curves for various load resistors.

and D'. It should be noted that a different value of R_L would result in a different transfer curve. Using this technique, Fig. 4.15(b) shows the transfer curves for several values of R_L. As can be seen, the greater the R_L, the more linear is the transfer curve. This is due to the fact that as R_L increases, the resistance of the nonlinear diode becomes small compared with R_L. Also, the original diode characteristic curve becomes a part of the family of transfer curves. It is the transfer curve for $R_L = 0$.

The use of the transfer curve in predicting output voltages will now be shown. Consider the circuit shown in Fig. 4.12, and assume that $e_{in} = 20 \sin \omega t$ volts, $R_L = 1$ K, and the tube characteristics are those of Fig. 4.14(b). The resulting transfer curve is shown in Fig. 4.14(c). Superimposing the variable input voltage on this diagram allows one to construct the output current, and thus the output voltage. This procedure is shown in Fig. 4.16.

The points on the input wave form e_{in}, (A, B, C, D, and E) are projected into the output-current wave form points A', B',

C', D', and E'. From this, the output voltage waveform, the voltage drop across R_L can be obtained from the relation

$$e_{\text{out}} = i_b R_L \tag{4.4}$$

As can be seen, the input voltage is distorted by the diode circuit. Only every other half-cycle is allowed to produce current flow in the circuit, since it is only during these periods that the diode is forward biased. This process is called *rectification*.

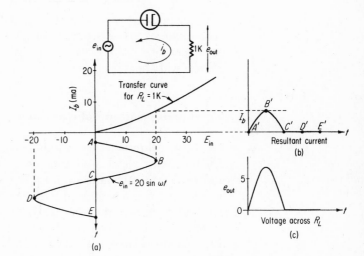

Fig. 4.16 Dynamic solution utilizing the transfer curve.

Example 4.2 Using the circuit and characteristics shown below, obtain the output waveform for $R_L = 10$ K and $e_{\text{in}} = 20 \sin \omega t$.

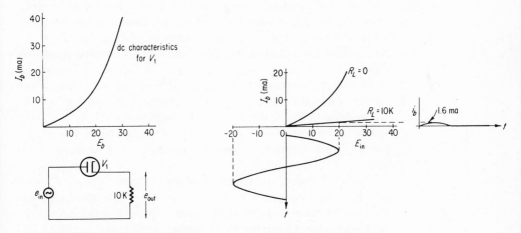

1.   The transfer curve for $R_L = 10$ K is drawn first. Then the input voltage is projected to find i_b.

2. The maximum value of i_b is 1.6 ma. Thus the maximum value of e_{out} is given as $e_{out} = i_b R_L = 16$ volts.

3. The output voltage becomes

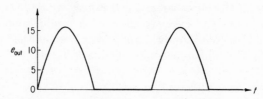

4. Comparing this with the results for $R_L = 1$ K, where the peak value of e_{out} was only about 7 volts for the same input as used here, one can see that as R_L gets larger, more of the input voltage will appear across R_L.

4.9 Conduction in Solids—Energy-Band Picture

The electrical resistance of solids varies over a tremendous range. The resistivity, which is defined in Eq. (1.5), $dR = \rho \ (dl/S)$, where ρ is the resistivity, varies from 1.6×10^{-8} ohm per meter at 20 °C for silver to over 5×10^{15} ohm per meter at 20 °C for fused quartz. This range is illustrated in Table 4.2. The metals are characterized by relatively low resistivity. On the other end of the scale, the insulators have extremely high resistivities. This range in resistivity is succinctly explained by the energy-band concept.

An isolated atom has electrons ordered in discrete energy levels, as shown in Fig. 4.17(a). In this diagram the energy levels are shown in two ways. The first is an energy-level diagram. It has only one axis, vertical. The horizontal extension is for convenience only. In the energy-level diagram, the electrons are shown to have discrete energies. The most energetic electron, in the $n = 4$ level, is still a few electron volts below the ionization level. The lines $n = 1, n = 2, \cdots, n = 4$, represent the permissible energy levels for the electrons in the atom shown, for the ground state. In this example, there would be 2 electrons in the $n = 1$ level, 8 electrons in the $n = 2$ level, 18 electrons in the $n = 3$ level, and the remaining electrons would be contained in the $n = 4$ level. These outermost electrons determine the valence of the atom.

There are other permissible discrete levels beyond $n = 4$, which lie between the $n = 4$ and $U = 0$ levels. These levels correspond to excited states of the atom. Given the proper amount of energy, electrons in the lower levels can be raised to these other levels, producing an excited atom that usually reverts back to the ground state (after an excited lifetime of about 10^{-8} sec) with the subsequent radiation of the excess energy. Energies above $U = 0$ symbolize electron energies that are above the escape energy for the particular atom. If the minimum energy required for escape from the atom is arbitrarily taken as zero, then trapped or bound electrons will have negative energies and free electrons will have positive energies. In the second type of diagram, the potential-energy diagram, both the energy levels and their radial extension are shown. The potential energy diagram uses the same vertical scale, but adds the radial dimension along the abscissa. This diagram shows not only the energy levels, but also their

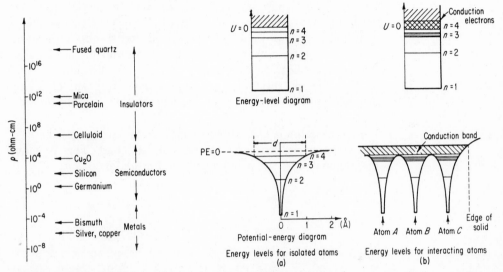

TABLE 4.2 The resistivity of several materials.

Fig. 4.17 Energy levels for the isolated atom. (b) Energy levels for interacting atoms.

radial extent that is, the radial dimensions of the electron orbits. Thus, the dimension d in Fig. 4.17(a) represents the physical size of the atom. Each diagram is useful in its own way. The potential-energy diagram is the more complicated and is usually not too well known for a given atom. Energy transitions, however, can be measured precisely from optical and X ray spectra data, so that energy levels are well known for a given atom. Thus the energy-level diagram is the more frequently used explanatory device.

If the energy-level diagram shown represents a copper atom, the $n = 1$ level would be at -8047 eV, the $n = 2$ level at -928 eV, and the $n = 4$ level at -7.7 eV. If a correct vertical scale were used, the $n = 4$ level would be too close to the $U = 0$ level to be seen. The levels shown in Fig. 4.17(a) are what would be expected in a copper gas, where the atoms are separated by five or more radii. When the atoms are compacted to form a solid, the interactions between the outermost electrons and the surrounding atoms produce a change in the energy-level structure for each atom. This is shown in the energy-level diagram of Fig. 4.17(b) as a broadening and splitting of the energy levels. For copper, the uppermost level (the $n = 4$ level) broadens until it overlaps the conduction band and the lower level, $n = 3$. This means that the single electron in the $n = 4$ level is now free of the electrostatic force binding electrons to nuclei. The electron is still bound to the lattice and its motion is not independent of the motion or position of the other electrons.

The potential-energy diagram shows this much better. Not only are the energy levels broadened, but the potential energy level between atoms is also reduced. Thus electrons acquiring sufficient energy in the conduction band find turning points only at the lattice termination. To be precise, however, the situation is more complicated than shown. A lattice is three-dimensional, so that the potential-energy diagram is a function of spatial orientation with respect to each nucleus. But, the simple approach will suffice here.

The insulator is characterized by a large energy gap between the valence electrons and the conduction band. This is shown diagrammatically for diamond, a good insulator, in Fig. 4.18(a).

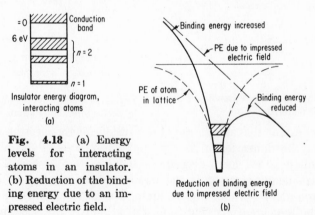

Insulator energy diagram, interacting atoms

(a)

Reduction of binding energy due to impressed electric field

(b)

Fig. 4.18 (a) Energy levels for interacting atoms in an insulator. (b) Reduction of the binding energy due to an impressed electric field.

The carbon atom, the constituent of diamond, has two electrons in the $n = 1$ level and four electrons in the $n = 2$ level. When the atoms interact in the diamond lattice, the energy levels for the $n = 2$ level split into two bands. However, the upper band is still 6 eV below the conduction band. Thus at room temperatures, where the average thermal energy that can be imparted to an electron is of the order of 0.025 eV, the electrons are tightly bound in their respective levels and cannot participate in any current when an electric field is applied. If the electric field is made strong enough, the potential energy (PE) diagram for an atom can be distorted sufficiently so that the binding energy is reduced to the point where electrons from the uppermost $n = 2$ level can escape. The electric field, in this case will be so great that the free electrons gain enough kinetic energy between atoms to ionize these atoms by collision. These electrons in turn liberate others, and an avalanche of electrons is produced.

The final result is a complete breakdown of the insulator. As this initial current passes through the insulator, it heats up the atoms adjacent to the current path, liberating more electrons. The current builds up very quickly (in microseconds) to a sizable value, literally punching a hole through the insulator by vaporization. Ceramic types of insulators are actually split apart when this breakdown field is reached.

Semiconductors are characterized by small energy gaps between the valence band and the conduction band. The energy level diagrams for silicon and germanium, two well-known semiconductors are shown in Fig. 4.19(a). Silicon has an energy gap of

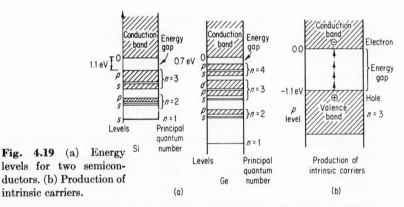

Fig. 4.19 (a) Energy levels for two semiconductors. (b) Production of intrinsic carriers.

about 1.1 eV and germanium has an energy gap of 0.72 eV. These energy gaps are small enough so that there is a finite probability that some of the electrons in the valence band will acquire enough thermal energy to jump to the conduction band. This leaves behind a net positive charge, a hole.

The electron now has enough kinetic energy to stay free and can be accelerated by an externally applied electric field, thus producing a current. The hole can also move under the influence of the externally applied field. With no external electric fields, the hole and electron drift about in random motion, producing no net current. At this point a physical picture of what is happening may be of help. Both germanium and silicon form solids held together by covalent bonds, that is, neighboring atoms share electrons. Such a two-dimensional lattice is shown in Fig. 4.20(a). Each germanium atom has four electrons in its outer shell. These are shared with four others from neighboring atoms, forming covalent bonds. The thermal energies of vibration are sufficient to free some of these electrons. When this happens, the site formerly occupied by the electron is now positively charged (the atom is ionized) and the positive site is called a *hole*, Fig. 4.20(b). Therefore, whenever a conduction electron is formed, a hole is also formed. Both the hole and electron wander about the lattice.

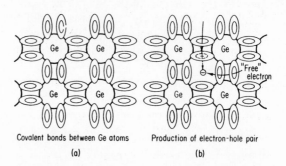

Covalent bonds between Ge atoms   Production of electron-hole pair

(a)   (b)

Fig. 4.20 (a) Covalent bands between germanium atoms. (b) The production of an electron-hole pair.

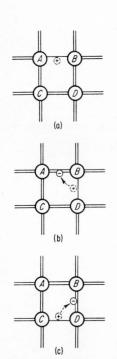

Fig. 4.21 Motion of holes in a lattice.

The electrons behave much like conduction electrons in metals, but the wandering of the holes is due to a different mechanism. This is illustrated in Fig. 4.21. A hole existing between atoms A and B is due to the ejection of an electron from this bond. The electron will be omitted, for simplicity. If the conditions are correct, at some later instant of time the electron in the bond between atoms B and D will find it possible to jump to the vacant bond, Fig. 4.21(b). This effectively moves the hole to a position between atoms B and D. Another transfer of a bond electron moves the hole to a position between atoms C and D. Instead of describing this complicated process each time, it is more convenient to say simply that the hole has wandered from a position between atoms A and B to a position between atoms C and D. The effect an external electric field will have, when applied to this semiconductor, is to cause the electrons to migrate in one direction and the holes to migrate in the other direction, Fig. 4.22. When the holes reach the negative end of the semiconductor, they are filled by electrons from the external conductor. Also, the electrons reaching the positive end of the semiconductor enter the external conductor. This produces a current in the external conductor, which can be measured by the ammeter. This current in the semiconductor is called *intrinsic conduction*. If no more electron-hole pairs were formed, this current would soon stop. However, the semiconductor is in thermal equilibrium, which means that as soon as an electron-hole pair vanishes at the edges of the semiconductor, on the average, another electron-hole pair forms in the semi-conductor. At a particular temperature the number of electron-hole pairs is a constant for a given material. At room temperature germanium has about 2.5×10^{13} electron-hole pairs per cubic centimeter and silicon has about 1.4×10^{10} electron-hole pairs per cubic centimeter.

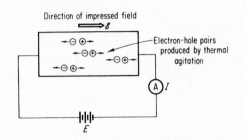

Fig. 4.22 Intrinsic conduction in a semi-conductor.

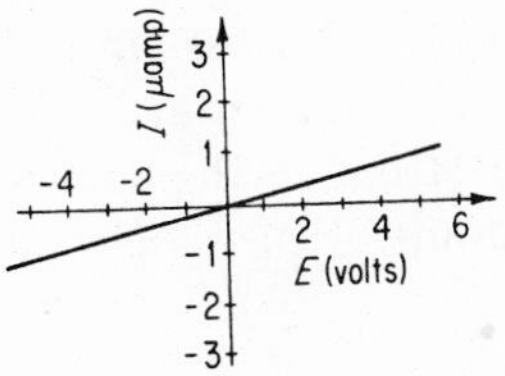

Fig. 4.23 E–I characteristics for intrinsic conduction.

At higher temperatures, more electron-hole pairs are formed per unit volume, and therefore the current will increase at higher temperatures. Voltage-current characteristics show that the resistance of a typical semiconductor is rather high, Fig. 4.23. This is shown by the fact that for an applied voltage of 5 volts, for example, the resultant current is only about 1 μamp. This is equivalent to a resistance of about 5 million ohms. The intrinsic current, therefore, is usually very small. Figure 4.24 shows the same characteristics for the case in which not only the voltage but also the temperature is changed. The temperature is held constant for each curve, but $T_3 > T_2 > T_1$. It should be noted that each point on the curves stands for three quantities I, E, and T.

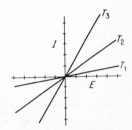

Fig. 4.24 Temperature dependence of intrinsic conduction.

4.10 Doping of Semiconductors

In effect, the semiconductor acts like a high-resistance material. In order to make the semiconductor a useful material, it is necessary to place other atoms (impurity atoms) into the semiconductor lattice. These atoms are mixed in with the semiconductor atoms in the ratio of 1 impurity atom to about 10^6 semiconductor atoms.

Two important semiconductor materials are germanium and silicon. Two much used impurities are arsenic and indium. Phosphorus, antimony, boron, aluminum, and gallium are also used as impurities. The controlled addition of these foreign atoms to the semiconductor is called *doping*. Although the ratio sounds large, a million to one, there is still an appreciable number of impurity atoms in a cubic centimeter of the semiconductor (about 10^{16} per cubic centimeter) simply because there are so many atoms per cubic centimeter in the solid (about 10^{22} per cubic centimeter). The effect of doping germanium with arsenic is entirely different from the reaction when germanium is doped with indium. This difference is what makes possible the manufacture of useful electric devices out of semiconductors.

If germanium is doped with arsenic, then an atom with five electrons in the outer shell, arsenic, is placed between germanium atoms, Fig. 4.25. Four of the arsenic electrons will bond with

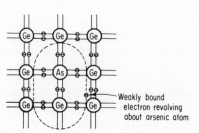

Weakly bound
electron revolving
about arsenic atom

Fig. 4.25 Introduction of arsenic to produce a weakly bound electron.

neighboring germanium atoms, leaving an excess electron in the outer shell. This electron will not be bound tightly to the arsenic atom because of shielding. This shielding arises for two reasons: first, the inner shells in the arsenic atom screen the nucleus, and second, the bonding electrons also produce screening that was not there before the bonds were established. The excess electron attempts to describe an orbit about the parent arsenic atom, but is converted into a free, or conduction, electron because the ionization energy is only 0.05 eV. The introduction of arsenic, therefore, converts the semiconductor into a relatively low-resistance material that contains conduction electrons. If an external electric field is applied to this material, a current will result as these conduction electrons migrate unidirectionally.

The current produced by the introduction of impurity atoms is called *extrinsic conduction*. There is conduction in this material, due to two sources: first, the major conduction due to free electrons, the extrinsic conduction; and second, the electron-hole pairs formed at random by the thermal breaking of bonds, the intrinsic conduction. The extrinsic conduction is greater than the intrinsic conduction by a factor of at least 10^3. Therefore, in the arsenic-doped material, the electrons are the major carrier. This type of doped material is known as an N-type semiconductor (N for *n*egative majority carrier).

If the germanium is doped with indium, an atom having only three electrons in its outer shell, then when the covalent bonds are formed between neighboring germanium atoms, one bond remains devoid of an electron. The introduction of indium into the semiconductor forms excess holes. These holes wander about for exactly the same reason given in Sec. 4.9. Under the influence of an external electric field, the holes migrate in the direction of the electric field and constitute a current. The conductor thus formed is known as a P-type material because the majority carrier is a *p*ositively charged hole.

The introduction of impurity atoms in controlled amounts into semiconductors produces two types of conductors, N-type and P-type. The resistance of the semiconductor is reduced with the introduction of these impurities.

4.11 Junction Diodes

When P and N types of materials are joined to form a thin interface, a junction diode is formed. For efficient operation, the interface, or junction, must be microscopically smooth. This is accomplished by several methods such as crystal pulling, the alloy process, and the diffusion method.

The important property of a junction diode is that of rectification. The diode will act like an electronic switch that is either open or closed, depending upon the polarity of the voltage impressed across it. This property of unidirectional current flow will be explained qualitatively by looking at the motion of the holes and electrons in the P and N types of semiconductors composing the diode.

To begin with, when the two materials are joined electrically, a contact potential is produced at the junction. This is shown diagrammatically in Fig. 4.26. There will be a drift of electrons and holes across the junction due to diffusion. This will cause a number of electrons to fall into the holes. These carriers are then in the valence band and cannot contribute further to current flow. This will produce on each side of the junction a charge layer due to the impurity ions, which cannot move. The dipole layer produces an electric field that stops further diffusion of holes and electrons, and equilibrium is established.

The electric field $\mathcal{E}'$ established across the junction produces a voltage drop called the *contact potential*. For germanium, this is of the order of 0.6 volt. The field is established over a very small

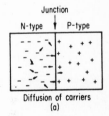

Diffusion of carriers
(a)

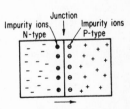

Equilibrium field established
(b)

Fig. 4.26 (a) Diffusion of carriers across a junction. (b) Establishment of a contact potential field.

distance, of the order of microns. Holes and electrons physically removed from the junction do not "see" this field until they are very close to the junction. In the field-free region, the motion of the carriers is essentially governed by diffusion processes. If a battery is connected across the diode so that the positive end is connected to the P type of material and the negative end is connected to the N type of material, Fig. 4.27(a), a current will be produced. It should be noted that no current can flow until the contact-potential field ε' is overcome. Thus, no appreciable current occurs in the forward direction until the external battery voltage exceeds about 0.6 volt. Any further increase in externally applied voltage produces a large increase in current as the electrons and holes drift toward the junction, "see" the electric field, are swept across the junction, and combine. Holes and electrons thus united can no longer participate in conduction. At the same time more electrons are entering the N type of material from the external circuits and holes appear at the other terminal as electrons are pulled out of the P type of material into the external circuit. There is an overall neutrality throughout the circuit as the carriers move within the semiconductor and connecting wire under the influence of the impressed field due to the battery.

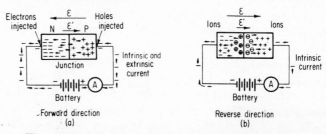

Fig. 4.27 (a) Forward direction. (b) Reverse direction.

In the reverse connection, the negative end of the battery is connected to the P type of material and the positive end is connected to the N type of material, Fig. 4.27(b). The impressed electric field is in such a direction as to drive the electrons and holes away from the junction, enhancing the field due to the contact potential. The majority carriers can no longer drift toward the junction. The current decreases by a factor of 10^{-3} to 10^{-5} compared to that in the forward direction. This small reverse current is due to the presence of the intrinsic carriers, the electron-hole pairs that appear at random throughout the semiconductor as a result of thermal agitation. These minority carriers were neglected in the explanation of forward current because their number was so small compared with the majority carriers. Now, however, they are the dominant carriers. Holes formed in the N type of material and electrons formed in the P type of material drift toward the junction, are swept across, and recombine. Holes and electrons are injected to keep the number of minority carriers constant in the semiconductor. The result is a small current in the external circuit, in the order of 10^{-6} to 10^{-7} amp.

The dc characteristics can be summed up in the $E-I$ characteristics shown in Fig. 4.28. In the forward direction, it can be seen that no appreciable current flows until the voltage has reached about 0.5 to 0.6 volt (point C), then the current increases rapidly, with further increase in voltage (C to D). This shows that when the diode conducts, it has a small internal resistance. In the reverse direction, the current will appear to be zero on the scale shown because it is in the microampere range (B to A). The overall effect, then, is that when the diode is forward-biased, it acts like a small resistance, of the order of 10 ohms, and when it is biased in the reverse direction, it acts like a very large resistance, of the order of 10^6 or 10^7 ohms.

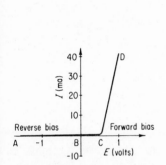

Fig. 4.28 $E-I$ characteristics for the PN junction diode.

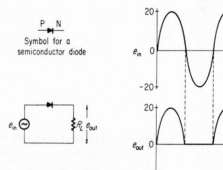

Fig. 4.29 The behavior of a junction diode in a dynamic circuit.

If used in a rectifier circuit, as shown in Fig. 4.29, the diode will produce a pulsating dc-voltage output similar to the output produced by a thermionic vacuum diode. The input voltage is given by $e_{in} = 20 \sin \omega t$ volts. At alternate cycles, the diode is forward-biased and will conduct. Since in most cases $R_L \gg R_{diode}$ during the forward half-cycle, the voltage across R_L is equal to e_{in}. During the alternate half-cycles, the diode is reverse-biased and will not conduct, for all practical purposes, and the voltage across R_L will be zero.

The symbol for a semiconductor diode is also shown in Fig. 4.29, that is, an arrow pointing into a line. The arrow represents the P type of material and the vertical line represents the N type of material. The arrow points in the direction of positive current flow when forward-biased.

4.12   The Gas Diode

Gas tubes are used in many applications, such as rectification, voltage regulation, particle detection, photon detection, and switching. The overall characteristics of gaseous conduction are very complicated and lengthy to relate. The following is a brief introduction, sufficient to help describe the use of a gas diode in the applications mentioned above. In essence, the description will be operational, that is, determining the dc E–I characteristics and, with the use of these characteristics, predicting the operation of the device in a circuit.

Consider the device shown in Fig. 4.30(a), a gas diode consisting of a thin wire, the cathode, and a concentric cylinder, the plate.

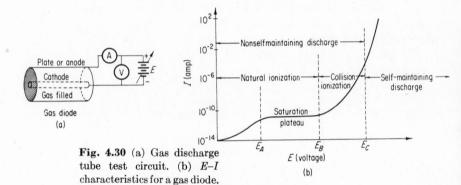

Fig. 4.30 (a) Gas discharge tube test circuit. (b) E–I characteristics for a gas diode.

The intervening region contains a gas, usually at a reduced pressure. The ammeter measures the current from the cathode to plate and the voltmeter indicates the voltage across the diode. It is assumed that the voltage drop across the ammeter is negligible. The voltage supply E_{bb} can be varied from zero to several hundred volts. The result of varying E_{bb} and measuring the resultant current is shown in Fig. 4.30(b). In general, as the voltage E is increased, I increases until a potential E_A is reached, and then the current saturates, that is, even though E is increased, I does not increase. A plateau of current is established from E_A to E_B, wherein a change in applied voltage produces no appreciable change in current. The current saturates because, above the voltage E_A, all available charge carriers are being collected.

Thus a further increase in voltage (up to E_b) produces no increase in current. The charge carriers are electrons and ions produced by the ionization of some of the gas atoms. This ionization is due to natural radioactivity and cosmic radiation. A condition of equilibrium exists between the number of gas molecules being ionized and the number of ions recombining. Because of considerations such as conservation of momentum and energy, most of the ions effect recombination at the cathode or walls of the gas diode. This ionization current is represented as a smooth curve in Fig. 4.30(b). This is the result that would be obtained with a dc meter or other current-measuring device having an averaging effect.

In reality the current consists of pulses, as units of charge are collected at the plate and cathode. Each ionizing event, such as a fast-moving cosmic ray, produces a track of ionized gas molecules in the diode. The electrons drift rapidly to the plate and produce a pulse of current. The slower-moving ions drift toward the cathode, recombining there. Thus, if the current in the diode were resolved timewise, it would appear somewhat as shown in Fig. 4.31. The dc meter has indicated the average value shown by the dashed line and labeled I_{av}. This current is of the order of 10^{-12} amp.

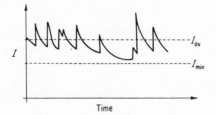

Fig. 4.31 Current as a function of time in a gas diode.

A device such as an oscilloscope would show the pulsating, random shape of the current. Each pulse is due to an ionizing event in the gas diode. There is a nonzero baseline due to the slowly drifting ions that are constantly recombining at the cathode. This value of current has been labeled I_I. A gas diode operating in the plateau region, then, can be used to detect ionizing radiation.

If the potential is increased beyond E_b, there is a region of potential where multiplication of the incident ionization is produced because of collision ionization. This is called the *proportional region*. In Fig. 4.30(b), this region is between potentials E_b and E_c. The current scale shows that there is a considerable increase in current over the plateau region. Again the curve indicates that the current is a smooth function. This is due to the fact that only average current has been plotted as a function of voltage. The current is still made up of pulses, as in the preceding case. However, for each electron-ion pair produced by an incoming cosmic ray, for example, there may be 10^4 to 10^6 electrons arriving at the anode. This multiplication of initial ionization occurs because the electric field is now high enough to accelerate the electrons between collisions with gas molecules to kinetic energies high enough to produce ionization upon impact. Each of the released electrons in turn gains enough kinetic energy between collisions to produce ionization.

This procedure produces a cascade of electrons as they journey through the gas toward the positive plate, where they are collected. The procedure is shown schematically in Fig. 4.32. A single ion-electron pair is shown created by a cosmic-ray ionization of a gas molecule in Fig. 4.32(a). By the avalanche process, the electrons multiply in number as they quickly move to the plate. The resulting ions drift slowly to the cathode. The ions move rather slowly compared to the electrons because they are more massive.

In Fig. 4.32(c), the electrons have reached the plate and are being absorbed as they enter the external circuit. In Fig. 4.32(d)

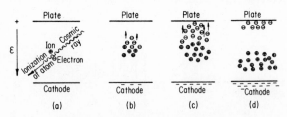

Fig. 4.32 Collision ionization process.

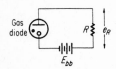

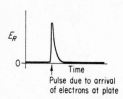

Fig. 4.33 A simple gas diode circuit and response in the proportional region.

the electrons have been almost wholly absorbed by the plate while the ions still drift slowly toward the cathode. There is now a capacitor-like effect at the cathode because of the space charge of ions approaching the cathode. When they strike the cathode, they recombine. Thus the most probable place for a momentary glow to appear is at the cathode. In the external circuit of Fig. 4.33, the avalanche of electrons collected by the plate appears as a pulse of current through resistor R as the electrons move through the external circuit to the cathode of the gas diode. The current lasts for a brief time because the flow of electrons is nonself-sustaining. In order for current to continue, the supply of electrons must be replenished by some mechanism. In the proportional region of collision ionization, as soon as all the electrons are collected, ionization stops. The ions are moving too slowly to produce ionization.

With respect to the E–I characteristics shown in Fig. 4.30, if the voltage is raised above the critical voltage E_c, a self-sustaining discharge can be produced. This can come about for several reasons, such as secondary emission at the cathode, photoelectric emission by ultraviolet or soft X rays, and thermionic emission at the cathode due to ion impacts heating up the cathode. In any event, to sustain the discharge, electrons must be liberated from the cathode. In the secondary emission or thermionic emission methods, the field must be high enough to accelerate the ions to the energies required. In the photoelectric method, the gas chosen must be such that it produces ultraviolet or soft X radiation upon recombination. Since the recombination takes place near the cathode, the X-ray or ultraviolet photons can produce electrons by photoelectric emission. These electrons then avalanche to the plate, creating more ions, which drift toward the cathode and recombine. Characteristics of this self-sustaining discharge are a characteristic glow pattern in the tube and a markedly reduced resistance to current flow. The voltage at which this occurs will be called the *firing potential*.

The glow pattern is shown in Fig. 4.34. Close to the cathode there is a bright glow called the *sheath*. In this region the electrons

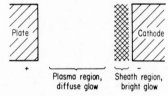

Fig. 4.34 Glow discharge characteristics.

produced at the cathode are accelerated toward the plate. It is also the region of greatest ion density. Hence, the sheath is the region of greatest recombination and therefore the region of brightest glow. Between the sheath and the plate is a volume of diffuse glow called the *plasma region*. It contains a large number of positive ions (produced by collision with the electrons that pass through the sheath region) and a large number of electrons (avalanching to the plate). The diffuse glow in the plasma region is due to the small number of recombinations that occur.

The firing potential can be found experimentally with the aid of the circuit shown in Fig. 4.35(a). The resistor R is placed in the circuit as a current-limiting resistor. When the gas diode ionizes, its resistance drops drastically, in a matter of microseconds. The voltage supply E_{bb} is started at zero and raised slowly; the voltmeter V will indicate the voltage across the gas diode. This is indicated in the graph of voltage across the tube versus current through the tube as the line OA. When the ionization voltage E_I is reached, the tube suddenly enters the glow discharge region. The resistance of the tube drops from an open circuit to a few thousand ohms. The resulting current flow through R and the gas diode causes the voltage across the tube to drop (some of the applied voltage is dropped across R). This jump is shown as the dashed line AB in Fig. 4.35(b). Further increase in E_{bb} produces

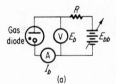

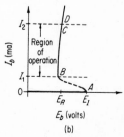

Fig. 4.35 Test circuit and $E–I$ characteristics in the glow discharge region.

an increase in the voltage across the tube and a rapid increase in resulting current. This is shown as the near-vertical line BCD. A small change in applied voltage produces a large change in current.

Characteristics like those described above are used in the design of voltage regulator tubes. Voltage regulator tubes and associated circuits are described in Chapter 5. In the $E-I$ characteristics shown here, the range of operation lies between I_1 and I_2. Below I_1, the operation becomes spotty and irregular, while above I_2, the current would be too great for the tube used and would produce dangerous overheating.

In summary, a gas diode can be used in one of three regions: the plateau region, the proportional region, and the glow discharge region. The first two regions are used for photon and particle detection, and the glow discharge region is used in voltage regulator tubes. Normally, a gas diode is designed for specific use in one of the three regions.

4.13   The Thyratron

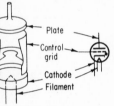

Fig. 4.36 Thyratron construction and schematic symbol.

Although the thyratron is not a diode, it is used as a voltage-control switch. Its mode of operation is such that it is either conducting or nonconducting, depending upon the voltages applied to it. The construction of a thyratron and its schematic symbol are shown in Fig. 4.36. The control grid is a cylinder with a small aperture to allow electrons to flow from cathode to plate. The tube operates in the glow region. The source of electrons for the self-sustaining discharge is thermionic emission from the filament-heated cathode.

In normal operation the plate is positive with respect to the cathode and the control grid is negative with respect to the cathode. There is a relationship between the firing potential on the plate and the grid potential. Once the tube has ionized and is in the glow region, the control grid has no further control on the thyratron because the grid is immediately surrounded by a positive-ion sheath, and the tube will remain ionized until the plate voltage of the tube is reduced to a value of about 5 to 10 volts. At this point the tube deionizes.

The characteristics of importance for a thyratron are the firing voltage and grid voltage. The firing characteristics of a typical thyratron are shown in Fig. 4.37. The circuit used to determine the characteristics is also shown. These characteristics are obtained by starting with the plate voltage initially at zero and setting the grid to a particular value; for example, $E_c = -10$ volts. Then, the plate voltage is increased until the tube fires. The voltage recorded just before the tube fires is the firing potential. In this case, for $E_c = -10$, the firing potential will be recorded as $E_F = 100$ volts, and point B on the curve is recorded. Then the plate voltage is reduced to zero, to deionize the tube, E_c is changed, and E_{bb} is raised to determine the new firing potential. With this procedure, the curve of firing potential, E_F versus grid voltage E_c can be determined.

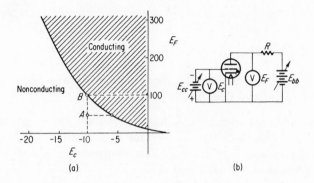

(a) (b)

Fig. 4.37 Test circuit and on-off characteristics for a thyratron.

In general, as the grid voltage is made more negative, the firing potential is increased. The curve indicates whether the tube will be ionized or nonconducting for a given set of conditions. For example, if the voltage on the plate is 50 volts and the grid voltage is -10 volts, the tube will be in the nonconducting state. The tube will fire if one of two things is done. Starting at point A, the tube will fire if the plate voltage is raised to 100 volts. Also, starting at point A, the tube will fire if the grid voltage is reduced sufficiently to -6 volts. Once the thyratron fires, it remains in that condition until the plate voltage is reduced to the deionization point. The use of the thyratron in an oscillator circuit is shown in Chapter 9.

Problems

4.1 Calculate the saturation current for a 5-cm long, 1-mm diameter tungsten wire heated to 2000°C.

4.2 Find the dc resistance for a vacuum diode, whose characteristics are shown, at the points $E_b = 5$, 10, and 15 volts.

4.2

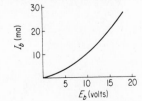

4.3 Find the dc resistance for the diode characteristics shown for the points $I_b = 20, 40,$ and 60 ma.

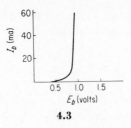

4.3

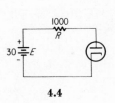

4.4

4.4 Find I_b and E_b for the diode circuit shown. Use the dc characteristics from Problem 4.2.

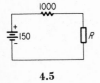

4.5

4.5 A nonlinear component has dc characteristics given by $I = 5E^2$ for $E > 0$ and $I = 0$ for $E < 0$, where I is in milliamperes and E is in volts. This component is placed in the circuit shown. Find I and E.

4.6 Given: An I versus E curve for a nonlinear device. If the applied voltage is a sinusoidal one, $e_D = 10 \sin \omega t$, find the instantaneous current i_D graphically. For your plot, use intervals of 30 deg.

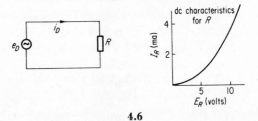

4.6

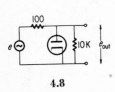

4.8

4.7 For the characteristics shown in Problem 4.3, plot the dynamic transfer curves for $R = 10, 50,$ and 100 ohms.

4.8 In the circuit shown, the input voltage is given by $e = 100 \sin \omega t$. Determine e_{out}. Graph it for several cycles. Assume the diode resistance is 1 K when conducting.

5.1 Introduction

Power supplies are used in electronic equipment to replace batteries. Most circuits require dc voltages to operate properly. Also, if vacuum tubes are being used, a low voltage, high-current source is necessary to heat the filaments. Therefore the purpose of a power supply is to take the available 115 volt, 60 cycle ac and convert it into dc voltages and into filament voltages. Vacuum tubes are usually designed to utilize an ac voltage source for the filament supply.

Power Supplies

Figure 5.1 shows two block diagrams of the typical requirements for a vacuum tube circuit and for a transistor circuit. Vacuum tubes require a relatively high dc voltage supply (100 to 300 volts) and an ac filament voltage supply (6.3 volts). The transistor circuits do not require filament voltages, but only a dc voltage source at a relatively low value (4 to 24 volts). In order to obtain the voltages necessary, a power transformer is often used.

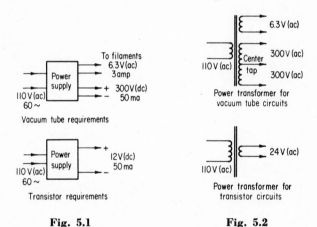

Fig. 5.1   Fig. 5.2

The power transformers shown in Fig. 5.2 could supply the voltage requirements for a vacuum tube circuit and for a transistor circuit. For the vacuum tube circuit, the secondary outputs are a stepdown winding for the 6.3-volt ac filament supply, and a step-up winding to produce the 300-volt supply. This high-voltage output is really a 600-volt center-tapped winding. Thus, with respect to the center tap, there is 300 volts across each half of the high-voltage winding. For the transistor circuit power supply, a simple stepdown transformer is all that is required. The one shown in Fig. 5.2 steps the 110-volt ac down to 24 volts ac. It should be noted that all ac voltages are given in terms of their rms values. The ac voltages produced by the secondary of the power transformer must now be transformed into dc voltages, with the exception of the filament voltage.

5.2 Rectification

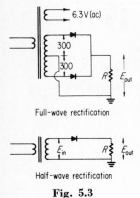

Full-wave rectification

Half-wave rectification

Fig. 5.3

In order to transform the ac voltages to constant dc voltages, the ac voltages must first be transformed to pulsating dc. This is done by rectifying the output from the power transformers. The on-off switching characteristic of a diode is used for this purpose.

Two rectifier circuits having resistive loads are shown in Fig. 5.3: a half-wave rectifier and a full-wave rectifier. The voltage developed across R is the output voltage for the rectifier circuit. To explain the action of the half-wave circuit, look at Fig. 5.4. All voltages will be given with respect to the ground symbol on a schematic diagram, unless otherwise specified. During the first half-cycle, the voltage developed across the power transformer winding is of such a polarity that the diode is biased in the forward direction; hence it conducts. Since the resistance of a diode when

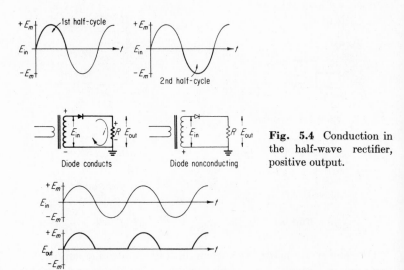

Diode conducts   Diode nonconducting

Fig. 5.4 Conduction in the half-wave rectifier, positive output.

conducting is small, almost all the voltage E_{in} appears across R. The current is in a direction such as to produce a positive voltage across R (with respect to ground). During the second half-cycle, the voltage across the transformer has reversed polarity and the diode is biased in the reverse direction. Thus it becomes nonconducting. For this half-cycle there is no voltage drop across R because the current through R is zero. The output voltage, E_{out}, for such a circuit is shown to be a pulsating dc voltage whose maximum value is equal to E_{in}.

Negative voltages can also be obtained from such a circuit. All that need be done is to reverse the diode connections. This is illustrated in Fig. 5.5. In this case, the diode is reverse-biased during the first half-cycle, and no current flows. During the second half-cycle, the bias is in the forward direction, and current will flow. However, the direction of the current through R is opposite to that in Fig. 5.4. In this case, a negative, pulsating dc voltage is produced across R. So, all that is required to produce a positive or negative voltage output is to connect the diode in the proper direction.

In the full-wave rectifier circuit of Fig. 5.3, two diodes are connected in such a way that there is current flowing through resistor R during every half-cycle. This action is explained graphically in Fig. 5.6. During the first half-cycle, the voltage across the center-tapped secondary is as shown in Fig. 5.3, that is, point A is positive with respect to the center tap (CT), and point B is negative with respect to the center tap. This means that during the first half-cycle, diode I is biased in the forward direction and will conduct, while diode II is biased in the reverse direction and will not conduct. Current will flow through resistor R, as shown, producing a positive voltage drop.

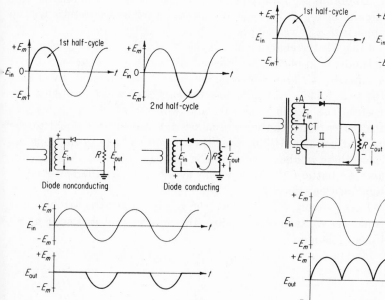

Fig. 5.5 A half-wave rectifier to produce a negative output.

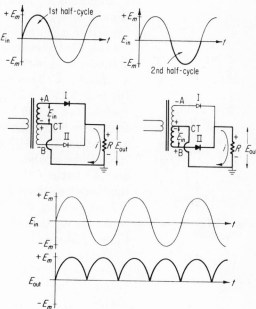

Fig. 5.6 Conduction in the full-wave rectifier.

During the second half-cycle, point A is negative with respect to the center tap CT, and point B is positive with respect to the center tap. Thus diode I is now biased in the reverse direction and diode II is biased in the forward direction. The current will flow through diode II and R, as indicated, that is, in the same direction through R as during the first half-cycle. As a result, the voltage E_{in} is transformed into a pulsating dc, E_{out}, as shown in Fig. 5.6. This action, where current flows through R every half-cycle, is called *full-wave rectification*. Where the recurrence frequency for a half-wave rectified-voltage wave form was 60 cps, the full-wave rectified-voltage wave form will have twice the applied frequency, 120 cps in this case.

Thus full-wave or half-wave rectification can change an ac voltage into a pulsating dc voltage. Now some method of smoothing this pulsating dc must be found.

5.3   Transients in RC Circuits

Before anything is done to smooth out the pulsating dc produced be rectifying an ac signal, the general principles of the charging and discharging of a capacitor should be reviewed. It will turn out that a resistor-capacitor (RC) combination can be used to transform a pulsating dc voltage into a fairly constant dc voltage.

To obtain the principles for capacitor charge and discharge, the circuits in Fig. 5.7 will be discussed, and then the results will be applied to a power supply.

Consider the charging of a capacitor. In Fig. 5.7(a), a voltage E is suddenly applied to a series circuit containing a resistor R and a capacitor C. The sudden application is made possible by closing the switch. The voltage across the capacitor will be initially zero (before the switch is closed) and will rise as charge flows into the plates, until the voltage across the capacitor is equal to the applied voltage. At this time the current in the circuit stops because there will be no difference in potential between the capacitor and the battery E. To show this analytically, a differential equation must be solved.

In the circuit of Fig. 5.7(a), the voltage across the battery must equal the voltage drops across the resistor and capacitor at all times. Therefore

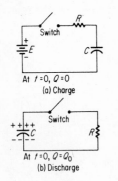

At $t=0$, $Q=0$
(a) Charge

At $t=0$, $Q=Q_0$
(b) Discharge

Fig.   5.7   (a) Charge. (b) Discharge.

$$E = e_R + e_C$$

$$= iR + \frac{q}{C} \tag{5.1}$$

where i is the instantaneous current through the resistor and q is the instantaneous charge on the capacitor plates. Both i and q are functions of time. Since it would be of interest to find i and q as functions of time, the following definition is used:

$$i = \frac{dq}{dt} \tag{5.2}$$

Eliminating i between Eq. (5.1) and Eq. (5.2),

$$E = \frac{dq}{dt} R + \frac{q}{C} \tag{5.3}$$

Multiplying through by C and solving for dq/dt,

$$RC \frac{dq}{dt} = EC - q$$

This can be rearranged as

$$\frac{1}{EC - q} \frac{dq}{dt} = \frac{1}{RC} \tag{5.4}$$

Integrating both sides with respect to t produces the following equation:

$$\int \frac{dq}{EC - q} = \int \frac{dt}{RC} \tag{5.5}$$

The left-hand side of the equation is a function of q only, and the right-hand side of the equation is a function of t only. The variables have been separated and the equation can be integrated as it stands.

$$\ln (EC - q) = -\frac{t}{RC} + \text{const} \tag{5.6}$$

To evaluate the constant of integration in Eq. (5.6), the initial conditions are inserted into the equation. These initial conditions are

$$q = 0 \quad \text{at} \quad t = 0 \tag{5.7}$$

Equation (5.6) becomes

$$\ln EC = \text{const} \tag{5.8}$$

Putting this value of the constant into Eq. (5.6),

$$\ln (EC - q) - \ln (EC) = -\frac{t}{RC}$$

$$\ln \left(\frac{EC - q}{EC} \right) = -\frac{t}{RC} \tag{5.9}$$

Raising the number e to the power given by both sides of the equation allows one to solve for q:

$$e^{\ln [(EC-q)/EC]} = e^{-t/RC} \tag{5.10}$$

$$\frac{EC - q}{EC} = e^{-t/RC} \tag{5.11}$$

$$q = EC(1 - e^{-t/RC}) \tag{5.12}$$

This equation for q gives the value of the charge on the capacitor for any time t. From this equation, all other facts about the circuit in Fig. 5.7(a) can be obtained. For example: The voltage across the capacitor is given by

$$e_C = \frac{q}{C} = E(1 - e^{-t/RC}) \tag{5.13}$$

The current in the circuit is given by

$$i = \frac{dq}{dt} = \frac{E}{R} e^{-t/RC} \tag{5.14}$$

and the voltage across the resistor is given by

$$e_R = iR = Ee^{-t/RC} \tag{5.15}$$

As a check:

$$E = e_R + e_C = Ee^{-t/RC} + E(1 - e^{-t/RC}) = E$$

The time represented by RC is called the *time constant* of the circuit, τ. The voltage across the capacitor, Eq. (5.13), can be plotted by using increments of t in multiples of RC. This produces a curve that can be used for any value of R and C. The curve is plotted in Fig. 5.8. After a time equal to 5 RC seconds (5 time constants),

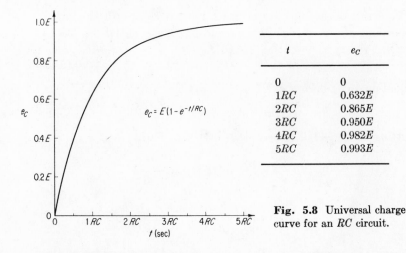

t	e_C
0	0
1RC	0.632E
2RC	0.865E
3RC	0.950E
4RC	0.982E
5RC	0.993E

Fig. 5.8 Universal charge curve for an RC circuit.

the capacitor can be assumed to be fully charged, for all practical purposes. Initially the capacitor charges quickly, and then the voltage across the capacitor approaches E asymptotically.

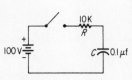

Example 5.1 In the circuit shown at left, find the time required to charge the capacitor to 50 percent of the initial voltage.

1. The *RC* time constant $\tau = RC = 0.001$ sec.

2. Using Eq. (5.13), it is rearranged to express the percentage as a fraction:

$$\frac{e_C}{E} = 1 - e^{-t/0.001} = 0.50$$

3. Solving for t:

$$e^{-t/0.001} = 0.50$$

$$-\frac{t}{0.001} = \ln\,(0.50)$$

$$t = 0.001\,\ln 2$$

$$\cong 6.93 \times 10^{-4}\ \text{sec}$$

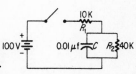

4. This can be verified graphically by using the curve in Fig. 5.8. Find t for the case $e_C = 0.5\,E$. This turns out to be about 0.7 *RC*. Since $RC = 10^{-3}$ sec, $0.7\,RC = 7.0 \times 10^{-4}$ sec. This is in good agreement with the analytical solution, which is more accurate.

Example 5.2 In the diagram shown at right top, find the voltage across the capacitor as a function of time.

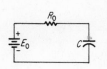

1. The important parameter is the time constant. This should be determined. To do this, change the circuit, by Thévenin's theorem, into the next diagram.

Then the time constant can be read off as $\tau = R_0 C$ seconds, and the voltage to which the capacitor charges is E_0. The result can then be plotted.

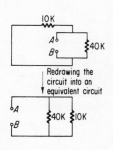

Redrawing the circuit into an equivalent circuit

2. To find R_0, remove the load C, short-out the voltage sources, and find R_{AB}. It can be seen that

$$R_0 = \frac{R_1 R_2}{R_1 + R_2} = 8\ \text{K}$$

Thus

$$\tau = R_0 C = 80\ \mu\text{sec}$$

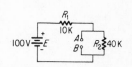

3. To find E_0, remove C, and then find E_{AB}.

From this, the voltage E_{AB} can be seen to be

$$E_{AB} = E\left(\frac{R_2}{R_1 + R_2}\right) = 100\left(\frac{40}{10 + 40}\right)$$

$$= 80\ \text{volts}$$

4. Thus the maximum voltage to which the capacitor will charge is 80 volts; the time constant is given by R_0C, where R_0 is effectively the resistance of R_1 and R_2 in parallel.

5. The curve of e_C versus t can be plotted, as shown below.

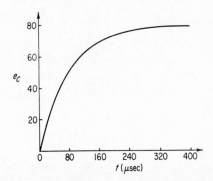

Once a capacitor is charged, it can be discharged by placing a resistance across its terminals. As might be expected, the smaller the resistance across C, the faster the capacitor will discharge. To obtain the analytical solutions, the circuit in Fig. 5.7(b) will be discussed. In this circuit the initial conditions are that, at $t = 0$ (just before the switch closes), there is a charge Q_0 on the capacitor plates. The sum of the voltage drops around the loop must equal the applied voltage:

$$e_C + e_R = 0 \tag{5.16}$$

In differential form,

$$\frac{q}{C} + iR = 0 \tag{5.17}$$

Utilizing Eq. (5.2), Eq. (5.17) becomes

$$\frac{q}{C} + \frac{dq}{dt} R = 0 \tag{5.18}$$

The variables can be separated to give

$$\frac{dq}{q} = -\frac{dt}{RC} \tag{5.19}$$

Integrating,

$$\int \frac{dq}{q} = -\int \frac{dt}{RC} \tag{5.20}$$

$$\ln q = -\frac{t}{RC} + \text{const}$$

Utilizing the initial conditions at $t = 0$, $q = Q_0$:

$$\ln Q_0 = \text{const} \qquad (5.21)$$

Putting this value of the constant of integration into Eq. (5.20) and solving for q,

$$q = Q_0 e^{-t/RC} \qquad (5.22)$$

To obtain the voltage across the capacitor as a function of time,

$$e_C = \frac{q}{C} = \frac{Q_0}{C} e^{-t/RC} = E_0 e^{-t/RC} \qquad (5.23)$$

where E_0 is the initial voltage across the charged capacitor. Equation (5.23) shows that the capacitor discharges in such a way that its voltage decreases exponentially with time. The time constant, $\tau = RC$, is again an important parameter. A plot of e_C in increments of τ is shown in Fig. 5.9. As can be seen, it is safe to say that after 5 RC time constants, the capacitor is discharged.

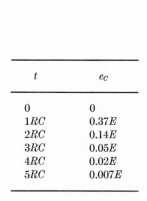

t	e_C
0	0
$1RC$	$0.37E$
$2RC$	$0.14E$
$3RC$	$0.05E$
$4RC$	$0.02E$
$5RC$	$0.007E$

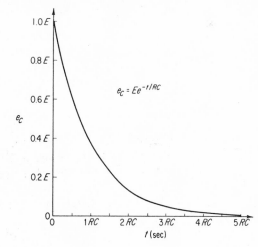

Fig. 5.9 Universal discharge circuit for an RC circuit.

Example 5.3 Show on the same graph the discharge curves for similar capacitors (0.1 μf), where each is initially charged to 100 volts, but where each is connected across a different resistance. This is illustrated at the right.

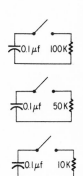

1. The time constants must be calculated first:

$$\tau_A = 10^5 \times 0.1 \times 10^{-6} = 0.01 \text{ sec}$$

$$\tau_B = 5 \times 10^4 \times 0.1 \times 10^{-6} = 0.005 \text{ sec}$$

$$\tau_C = 10^4 \times 0.1 \times 10^{-6} = 0.001 \text{ sec}$$

2. The following functions can now be plotted:

$$e_A = 100 \ e^{-t/0.01}$$

$$e_B = 100 \ e^{-t/0.005}$$

$$e_C = 100 \ e^{-t/0.001}$$

3. The plots become

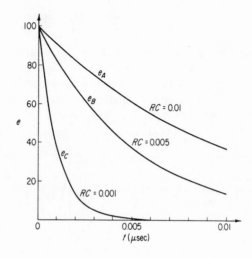

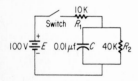

Example 5.4 At $t = 0$, the switch is closed. At $t = 500$ μsec, the switch is opened again. Plot the voltage across the capacitor as a function of time for at least 1600 μsec.

1. In the charge cycle, the time constant is given as

$$\tau = \left(\frac{R_1 R_2}{R_1 + R_2} \right) C = 80 \ \mu\text{sec}$$

and the voltage to which C can charge is given by

$$E = 100 \left(\frac{R_2}{R_1 + R_2} \right) = 80 \text{ volts}$$

Thus the charge equation is given by

$$e_C = 80 \ e^{-t/(80 \times 10^{-6})} \text{ volts}$$

2. Since the switch is held closed for 500 μsec, the capacitor charges up to 80 volts, that is, $5\tau < 500$ μsec.

3. During the time the switch is open, the capacitor discharges through R_2 only. The new time constant for the discharge is

$$\tau' = R_2 C = 400 \ \mu\text{sec}$$

and the initial voltage is 80 volts. Thus the discharge curve is given by

$$e_C = 80e^{-t'/(400\times10^{-6})} \text{ volts}$$

where t' is the time measured after the switch is closed.

4.   The plot of these two equations becomes

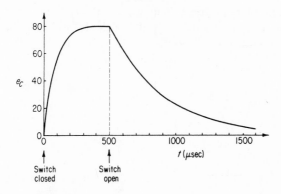

The second equation can be put in terms of t rather than t' by using a simple transformation

$$t' = (t - 500 \times 10^{-6})$$

That is, when $t = 500 \times 10^{-6}$ sec, $t' = 0$.

5.4   Simple *RC* Filter

As a first-order attempt at changing the pulsating dc of a rectified ac voltage into a constant dc voltage, consider the filter in Fig. 5.10. It consists of a capacitor C and resistor R_L. The power transformer is represented by the sinusoidal voltage generator. Without C across R, the circuit is a simple half-wave rectifier and the wave form for the output voltage would look like the curve labeled "without C," in Fig. 5.10. With the introduction of C, however, the output voltage changes drastically. The voltage output with C in the circuit is shown in the curve labeled "with C." This output voltage is a definite improvement and is nearly a constant dc voltage. The qualitative reason for this circuit action is as follows: When the diode is conducting, the circuit is essentially that of Fig. 5.7 (a), where R is the internal resistance of the diode. The internal resistance is low, so that the value of the resistance in the charging time constant is essentially R, that is (from Example 5.2), $R_0 = RR_L/(R + R_L)$; if $R_L > R$, then $R_0 \cong R$.

Thus the capacitor C charges through the diode when it is conducting. This occurs during time t_1 in Fig. 5.10. When the diode is not conducting, during the interval t_2 in Fig. 5.10, then the capacitor must discharge through R_L. This time constant, $R_L C$, is much larger than the charge time constant RC. The capacitor discharges slowly through R_L until, at point A in Fig. 5.10, the diode becomes biased in the forward direction as the input voltage exceeds the voltage across the capacitor, and the capacitor recharges quickly. At point B in Fig. 5.10, the input voltage goes below the capacitor voltage, and the diode is reverse-biased and ceases to conduct until point A is reached again. Therefore the diode conducts in pulses only, for the time t_1.

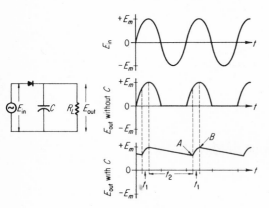

Fig. 5.10 Filtering action due to C.

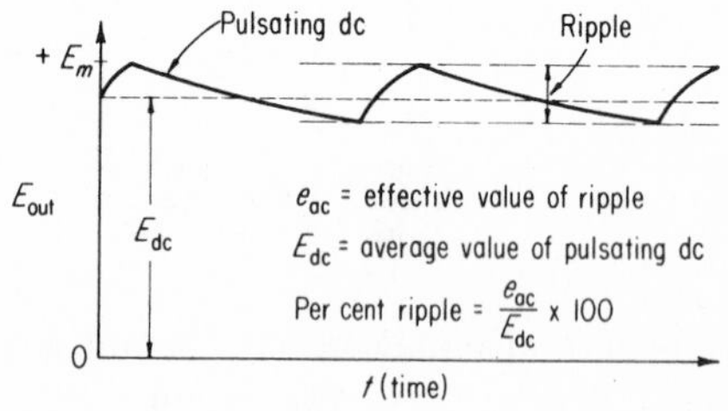

Fig. 5.11 Ripple, due to the charge and discharge of C.

The rapid charge and slow discharge of the capacitor produces an almost constant voltage across R_L. This type of simple filtering can reduce the pulsating dc amplitude to at least 5 percent of its original value by a proper choice of C. The peak-to-peak voltage variation is called a *ripple*. To give some quantitative value for the effectiveness of filtering, percent ripple is introduced. This is defined as the effective value of the ripple divided by the average value of the dc voltage. The quantities are illustrated in Fig. 5.11.

$$\text{Percent ripple} = \frac{e_{ac}}{E_{dc}} \times 100 \qquad (5.24)$$

This quantity is so defined that the better the filtering, the smaller the factor. For perfect filtering, that is, a constant dc voltage, the percent ripple would be zero.

5.5 Choke Input Filter

To produce more effective filtering, a filter more complex than the RC filter described in Sec. 5.4 must be used. There are two types in common usage, the choke input filter and the capacitor input

filter. The choke input filter is shown in Fig. 5.12(a). The inductor
L is called a *choke* when used in power-supply filters. It usually
has an inductance of the order of 1 to 20 henrys in this application.

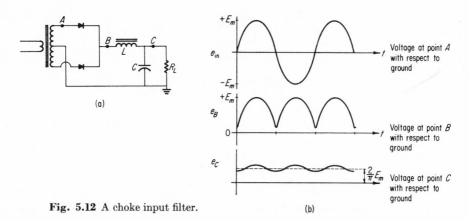

Fig. 5.12 A choke input filter.

Typical wave forms for a choke input filter are shown in Fig.
5.12(b). Point A shows the transformer secondary voltage as a
sine wave. The output of the full-wave rectifier is shown at point B.
The sharp peaks are rounded off because of the effect of the L–C
load. This voltage e_B has twice the transformer frequency because
of the full wave rectifier. At point C, the ripple has been reduced
considerably because of the filtering action of the choke L and the
capacitor C. The ripple is superimposed on the dc component. The
dc voltage drop across the resistance of the choke was neglected
here.

The rectifier circuit used with a choke input filter is always a
full-wave circuit. This helps ensure that current will always flow
through the choke. If a half-wave rectifier were inadvertently
hooked up to a choke input filter, then for one half-cycle, there
would be no current through the choke because there is no input
capacitor. The sudden cessation and beginning of current through
the choke would produce large voltage transients across the choke,
which could break down the insulation between windings on the
choke and short it out. The voltage across an inductor is given by

$$e_L = L \frac{di}{dt} \tag{2.24}$$

and if the current is stopped suddenly (small dt), a very large
voltage can be produced across the choke (of the order of 1000
volts).

In order to deal with this circuit quantitatively, ac network
theory developed in Chapter 2 must be used. In this case, a com-
plex-looking voltage is passed through a filter network and the

resultant output voltage is the one desired. This is illustrated in Fig. 5.13. The input to the filter, due to the power-transformer diode circuit, is a full-wave rectified signal. In order to calculate

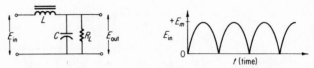

Fig. 5.13 The input voltage for a choke input filter.

the output voltage, the voltage E_{in} must be represented analytically. This can be done with aid of an infinite series:

$$E_{in} = \frac{2}{\pi} E_m - \frac{4}{3\pi} E_m \cos 2\omega t - \frac{4}{15\pi} E_m \cos 4\omega t + \cdots \quad (5.25)$$

This equation can be obtained from a Fourier analysis of the full-wave rectified voltage. It tells us that the voltage E_{in} can be represented by a converging, infinite series of voltages. The first term is the average value of the pulsating dc voltage, and the other terms are sinusoidal voltages of increasing frequency (2ω, 4ω, 6ω, etc.) and decreasing amplitude ($4/3\pi$, $4/15\pi$, etc.). If all these voltages are added together, they will reproduce the input voltage.

A simplification can be effected here because of the choices of L and C in the filter. The inductance is chosen relatively large so that X_L will be large, and the capacitor is made large (10 to 80 μf) so that X_C will be small. Thus all the higher-frequency terms in Eq. (5.25) will be greatly attenuated in passing through the filter. For this reason, only the first two terms are used in calculating the effect of the filter. The voltage E_{in} can then be represented by

$$E_{in} \cong \frac{2}{\pi} E_m - \frac{4}{3\pi} E_m \cos 2\omega t \quad (5.26)$$

for a full-wave rectifier. This means that the equivalent circuit for the full-wave rectifier can be represented as the sum of a dc voltage (a battery) and an ac generator of twice the applied frequency (2ω). The equivalent circuit that results is shown in Fig. 5.14. This is the circuit that will be used to calculate the output voltage E_{out}. One of the conditions that results from making C large is that $X_C \ll R_L$, so that the parallel combination of R_L and X_C becomes

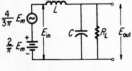

Fig. 5.14 The equivalent circuit for the input voltage.

$$Z_{11} = \frac{R_L(-jX_C)}{R_L - jX_C}$$

Since $X_C \ll R_L$,

$$Z_{11} \cong \frac{R_L(-jX_C)}{R_L} = -jX_C \qquad (5.27)$$

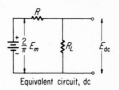

Equivalent circuit, dc

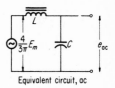

Equivalent circuit, ac

Fig. 5.15

So, for ac calculations, R_L is effectively out of the circuit.

To find the resultant voltage E_{out}, the problem is now broken up into two parts, an ac problem and a dc problem. Each part is solved separately and then the results are added to give the resultant voltage E_{out} by the principle of superposition.

To form the dc problem, start with the equivalent circuit of Fig. 5.14 and draw in only those components that will produce a voltage drop due to the battery current. The equivalent dc circuit for Fig. 5.14 is shown in Fig. 5.15. The dc resistance of the inductor is indicated by R. The capacitor does not pass a dc current, so it acts like an open circuit and does not appear. For the ac equivalent circuit, only the components producing voltage drops due to the ac current are shown, with the generator representing the input ac voltage. The resistance of the inductor is not shown, since it will be assumed that $X_L \gg R$, nor is the resistor R_L shown, for reasons given in Eq. (5.27). Each circuit must now be solved and the voltages added to obtain the resultant output voltage.

In the dc circuit:

$$E_{dc} = E_{in} - IR$$

$$= \frac{2}{\pi} E_m - IR \qquad (5.28)$$

where I is the dc current through the choke. For many cases, $IR < (2/\pi) E_m$ and can be neglected.

In the ac circuit,

$$E_{ac} = |e_{ac}|_{max} = \left(\frac{4}{3\pi} E_m\right)\left(\frac{X_C}{X_L - X_C}\right)$$

$$= \left(\frac{4}{3\pi} E_m\right)\left(\frac{1}{4\pi^2 f^2 LC - 1}\right) \qquad (5.29)$$

and

$$\text{Ripple voltage} = \text{input voltage} \times \frac{1}{F}$$

where F stands for $4\pi^2 f^2 LC - 1$ and is called the *filter factor*. In general, the equation can be written as

$$E_{ac} = |E_{in}| \frac{1}{F} \qquad (5.30)$$

The results for this filter circuit are the sum of Eqs. (5.28) and (5.30).

$$E_{\text{out}} = \underbrace{\frac{2}{\pi} E_m - IR}_{\text{dc voltage}} + \underbrace{\frac{4}{3\pi} \frac{E_m}{F} \cos 2\omega t}_{\text{ac voltage (ripple)}}$$

$$= \quad E_{\text{dc}} \quad + \quad E_{\text{ac}} \cos 2\omega t \qquad (5.31)$$

Remember that E_{ac}, as expressed here, is a peak value, not an effective value.

Example 5.5 Calculate the value of the ac ripple, and percent ripple for the following power supply:

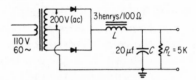

1. The maximum voltage E_{in} is given by

$$\sqrt{2}\, E_{\text{eff}} = E_m$$
$$\sqrt{2} \times 200 = E_m$$
$$283 = E_m$$

2. The filter factor F is given by

$$F = 4\pi^2 f^2 LC - 1$$

where f is twice the applied frequency, since full-wave rectification is being employed, and $F = 33$.

3. The effective value of the ripple voltage is

$$E_{\text{ac}}(\text{eff}) = \frac{4E_m}{\sqrt{2}\,3\pi F} = 2.54 \text{ volts}$$

4. The percent ripple is given by

$$\text{Percent ripple} = \frac{E_{\text{ac}}(\text{eff})}{E_{\text{dc}}}$$

where

$$E_{\text{dc}} = \frac{2E_m}{\pi} - IR = 176 \text{ volts}$$

Thus

$$\text{Percent ripple} = 1.4 \text{ percent}$$

5. The output voltage, therefore, is a dc value of 176 volts, with a ripple having a peak value of 3.6 volts that is superimposed upon it.

5.6 Capacitor Input Filter

In the capacitor input filter, a capacitor is placed before the L–C filter to form what is known as a pi filter (π). This is shown in Fig. 5.16. Either a half-wave or a full-wave rectifier can be used with a capacitor input filter. C_1 is the input capacitor, and L and C_2 form a filter like that discussed in Sec. 5.5. To perform a quantitative analysis of this circuit, Fourier analysis shows that the combination of the power transformer, diode, and C_1 can be replaced by an equivalent circuit consisting of a dc battery and an ac generator, where

$$E_{dc} \cong E_m \tag{5.32}$$

$$E_{ac}(\max) \cong \sqrt{2}IX_{C_1} \tag{5.33}$$

where I is the dc current through L.

$$E_{in} = E_{dc} + E_{ac} \cos \omega t \tag{5.34}$$

The input capacitor tends to charge to the maximum voltage E_m and maintain that value. The ac voltage is more complicated and is in reality composed of an infinite series, of which Eq. (5.33) is just the first term. The other terms decrease in amplitude and are neglected in this approach. Thus, the equivalent circuit for Fig. 5.16 becomes as shown in Fig. 5.17. This is the same

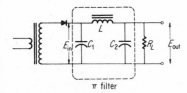

Fig. 5.16 Capacitor input filter and half-wave rectifier.

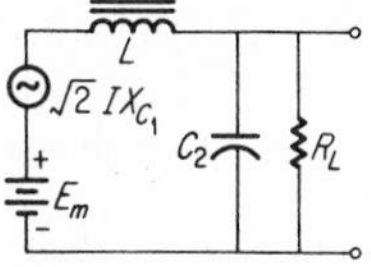

Fig. 5.17 The equivalent circuit for the half-wave rectifier and C_1.

diagram as that obtained in Fig. 5.14, except for the values of E_{ac} and E_{dc}. Therefore the solutions for the capacitor input filter have the same forms as Eqs. (5.28) and (5.30).

$$E_{dc} = E_m - IR \tag{5.35}$$

$$E_{ac} = \frac{\sqrt{2}IX_{C_1}}{F} \quad \text{(peak voltage)} \tag{5.36}$$

Example 5.6 For the circuit shown, calculate the percent ripple. Let $R = 100$ ohms.

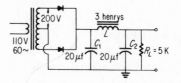

1. $E_m = \sqrt{2}\, E_{\text{eff}} = \sqrt{2} \times 200 = 283$ volts

2. $X_{C_1} = \dfrac{1}{2\pi f C_1} = 66$ ohms

where $f = 120$ cps, since full-wave rectification is being used.

3. I is given by

$$I = \frac{E_m}{R + R_L} \cong 55 \text{ ma}$$

4. $E_{\text{dc}} = E_m - IR = 277.5$ volts

$$E_{\text{ac}}(\text{eff}) = \frac{I X_{C_1}}{F} = \frac{3.63}{33} = 0.11 \text{ volt}$$

5. Percent ripple $= \dfrac{E_{\text{ac}}(\text{eff})}{E_{\text{dc}}} \times 100$

$$\cong 0.04$$

In comparing Example 5.6 with Example 5.5, a capacitor input filter and a choke input filter are being compared, since the transformer and R_L are the same in both examples. The comparison shows that the output dc voltage is higher for a capacitor input filter and that the percent ripple is lower for a capacitor input filter (better filtering). Thus, with respect to filtering, the capacitor input filter does a better job.

A choke input filter is usually resorted to when high-current power supplies are needed (output currents of the order of amperes). In this case, mercury vapor diodes are usually used. The mercury vapor reduces the internal resistance of the diode so that high currents can be conducted without producing excessive $I^2 R$ power losses in the tube. These gas diodes cannot take large transient currents. The choke input filter has a relatively large time constant, which helps keep the current transients small.

Most of the power supplies encountered in normal use will

utilize capacitor input filters. They produce good to excellent filtering over a gamut of power supply requirements.

Some applications require that a power supply maintain a constant voltage output even though the output current changes considerably. A measure of this type of regulation is given by the following definition:

$$\text{Percent regulation} = \frac{E \text{ (min load)} - E \text{ (max load)}}{E \text{ (min load)}^{NL}} \times 100$$

$$(5.37)$$

where E (min load) and E (max load) are the dc output voltages of the power supply under both conditions. The dc-voltage output for a choke input filter is given by Eq. (5.28) as

$$E_{dc} = \frac{2E_m}{\pi} - IR$$

and is independent of load current except for the term IR. $2E_m/\pi$ is not a function of load current.

For a capacitor input filter, the output voltage is given by Eq. (5.35), but this is not exactly correct, for if the load resistance is relatively low, the capacitor C_1 can discharge appreciably between cycles, producing a lower dc-voltage output. This is illustrated in Fig. 5.18. For the case where R_L is large, C_1 does not

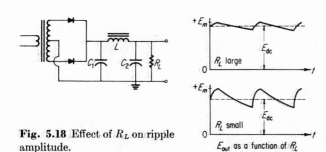

Fig. 5.18 Effect of R_L on ripple amplitude.

discharge much between cycles and $E_{dc} \rightarrow E_m$. For large current drains, that is, R_L small, C_1 discharges appreciably during cycles, and E_{dc} drops as a consequence. A more accurate representation for E_{dc} for a capacitor input filter, when large dc currents are drawn, would be

$$E_{dc} = (E_m - \sqrt{2}IX_{C_1}) - IR \qquad (5.38)$$

where the term $\sqrt{2}IX_{C_1}$ has been inserted to take into account the amount the dc level drops for small R_L.

Example 5.7 Find the percent regulation for a choke input filter when the load resistance is reduced to one-fifth its original value. Use the circuit of Example 5.5.

1. In the choke input filter, the voltage under minimum load is given by

$$E_m = \frac{2E_m}{\pi} - I_{min}R$$

where

$$I_{min} = \frac{2E_m}{\pi} \frac{1}{(R + R_L)}$$

and from Example 5.5

$$E \text{ (min load)} = 176 \text{ volts}$$

2. For E (max load) calculations, $R_L = 1000$ ohms, and I increases as a consequence.

$$E \text{ (max load)} = \frac{2E_m}{\pi} - \frac{2E_m}{\pi}\left(\frac{R}{R + R_L}\right)$$

$$= \frac{2E_m}{\pi}\left(1 - \frac{R}{R + R_L}\right) = \frac{2E_m}{\pi}\left(\frac{R_L}{R + R_L}\right)$$

$$\cong 164 \text{ volts}$$

3. For the choke input filter,

$$\text{Percent regulation} = \frac{176 - 164}{176} \times 100$$

$$\cong 6.8 \text{ percent}$$

5.7 Voltage Dividers

In many cases the power supply must furnish several dc voltages, both positive and negative. Also, the current requirements for each voltage will be different. These varied voltages can be supplied by one power supply. Typical power-supply requirements are shown in Fig. 5.19, a positive 270 volts at 90 ma and a negative 30 volts at 5 ma. The currents represent the anticipated currents that will be produced when the power supply is connected to the circuit for which it was designed.

The voltages desired are obtained by using a voltage divider, as shown in Fig. 5.20. It is the action of the voltage divider R_1 and R_2 that establishes the division of the 300-volt output of the power supply into 270 volts and 30 volts. In this case, the center tap of the power-transformer secondary is not grounded. The junction of the voltage divider R_1 and R_2, point B, is grounded, and the output of the power supply is measured with respect to point B. Point A in Fig. 5.20 will then be positive with respect to point B, and point C will be negative with respect to point B. Thus positive and negative voltages are obtained by moving the reference point to some point between A and C.

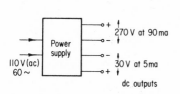

Fig. 5.19 Typical power-supply requirements.

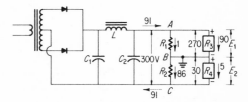

Fig. 5.20 A voltage divider used to produce positive and negative output voltages.

The only problem remaining is to obtain the correct values for R_1 and R_2 so that the total voltage is divided correctly into 270 volts and 30 volts. To do this, the current requirements of the two loads, R_3 and R_4, are needed. Since, for this example, the current drawn by load R_3 is 90 ma when placed across 270 volts, and the current for load R_4 is 5 ma when placed across 30 volts, the effective resistances for R_3 and R_4 are 3000 ohms and 6000 ohms, respectively. Refer to Fig. 5.20; 90 ma flows down through R_3 (positive current is being used) and 5 of this 90 ma flows through R_4. The remainder must flow through R_2. Since R_1 and R_2 have not been determined yet, a decision must be made as to how much current must flow through R_1. This current will also flow through R_2. As a rule of thumb, a choice of 1 ma will be made for the current flow through a resistor when the current is arbitrary. Looking at the parallel circuit of R_1, R_2, R_3, and R_4, it can be seen that the current through R_3 is fixed as 90 ma. The current through R_4 is fixed as 5 ma, and the current through R_2 is equal to the current through R_1 plus the current difference between R_3 and R_4. Thus the current through R_1 is arbitrary. The smaller R_1 is made, the more current will flow through R_2, and the smaller R_2 must be to maintain the proper ratio between E_1 and E_2. The reason for the 1-ma choice is twofold. First, an additional current of 1 ma will not load down the power supply significantly, and second, the resistor combination R_1 and R_2 is across the power supply and will discharge the capacitors when the on-off switch for the unit is turned to "off." This second use for the combination R_1 and R_2 is a safety factor. It ensures that the capacitors are discharged rapidly after the unit is turned off, so that one is not inadvertently shocked when repairing or modifying the power supply. The choice of 1 ma ensures that the capacitors will be discharged within seconds.

Thus, in this example, it is established that 1 ma flows through R_1 and 86 ma must flow through R_2. The voltage drop across R_1

must be 270 volts and the voltage drop across R_2 must be 30 volts. This fixes R_1 and R_2:

$$R_1 = \frac{E_1}{I_{R_1}} = \frac{270}{0.001} = 270 \text{ K}$$

$$R_2 = \frac{E_2}{I_{R_2}} = \frac{30}{0.086} \cong 352 \text{ ohms}$$

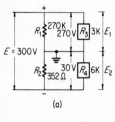

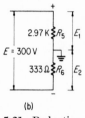

Fig. 5.21 Reduction of the voltage divider.

The voltage divider is shown in Fig. 5.21. These values for R_1 and R_2 will ensure that the 300 volts will be divided up so that there will be 270 volts across R_3 and 30 volts across R_4. To show this, reduce the parallel combination of R_1 and R_3 into R_5, given by

$$R_5 = \frac{R_1 R_3}{R_1 + R_3} \cong 2.97 \text{ K}$$

and reduce the parallel combination of R_2 and R_4 into R_6, given by

$$R_6 = \frac{R_2 R_4}{R_2 + R_4} \cong 333 \text{ ohms}$$

E_1 and E_2 can now be calculated:

$$E_1 = E \left(\frac{R_5}{R_5 + R_6} \right) = 270 \text{ volts}$$

$$E_2 = E \left(\frac{R_6}{R_5 + R_6} \right) = 30 \text{ volts}$$

The power requirements for R_1 and R_2 should also be calculated:

$$P_{R_1} = I_{R_1}{}^2 R_1 = 0.27 \text{ watt}$$

$$P_{R_2} = I_{R_2}{}^2 R_2 = 0.26 \text{ watt}$$

This is the required wattage for each resistor under ideal conditions. However, these resistors will be mounted with others under a chassis. Therefore they will have restricted ventilation. To ensure that they do not heat up appreciably, the wattage used is always a multiple of the minimum calculated by the equation $P = I^2 R$. The factor used is anywhere from 2 to 5, with the answer rounded off to the next highest wattage rating available for the resistor. Thus, using a factor of 2, R_1 would be a 0.54-watt resistor. Rounding this off to the nearest higher wattage, R_1 becomes a 1-watt resistor. R_2 will be 0.52 watt, and this also is rounded off to a 1-watt resistor.

Example 5.8 Find R_1, R_2, and R_3 and the wattage for each when used in a voltage divider to furnish the following voltages

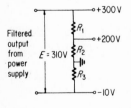

from a 310-volt supply: 300 volts at 100 ma; 200 volts at 50 ma; −10 volts at 10 ma.

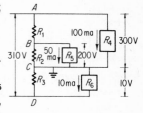

1. The loads would be connected as shown at the right (top).

2. R_1 would then pass current that would flow through R_5 and R_2, and R_3 would pass current that would flow through R_2, R_5, and R_4, minus the current flowing through R_6. From this it is seen that the current through R_2 is arbitrary, and it is chosen as 1 ma. The currents will then flow as shown at the right.

3. R_1, R_2, and R_3 can now be solved:

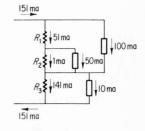

$$R_1 = \frac{E_{R_1}}{I_{R_1}} = \frac{100}{0.051} \cong 1.96 \text{ K}$$

$$R_2 = \frac{E_{R_2}}{I_{R_2}} = \frac{200}{0.001} = 200 \text{ K}$$

$$R_3 = \frac{E_{R_3}}{I_{R_3}} = \frac{10}{0.141} \cong 71 \text{ ohms}$$

4. The wattages calculated become

$$P_{R_1} = I_{R_1}{}^2 R_1 = 5.1 \text{ watts}$$

$$P_{R_2} = I_{R_2}{}^2 R_2 = 0.2 \text{ watt}$$

$$P_{R_3} = I_{R_3}{}^2 R_3 = 1.4 \text{ watts}$$

5. Using a factor of 2 on the power dissipation for the resistors,

$$R_1 = 1.96 \text{ K/10 watts}$$

$$R_2 = 200 \text{ K/0.5 watt}$$

$$R_3 = 71 \text{ ohms/5 watts}$$

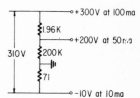

6. The voltage divider is as shown at the right.

Example 5.9 Using a power supply with an output voltage of 300 volts, construct a voltage divider for the following load requirements: 200 volts at 100 ma and −20 volts at 10 ma.

1. Since the voltage requirements, 200 and −20 volts, do not add up to the output voltage, 300 volts, a dropping resistor is also required. The circuit is as shown at the right.
The resistor R_1 is inserted in series with the voltage divider R_2 and R_3, simply to drop the 300 volts to the desired 220 volts.

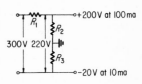

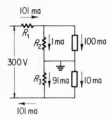

2. The current requirements for R_1, R_2, and R_3 are given at the left. The current through R_2 is the independent one and is arbitrarily chosen as 1 ma.

3. $$R_1 = \frac{E_{R_1}}{I_{R_1}} = \frac{80}{0.101} = 792 \text{ ohms}$$

$$R_2 = \frac{E_{R_2}}{I_{R_2}} = \frac{200}{0.001} = 200 \text{ K}$$

$$R_3 = \frac{E_{R_3}}{I_{R_3}} = \frac{20}{0.091} = 220 \text{ ohms}$$

4. The power calculations are

$$P_{R_1} = I_{R_1}^2 R_1 = 8 \text{ watts}$$

$$P_{R_2} = I_{R_2}^2 R_2 = 0.2 \text{ watt}$$

$$P_{R_3} = I_{R_3}^2 R_3 = 1.8 \text{ watts}$$

5. The requirements on R_1, R_2, and R_3, and the circuit used, are as shown at the left.

5.8 VR Tube Regulator Circuit

For critical circuits, the voltage regulation for a power supply can be increased by a factor of about 40 by using gas-filled voltage regulator (VR) tubes across the output of a power supply. VR tubes come in assorted physical sizes, operating voltage, and current limitations. The octal-based series has the parameters given in Table 5.1.

TABLE 5.1   VR Tube Characteristics

Tube Type	Operating Voltage, volts	Current Range, ma	Design Center, ma
0A3	75	5–40	20
0B3	90	5–40	20
0C3	105	5–40	20
0D3	150	5–40	20

Thus, the OD3, for example, will maintain the voltage across itself at 150 volts while the current through it varies from 5 to 40 ma.

For the seven-pin miniature tubes, the parameters given in Table 5.2 apply.

TABLE 5.2   Miniature Tube Parameters

Tube Type	Operating Voltage, volts	Current Range, ma	Design Center, ma
0A2	150	5–30	15
0B2	105	5–30	15
0C2	75	5–30	15
5651	87	1.5–3.5	2.5

The seven-pin miniatures are smaller, physically, and thus the operating current range is smaller because the smaller-size VR tube cannot dissipate as much heat. A typical circuit for a VR tube regulator is shown in Fig. 5.22.

Fig. 5.22 *VR* tube regulator circuit.

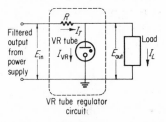

VR tube regulator circuit

The VR tube regulator circuit consists of a resistor R and the VR tube. To obtain a working equation for finding R, the voltage E_{in} is expressed as the sum of the voltage drop across R and across the VR tube:

$$E_{in} = I_T R + E_{out} \qquad (5.39)$$

and

$$I_T = I_{VR} + I_L \qquad (5.40)$$

Therefore

$$E_{in} = (I_{VR} + I_L)R + E_{out} \qquad (5.41)$$

Example 5.10   The 200-volt output of a power supply is to be regulated at 150 volts. The load will draw 50 ma at 150 volts. Design a VR tube circuit to accomplish this.

1.   The circuit and required voltages and currents are as shown in the following figure. The design center $I_{VR} = 20$ ma is used. Also, $I_L = 50$ ma; thus $I_T = 70$ ma.

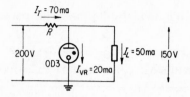

2. R can now be calculated:

$$R = \frac{E_{\text{in}} - E_{\text{out}}}{I_{\text{VR}} + I_L} = \frac{50}{0.070} = 715 \text{ ohms}$$

3.   The power requirements for R are

$$P_R = I_R{}^2 R = 3.5 \text{ watts}$$

4. $R = 715$ ohms/10 watts and is used in conjunction with an OD3 tube.

VR tubes can be used in series to regulate more than 150 volts. For example, two OD3 tubes can be put in series to regulate at 300 volts. VR tubes are not put in parallel to attempt to regulate for load-current changes larger than 40 ma unless special precautions are used to ensure that all the VR tubes fire before the load is applied. Since a VR tube operates at a lower potential than its firing potential, if two VR tubes were simply placed in parallel then only one would fire, immediately lowering the potential so that the other tube would never fire. This happens because no two VR tubes are exactly alike, and it would be quite unlikely to have two tubes with identical firing voltages. So, in general, VR tubes can be put in series but not in parallel for the circuit we are considering.

An advantage of VR tube regulation is that the effect is dependent upon a gaseous process and is therefore very fast acting. Response time for a VR tube circuit is on the order of microseconds. A VR tube circuit can thus smooth out very fast transients.

5.9   Bridge Circuit

A bridge circuit is very often used to produce a full-wave rectified signal from a transformer that has no center-tapped secondary. A circuit of this sort is most often used with semiconductor diodes to reduce the peak inverse voltage across each diode. This minimizes the probability of a failure in the diodes because of an excess reverse voltage across a diode unit. To make clear what is meant by a peak inverse voltage, refer to Fig. 5.23, which shows a simple capacitor input filter and half-wave rectifier. The voltage E_d across the diode is given by the difference in potential between points A and C.

Point C is held at $+E_m$ with respect to ground by capacitor C_1. The potential of point A varies sinusoidally from $+E_m$ to $-E_m$, with respect to ground. At the instant point A is at $-E_m$, the instantaneous voltage across the diode E_d becomes $2E_m$. The polarity is such that the diode is reverse-biased.

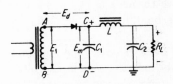

Fig. 5.23 Voltages encountered in a half-wave rectifier.

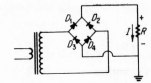

Fig. 5.24 A bridge rectifier circuit.

A diode used in this rectifier circuit must thus be able to withstand at least twice the peak applied voltage, or it will break down. To reduce this peak inverse voltage across a diode, the bridge circuit is often used. The circuit is shown in Fig. 5.24. The diodes are so connected that two diodes conduct each half-cycle, and the

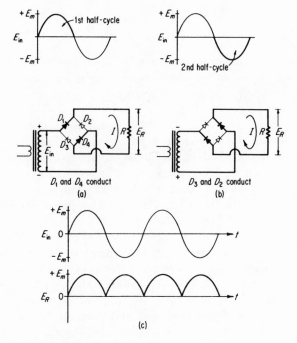

Fig. 5.25 Conduction processes in a bridge rectifier.

current flow through R is in the same direction each half-cycle. Thus the bridge circuit produces full-wave rectification. This is illustrated in Fig. 5.25. In the first half-cycle, diodes D_1 and D_4 will be biased in the forward direction and will conduct. The resulting current will flow through R in a direction such as to make the top of R positive with respect to the bottom, or ground point, Fig. 5.25(a). During the second half-cycle, diodes D_2 and D_3 are biased in the forward direction and will conduct as shown, Fig. 5.25(b). Again the current flows through R in the same direction. The voltage across R will be full-wave rectified, Fig. 5.25(c).

Since at all times there are two diodes in series, the maximum inverse voltage across each diode will be only the peak input voltage rather than twice the peak input voltage. This reduces the probability of damaging the diodes.

Also, the bridge circuit is useful in its own right because it is able to produce a full-wave rectified voltage from a source not having a center tap.

5.10   Voltage Doublers

In general, to secure a high-voltage output, all that is required is a step-up transformer giving the desired voltage output. However, the cost for such transformers goes up exponentially with respect to the voltage required.

So, in many cases, special circuits are utilized to get away from using an expensive transformer, or from using any transformer for that matter. These circuits are called *voltage doublers, voltage triplers,* etc.

There are several variations of these. For example, Fig. 5.26 illustrates the most common form of voltage doubler. This type of voltage doubler is used very often in table radios, many phono-amplifiers, and some television sets. The output across C_2 is equal to twice the peak input voltage. For a 115-volt ac line, this would be equal to about 320 volts. The operation is described with the aid of Fig. 5.27. During the first half-cycle, Fig. 5.27(a), the voltage input to the voltage doubler, E_{in}, is of the polarity shown. This biases diode D_1 in the forward direction and it conducts as shown, charging capacitor C_1 to the peak input voltage E_{in}. The capacitor C_1 charges up so that the side facing D_2 is positive. During the second half-cycle, Fig. 5.27(b), the polarity of the input voltage reverses, as shown, and diode D_2 is now biased in the forward direction and conducts, charging C_2. If the voltage input for diode D_2 is added up, it is seen to be equal to the voltage

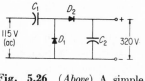

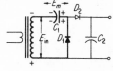

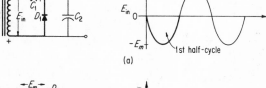

Fig. 5.26 (*Above*) A simple voltage doubler.

(a)

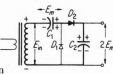

Fig. 5.27 (*Right*) Conduction processes in a voltage doubler.

(b)

across C_1 plus the voltage across the transformer secondary, E_{in}. Thus diode D_2 charges capacitor C_2 to $2E_{in}$. The load is connected across C_2 and tends to discharge C_2 between cycles, producing a ripple in the voltage across C_2, which can be taken out with additional filtering. The capacitor C_1 has a voltage applied in only one direction because of diode D_1. Thus capacitor C_1 can be an electrolytic type. It is usually of a large value, of the order of 100 μf, since it will charge C_2 through D_2 during alternate half-cycles.

This circuit can be used in tandem to produce a voltage quadrupler, or by adding other sections, to produce output voltages of $6E_{in}$, $8E_{in}$, etc. The circuit for a voltage quadrupler is shown in Fig. 5.28(b), with the circuit of Fig. 5.26 redrawn beside it for reference. To explain the operation of this circuit, refer to Fig. 5.29, which shows the flow of current for each half-cycle. For simplicity of explanation it will be assumed that one cycle has gone by and that C_1 and C_2 are charged as shown. Then, during the next half-cycle, Fig. 5.29(a), diodes D_1 and D_3 will be biased in the forward direction: diode D_1, from the voltage across C_1 and the generator, and diode D_3 from the voltage across C_2. Diode D_1 will conduct to keep C_1 charged to E_{in}, and D_3 will conduct to charge C_3 to $2E_{in}$. When E_{in} is at peak voltage and C_1 is charged, the voltage across C_1 and E_{in} at this time will be zero; thus the net voltage across C_3 will be $2E_{in}$, the voltage across C_2, and C_3 will charge through D_3 to a potential of $2E_{in}$ with a polarity as shown. During the next half-cycle, Fig. 5.29(b), the generator voltage E_{in} changes polarity, and now diode D_2 and diode D_4 are biased in their forward directions. C_2 charges to the peak voltage across the generator and C_1, that is, $2E_{in}$. C_4 charges to the sum of the voltages across the generator, C_1, C_3, and C_2, that is, $2E_{in}$ (observe the polarities

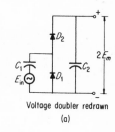

Voltage doubler redrawn

(a)

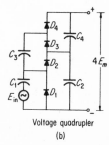

Voltage quadrupler

(b)

Fig. 5.28 Extension of a voltage doubler to produce a voltage quadrupler.

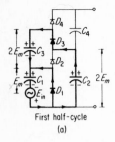

First half-cycle
(a)

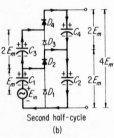

Second half-cycle
(b)

Fig. 5.29 Conduction processes in a voltage quadrupler.

when summing up these voltages). Thus the total voltage across C_4 and C_2 is $4E_{in}$.

A variation of the voltage doubler shown in Fig. 5.26 is that shown in Fig. 5.30. This voltage doubler has the advantage that it isolates the input generator from the output circuit. The operation is shown in Fig. 5.30(b) and Fig. 5.30(c). During one half-cycle, D_1 conducts, charging C_1 to the peak input voltage E_{in}. During the second half-cycle, D_2 conducts, charging C_2 to the peak input voltage E_{in}. The capacitors C_1 and C_2 are connected so that these voltages add, producing an output voltage equal to twice the peak input voltage, $2E_{in}$.

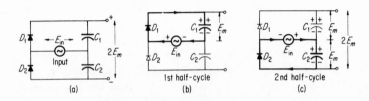

(a) 1st half-cycle (b) 2nd half-cycle (c)

Fig. 5.30 Another voltage doubler circuit.

Problems

5.1

5.1 Initially, capacitor C_1 is charged to a voltage E_0. At $t = 0$, switch S is closed. Find the voltage across C_2 as a function of time. What is the time constant for the circuit?

5.2 Initially, the voltage on capacitor C is zero. At $t = 0$, switch S is closed. Find the voltage across R as a function of time.

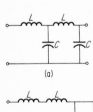

(a)

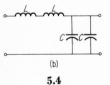

(b)

5.4

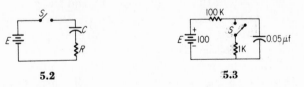

5.2   5.3

5.3 The switch S switches alternately open and closed, every 0.001 sec. Determine the shape of the voltage across C as a function of time for several cycles.

5.4 If one has two 10-henry chokes and two 20-μf capacitors, calculate the filter factor for the following filter circuits:

5.5   For the circuit shown:

(a) Determine the secondary voltage needed.

(b) Determine the ripple voltage.

The power frequency is 60 cps; $L = 10$ henry/100 ohms; $C_1 = C_2 = 20$ μf.

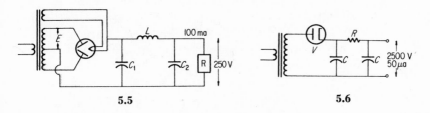

5.5   5.6

5.6   For the high-voltage supply shown, determine

(a) the value of C,

(b) the value of R,

(c) the secondary transformer voltage,

(d) the peak inverse voltage across V,

for a ripple voltage of 20 volts (effective).

5.7   A transistor power supply that will supply -9 volts at 5 ma, and a ripple of 0.01 volt, effective, is required. A transformer having a 12-volt secondary is available. Using a capacitor-input RC filter, design the power supply.

5.8   For the circuit shown (a) determine R and its wattage, and (b) determine the maximum and minimum values of E_{in} over which regulation will occur.

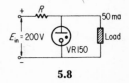

5.8

5.9   A 150-volt power-supply output is to be divided into 100 volts at 30 ma and -50 volts at 5 ma. Design a voltage divider that will do this.

5.10   The output voltage of a power supply is 300 volts. It is necessary to convert this into a $+300$ at 60 ma, $+200$ at 10 ma, and -30 volts at 10 ma output. Design a voltage divider that will produce these voltages.

5.11   Draw the circuit that the following block diagram suggests. Do not put in any component values.

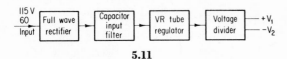

5.11

5.12 Determine the cutoff frequency for the L-C_2 filter in Problem 5.5. Neglect the resistance of the choke.

5.13 The secondary voltage is 150 volts, effective. Determine the following: (a) output voltage, (b) polarity of output, (c) peak inverse voltage for each diode.

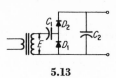

5.13

6.1 Triodes

In this chapter, triodes will be considered from an operational point of view, that is, the voltage-current characteristics for a triode in an amplifier circuit will be determined graphically. The triode is so named because it has three active elements. These are the plate and cathode, as in a diode, and a new element inserted between plate and cathode, called a *control grid*. The control grid is physically placed closer to the cathode than to the plate. The symbol for a triode is shown in Fig. 6.1(a). The filament is used simply to heat the cathode so that it can emit electrons, and is therefore not an active element in that it does not enter into any portion of the amplifier circuit.

In a typical physical construction, Fig. 6.1(b), the elements are composed of concentric cylinders. The plate surrounds the control grid and cathode. The control grid is made up of a fine wire mesh so that electrons can easily pass through. The cathode is a hollow cylinder. The external surface is coated with an oxide to enhance electron liberation; inside the cathode cylinder, coaxial with it, is the filament winding. All the leads (one for the plate, one for the control grid, one for the cathode, and two for the filament) pass through a glass envelope surrounding the elements and are soldered to the terminal pins at the base of the tube.

Vacuum Tubes — Graphical (Triodes and Pentodes)

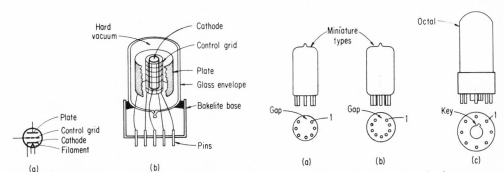

Fig. 6.1 Triode construction and schematic symbol.

Fig. 6.2 Miniature and octal tube types.

There are several bases and tube sizes in use. Three of the more common ones are shown in Fig. 6.2. Both the outline and tube base are shown. The miniature type in Fig. 6.2(a) has seven pins in the base. There is a numbering convention for these pins. To find pin number 1, hold the tube so that the gap is up, as shown, and pin number one is on the right-hand side of the gap. The numbers then proceed clockwise around the base. The same method is used for locating pin number 1 in the nine-pin miniature tube, Fig. 6.2(b). In the octal base tube, the pins are spaced equally around the base. To ensure that the tube can be put into the socket only one way, a key is molded into the plastic base. This must mesh with a similarly shaped hole in the tube socket in order to seat the tube into the socket. To locate pin number 1, hold the tube with the key up and pin number 1 will be to its right, as shown in Fig. 6.2(c). The reason the pins must be numbered is that different tubes, in general, have different base connections. For example, the plate is pin number 3 for a 6J5 triode, while for a 6SF5 triode it is number 5 .Thus the base connections for a tube must be looked up in a tube manual before a tube socket can be wired into a circuit.

The dc voltages applied to a triode are measured with respect to the cathode. In general, the plate is biased positively, so that a dc current will flow in the tube from cathode to plate. Also, the control grid is usually biased negatively so that it will tend to retard this current flow. The voltages, currents, and symbols are shown in Fig. 6.3.

In the plate circuit:

E_b = dc voltage between plate and cathode

I_b = dc current in plate circuit

E_{bb} = dc voltage source in plate circuit

In the control grid circuit:

E_c = dc voltage between control grid and cathode

I_c = dc current in grid circuit

E_{cc} = dc voltage source in control grid circuit

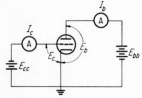

Fig. 6.3 The dc test circuit for a triode.

6.2   Triode Action

In a triode, the control grid is used to regulate the electron flow between plate and cathode. To establish this current flow in the tube, the plate is made positive with respect to the cathode. This is illustrated in Fig. 6.4(a). The battery E_{bb} biases the tube so that the plate is positive with respect to the cathode. Electrons thermionically emitted from the cathode are accelerated to the plate because of the electric field between plate and cathode. This produces a current I_b in the plate circuit. The plate circuit is defined by the circuital path $AGBCDA$. Because the control grid is made from a wire mesh, most of the electrons pass through. Occasionally one will strike the control grid and be absorbed. It will produce a current in the path $GFEA$, back to the cathode. This action maintains the control grid at the same potential as the cathode for the circuit in Fig. 6.4(a). This grid current can be anywhere from microamperes to milliamperes, depending upon the tube used and the potentials on the tube elements.

The dc characteristics for this circuit are shown in Fig. 6.4(b). This curve can be recognized as that for a diode. This is as it should be, since the control grid is at zero potential with respect to the cathode and does not interfere appreciably with the current between cathode and plate.

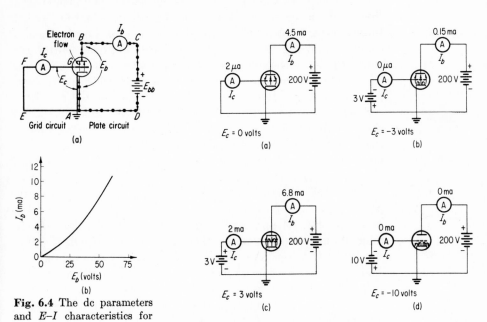

Fig. 6.4 The dc parameters and E–I characteristics for a triode.

Fig. 6.5 The effect of grid voltage on the plate current.

To show what effect the control grid has on plate current, the control grid can be made either positive or negative, with respect to the cathode, by placing a voltage source between the control grid and cathode (Fig. 6.3). Now, in additional to the electric field between plate and cathode, there will be an electric field between control grid and cathode. Since the control grid is very close to the cathode, a small voltage between control grid and cathode will produce a significant electric field. In fact, a difference in potential of a few volts between cathode and control grid can produce an electric field as great as that produced by a difference of potential of a few hundred volts between plate and cathode.

A typical example is shown in Fig. 6.5. In Fig. 6.5(a), the control grid is held at zero potential while the plate is held at +200 volts. In this case the plate current I_b will be 4.5 ma, and the control grid current I_c will be about 2 μamp. If the control grid is now made negative by 3 volts ($E_c = -3$ volts), the electric field in the region between control grid and cathode is weakened and not so many electrons are accelerated toward the plate. This is shown in Fig. 6.5(b). The net effect is that I_b drops in this example to about 0.15 ma, and I_c becomes zero, since the control grid is now negative and repels electrons. In Fig. 6.5(c), the control grid is made positive (E_c = +3 volts). Now the electric field in the vicinity of the cathode is increased and more electrons are accelerated toward the plate. The plate current increases as a result. Also, since the control grid is positive, it will attract electrons, producing an appreciable grid current. For the sake of this example, I_c is about 2 ma. If the control grid is made negative enough, as in Fig. 6.5(d), the electric field between control grid and cathode can become a retarding electric field, stopping even the fastest electrons emitted by the cathode. In this case, $I_b = 0$, $I_c = 0$, and the tube is cut off.

Summing up the action of a triode, if the control grid voltage goes up (becomes less negative), the plate current increases; if the control grid voltage goes down (becomes more negative), the plate current decreases. And if the control grid voltage becomes negative enough, the plate current can be cut off.

In practice, the control grid is usually operated at a negative potential with respect to the cathode. This is done for several reasons. First, no appreciable grid current can flow and I_c can then be taken as zero under these conditions. Second, the tube can be made to operate in the linear region of its characteristics. And third, if the ordinary tube were to operate with its control grid positive-biased for too long, the appreciable grid current (2 to 10 ma) that would result could heat up the control grid and severely damage it, if not melt it.

6.3 DC Characteristics of a Triode

To obtain the dc characteristics of a triode, not only is a positive voltage applied to the plate but also a negative voltage is applied to the control grid. These are both independent parameters. The dependent parameter becomes the plate current I_b. Thus, if either E_b or E_c is varied, I_b will vary. The question is: How do we plot these three parameters meaningfully on a graph?

The test circuit is similar to that shown in Fig. 6.3. To obtain the dc characteristics, E_{bb} and E_{cc} are varied. The plate characteristics turn out to be an informative way of plotting results.

In this method, I_b is plotted as a function of E_b for given increments of E_c. For example, E_c would be set for zero and then held at this value while E_b is varied from zero to several hundred volts. The resulting values of I_b are plotted versus E_b. Then the control grid voltage would be set for -1 volt and held at this value while E_b is again varied from zero to several hundred volts. The resulting I_b is also plotted versus E_b. This curve will be lower than the first one, since with $E_c = -1$ volt, less current, I_b, will flow for all values of E_b. E_c is then changed again (made more negative), and while held at this new potential, E_b is again varied through its range. This is done in increments of E_c until the tube is cut off or until enough curves are obtained.

The resulting curves for a typical triode are shown in Fig. 6.6. This set of curves is termed a *family* of curves. The fact that makes this family of curves so important is that not only will it give the dc parameters of the tube, such as resistance, but also, more important, the graph can be used to predict the behavior of the tube as an amplifier. To show how the dc resistance of the tube can be obtained from the graph, it must be remembered that each point on this graph stands for three values: E_b, I_b,

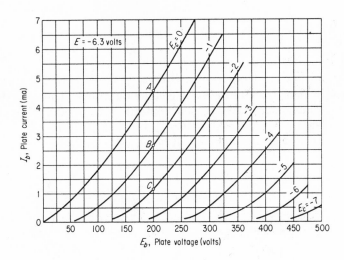

Fig. 6.6 Typical plate characteristics for a triode.

and E_c. Any two determine the other. For example, to find the dc resistance of the tube when the plate is biased at $+200$ volts, the control grid voltage must also be known. In Fig. 6.6, three points along the line $E_b = +200$ have been selected, one at $E_c = 0$, one at $E_c = -1$, and one at $E_c = -2$ volts. The dc resistance of the tube will be different for each point. Incidentally, the dc resistance described is between cathode and plate, since the plate voltage will be divided by the plate current to obtain the resistance.

At point A:

$$R_A = \frac{E_b}{I_b} = \frac{200}{4.5 \times 10^{-3}} = 44.4 \text{ K}$$

At point B:

$$R_B = \frac{E_b}{I_b} = \frac{200}{2.64 \times 10^{-3}} = 75.8 \text{ K}$$

At point C:

$$R_C = \frac{E_b}{I_b} = \frac{200}{1.13 \times 10^{-3}} = 177 \text{ K}$$

It is quite evident from this numerical example that even though the plate voltage remained constant, the dc current I_b decreased for more negative values of E_c. Thus the tube resistance can be made to vary by simply varying the grid-to-cathode potential. As the grid is made more negative, the resistance increases.

6.4   AC Characteristics of a Triode

To explain the action of the triode in an amplifier circuit, the characteristics must be defined. Ordinarily there are four parameters in a four-terminal network such as the triode. They are the input voltage and current, and output voltage and current, Fig. 6.7(a). For the triode, these become E_c, I_c, I_b, E_b, Fig. 6.7(b). Since the control grid will be biased negatively, I_c can be

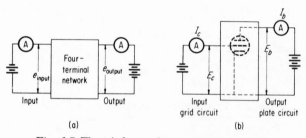

Fig. 6.7 The triode as a four-terminal network.

taken as zero. Therefore the current I_b is a function of E_b and E_c:

$$I_b = f(E_b, E_c) \tag{6.1}$$

The functional dependence is exactly that shown in Fig. 6.6. This is a complicated functional dependence and is not amenable to simple analytical forms. Since in an amplifier the voltage will be constantly changing, it is of interest to write Eq. (6.1) in terms of changes in current and voltage:

$$\Delta I_b = \left(\frac{\Delta I_b}{\Delta E_b}\right)_{E_c} \Delta E_b + \left(\frac{\Delta I_b}{\Delta E_c}\right)_{E_b} \Delta E_c \tag{6.2}$$

The total change in plate current can be summed up in terms of two changes: the change in plate current due to change in plate voltage, keeping the grid voltage constant,

$$\Delta I_{b_1} = \left(\frac{\Delta I_b}{\Delta E_b}\right)_{E_c} \Delta E_b \qquad (6.3)$$

plus the change in plate current due to a change in grid voltage, keeping the plate voltage constant,

$$\Delta I_{b_2} = \left(\frac{\Delta I_b}{\Delta E_c}\right)_{E_b} \Delta E_c \qquad (6.4)$$

Graphically (Fig. 6.8), this is equivalent to expressing the change in plate current, ΔI_b, in terms of the paths $AB + AC$. Along

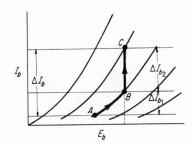

Fig. 6.8 The total plate current change can be expressed as the sum of two changes.

path AB, E_c = const; ΔI_b is given by Eq. (6.3) and is equal to ΔI_{b_1}. Along path BC, E_b = const; ΔI_b is given by Eq. (6.4) and is equal to ΔI_{b_2}. The quantities

$$\left(\frac{\Delta I_b}{\Delta E_b}\right)_{E_c} \qquad (6.5)$$

and

$$\left(\frac{\Delta I_b}{\Delta E_c}\right)_{E_b} \qquad (6.6)$$

are useful parameters and can be obtained graphically from the plate characteristics for a tube. The quantity (6.5) has to do with the change in plate current due to a change in plate voltage keeping the grid voltage constant. This ratio of changes has the units of ohms^{-1} and the reciprocal of quantity (6.5) is called the *ac plate resistance* (r_p) of the tube:

$$r_p = \left(\frac{\Delta E_b}{\Delta I_b}\right)_{E_c} \qquad \text{(plate resistance)} \qquad (6.7)$$

The quantity (6.6) relates the change in plate current due to a change in grid voltage keeping the plate voltage constant and also has the units of ohm^{-1}. This quantity is defined as the transconductance of the tube, g_m:

$$g_m = \left(\frac{\Delta I_b}{\Delta E_c}\right)_{E_b} \qquad \text{(transconductance)} \qquad (6.8)$$

In each definition the remaining variable is held constant during the change.

Inserting these definitions into Eq. (6.2),

$$\Delta I_b = \frac{1}{r_p} \Delta E_b + g_m \Delta E_c \qquad (6.9)$$

Another useful parameter can be defined for the case $I_b = \text{const}$. In this case Eq. (6.9) becomes

$$0 = \frac{1}{r_p} \Delta E_b + g_m \Delta E_c \qquad (6.10)$$

That is, if $I_b = \text{const}$, then $\Delta I_b = 0$. Equation (6.10) states that in order to hold I_b constant while the plate voltage is changing, the grid voltage must also be changed. If the plate voltage is increased, the current will tend to increase. To offset this, the control grid voltage must be decreased. Thus if ΔE_b is positive, ΔE_c will be negative, and Eq. (6.10) can be rewritten as

$$g_m r_p = -\left(\frac{\Delta E_b}{\Delta E_c}\right)_{I_b} \qquad (6.11)$$

The quantity

$$\mu = -\left(\frac{\Delta E_b}{\Delta E_c}\right)_{I_b} \qquad \text{(amplification factor)} \qquad (6.12)$$

is defined as the amplification factor μ of the tube. It relates how much the plate voltage must be changed in order to offset a change in grid voltage so as to keep the plate current unchanged. Combining Eqs. (6.11) and (6.12),

$$\mu = g_m r_p \qquad (6.13)$$

Thus, while the dc characteristics for a tube are not simple, analytically, the ac characteristics can be expressed in a simple equation.

6.5 Graphical Determination of AC Characteristics

The ac parameters defined in Eqs. (6.7), (6.8), and (6.12) can be obtained graphically from the plate characteristics. The three quantities μ, g_m, and r_p are defined at every point in the plate characteristics for a tube. Although they will not be used to any extent in this chapter, they are needed for the next chapter. The graphical method for determining μ, g_m, and r_p will be shown here.

The constructions for μ and g_m are shown in Fig. 6.9. The construction for r_p will be shown separately in Fig. 6.10. The values of μ, g_m, and r_p are to be determined at point Q in Figs. 6.9 and 6.10.

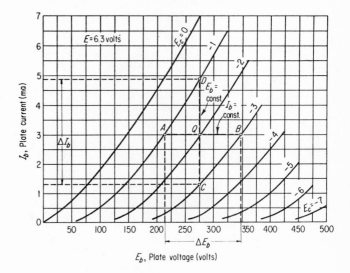

Fig. 6.9 Graphical construction for obtaining *mu* and g_m.

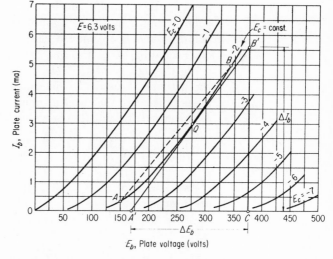

Fig. 6.10 Graphical construction for obtaining r_p.

To determine μ, the definition states that the ratio of ΔE_b to ΔE_c is taken for the condition $I_b = $ const. Thus μ is determined by the construction line AB in Fig. 6.9. Point A is located at $E_c = -1$, $E_b = 213$, and $I_b = 3$ ma. Point B is located at $E_c = -3$, $E_b = 348$, and $I_b = 3$ ma. Thus

$$\mu = -\left(\frac{\Delta E_b}{\Delta E_c}\right)_{I_b} = -\frac{348 - 213}{(-3) - (-1)} = 67.5$$

To determine g_m, the definition states that the ratio of ΔI_b to ΔE_c is taken for the condition $E_b = $ const. Thus g_m is determined by the construction line CD. At point C, the coordinates are $E_c = -3$, $E_b = 200$, and $I_b = 1.26$ ma. At point D, the coordinates are $E_c = -1$, $E_b = 200$, and $I_b = 4.82$ ma. Thus

$$g_m = \left(\frac{\Delta I_b}{\Delta E_c}\right)_{E_b} = \frac{(4.82 - 1.26) \times 10^{-3}}{(-1) - (-3)} = 1780 \ \mu\text{mho}$$

The units for g_m are ohm^{-1}, but to show that the transconductance for a tube is being discussed, the term "mho" has been adopted (ohm spelled backwards). Because the number is so small, the transconductance is usually given in micromhos.

For r_p, the ratio of ΔE_b to ΔI_b must be taken for the condition $E_c = $ const. As can be seen, the family of curves is not linear, that is, the curves are not equidistant and parallel to each other. Thus the definitions for μ, g_m, and r_p at any point are dependent upon the size of the changes taken about that point. The exact values are defined mathematically as the limit wherein the changes are infinitesimal. This fact enables the construction for r_p to be reduced to finding the sides of the triangle $A'B'C$ in Fig. 6.10, where the hypotenuse $A'B'$ is tangent to the $E_c = -2$ curve at point Q. In the definition for r_p, $E_c = $ const, which means that point A and B in Fig. 6.10 must be on the $E_c = -2$ curve. If the points A and B are moved closer to Q, to obtain a more precise determination of r_p, the line AB will approach the slope of the curve through point Q. This is the reason why the line $A'B'$ is the actual one used. To determine E_b and I_b, the line $A'B'$ is made tangent to the curve at point Q and thus has the same slope as the curve at point Q. The value of r_p at Q becomes

$$r_p = \left(\frac{\Delta E_b}{\Delta I_b}\right)_{E_c} = \frac{375 - 170}{(5.6 - 0) \times 10^{-3}} = 36.6 \ \text{K}$$

As a check on Eq. (6.13), the values determined graphically are inserted:

$$\mu = g_m r_p$$

(6.13)

$$67.5 = 65.3$$

The check is well within the precision inherent in a graphical determination.

Example 6.1 Find μ, r_p, and g_m graphically for the case $E_b = 275$

and for $E_c = -1, -2, -3,$ and -4 volts. Plot the resultant values of μ, r_p, and g_m as a function of I_b.

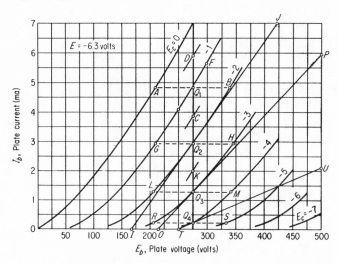

I_b, Plate current (ma)

E_b, Plate voltage (volts)

1.   Although this is an exercise in determining the three ac parameters for a triode, the results when plotted produce an important result.

2.   At point Q_1 the points A and B will give the value for μ_1:

$$\mu_1 = -\left(\frac{\Delta E_b}{\Delta E_c}\right) = -\frac{340 - 210}{-2 - 0} = 65$$

3.   At point Q_1 the points C and D will give the value for g_{m_1}. It should be noticed that points C and D were taken along the $E_c = -0.5$ and $E_c = -1.5$ curves, for more accuracy.

$$g_{m_1} = \left(\frac{\Delta I_b}{\Delta E_c}\right)_{E_b} = \frac{(5.90 - 3.88)10^{-3}}{-0.5 - (-1.5)} = 2020 \ \mu\text{mho}$$

4.   At point Q_1 the points E and F will give the value for r_{p_1}. The points are taken on the curve $E_c = -1$ because the curve is close to a straight line in the vicinity of Q.

$$r_{p_1} = \left(\frac{\Delta E_b}{\Delta I_b}\right)_{E_c} = \frac{300 - 250}{(5.65 - 4.08)10^{-3}} = 32.0 \ \text{K}$$

5.   As a check:

$$\mu_1 = g_{m_1} r_{p_1}$$

$$65 = 2.02 \times 10^{-3} \times 32 \times 10^3$$

$$= 64.6$$

A good check.

6.   For point Q_2, μ_2 is obtained by using points G and H; g_{m_2} is obtained by using points C and K; and r_{p_2} is obtained by using points I and J:

$$\mu_2 = -\left(\frac{\Delta E_b}{\Delta E_c}\right)_{I_b} = -\frac{346 - 211}{(-3) - (-1)} = 67.5$$

$$g_{m_2} = \left(\frac{\Delta I_b}{\Delta E_c}\right)_{E_b} = \frac{(3.88 - 2.04)10^{23}}{(-1.5) - (-2.5)} = 1840 \; \mu\text{mho}$$

$$r_{p_2} = \left(\frac{\Delta E_b}{\Delta I_b}\right)_{E_c} = \frac{425 - 168}{(7.00 - 0)10^{-3}} = 36.7 \text{ K}$$

As a check:

$$\mu_2 = g_{m_2} r_{p_2}$$

$$67.5 = 1.84 \times 10^{-3} \times 36.7 \times 10^3$$

$$= 67.5$$

A good check:

7.   For point Q_3, μ_3 is obtained by using points L and M; g_{m_3} is obtained by using points K and N; and r_{p_3} is obtained by using points O and P:

$$\mu_3 = -\left(\frac{\Delta E_b}{\Delta E_c}\right)_{I_b} = -\frac{342 - 207}{(-4) - (-2)} = 67.5$$

$$g_{m_3} = \left(\frac{\Delta I_b}{\Delta E_c}\right)_{E_b} = \frac{(2.04 - 0.70)10^{-3}}{(-2.5) - (-3.5)} = 1340 \; \mu\text{mho}$$

$$r_{p_3} = \left(\frac{\Delta E_b}{\Delta I_b}\right)_{E_c} = \frac{500 - 214}{(5.85 - 0)10^{-3}} = 48.8 \text{ K}$$

As a check:

$$\mu_3 = g_{m_3} r_{p_3}$$

$$67.5 = 1.34 \times 10^{-3} \times 48.8 \times 10^3$$

$$= 65.4$$

The check is still good. It should be realized that the graphical determinations, being in the nonlinear portion of the tube characteristics, will not be so precise.

8.   For point Q_4, μ_4 is obtained by using points R and S; g_{m_4} is obtained by using points N and Q; and r_{p_4} is obtained by using the points T and U.

$$\mu_4 = -\left(\frac{\Delta E_b}{\Delta E_c}\right)_{I_b} = -\frac{338 - 208}{(-5) - (-3)} = 65$$

$$g_{m_4} = \left(\frac{\Delta I_b}{\Delta E_c}\right)_{E_b} = \frac{(0.70 - 0.22)10^{-3}}{(-3.5) - (-4.0)} = 960 \ \mu\text{mho}$$

$$r_{p_4} = \left(\frac{\Delta E_b}{\Delta I_b}\right)_{E_c} = \frac{500 - 250}{(2.04 - 0)10^{-3}} \cong 123 \text{ K}$$

As a check:

$$\mu_4 = g_{m_4} r_{p_4}$$

$$65 = 0.960 \times 10^{-3} \times 123 \times 10^3$$

$$= 118$$

Poor agreement. This is due to the fact that all calculations are made in the nonlinear portion of the plate characteristics, and all that can be said is that the order of magnitude of the ac parameters has been obtained.

9. The values of μ, g_m, and r_p will now be plotted versus I_b. The values of I_b used are those for Q_1, Q_2, Q_3, and Q_4.

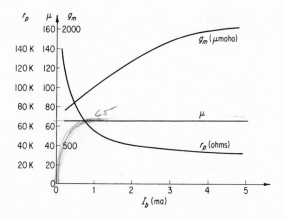

As can be seen from the figure, g_m increases as the current increases, while r_p decreases with increase in current. The surprising thing is that μ is essentially a constant for all plate currents involved in the example.

6.6   Amplifier with Resistive Load

To produce voltage amplification, a load resistor is inserted in the plate circuit, Fig. 6.11(a), and the voltage to be amplified is inserted in the grid circuit. The output voltage is developed across the tube as shown. Before this circuit can be discussed, the effect of R_L on the dc characteristics must first be determined, Fig. 6.11(b).

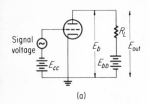

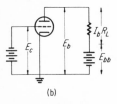

Fig. 6.11 Triode circuit with resistive load, R_L.

Adding up the voltages in the plate circuit,

$$E_{bb} = E_b + I_b R_L \qquad (6.14)$$

Solving for I_b,

$$I_b = \frac{E_{bb} - E_b}{R_L} \qquad (6.15)$$

This is called the *load-line equation*. It gives the dependence of I_b on E_b, the plate voltage. In general, both I_b and E_b are unknown. Thus there is one equation and two unknowns. To solve for I_b and E_b, another equation is needed, relating I_b and E_b. The graph in Fig. 6.6, the plate characteristics, is the other relationship needed. It cannot be put in analytical form, so the solution must be graphical. This is done by graphing Eq. (6.15) for the plate characteristics of the tube. The intersection of the curve represented by Eq. (6.15) with the plate characteristics curve represents the solution to the circuit of Fig. 6.11(b). From it, both E_b and I_b can be read off for any value of E_c. The intersection of the load-line equation with the tube characteristics is the load line itself. The tube can now take on only values of I_b and E_b that fall on the load line.

Example 6.2 Draw the load line for $R_1 = 20$ K and $E_{bb} = 350$ volts. Determine the voltage amplification of the circuit for the tube having the characteristic of Fig. 6.6.

1. The circuit to be solved is shown at the left.

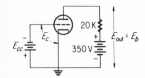

2. The load line equation becomes

$$I_b = \frac{350 - E_b}{20} \times 10^{-3} \text{ amp}$$

or

$$I_b = \frac{350 - E_b}{20} \text{ ma}$$

This equation for the tube characteristics is plotted at the top of page 187.

3. The operation of the circuit is such that only values of E_b and I_b that fall on the load line can be obtained. For example, if $E_c = -2$ volts, then the tube will be operating at point C, that is, $E_b = 285$ volts and $I_b = 3.22$ ma. The voltages in the circuit will be as shown at the left.

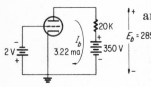

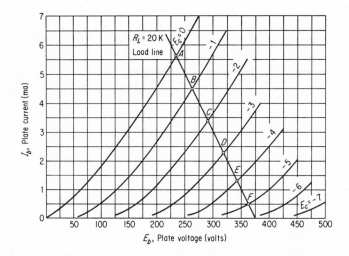

4.   If the grid voltage is changed to $E_c = -1$ volt, the corresponding values of E_b and I_b will become $E_b = 262$ volts and $I_b = 4.4$ ma, that is, point B on the load line.

5.   Voltage Amplification. In the previous parts of the example it was shown that a change of 1 volt in the grid circuit produces a change of 23 volts in the plate circuit. The voltage amplification for a circuit, A_v, is just the ratio of the output voltage change divided by the input voltage change. In terms of this example,

$$A_v = \frac{\Delta E_b}{\Delta E_c} \qquad \text{(along load line)}$$

$$= \frac{262 - 285}{(-1 - (-2))} = -23$$

6.7   Transfer Curve

To show quantitatively how a tube amplifies and in what region the amplification is linear, a transfer curve may be drawn, once the load line has been constructed. The transfer curve is a plot of plate current I_b as a function of grid voltage E_c.

Example 6.3   Using the circuit and load line of Example 6.2, plot the transfer curve for that circuit.

1.   Referring to the load-line graph of Example 6.2, the transfer curve would be composed of the values of E_c and I_b for the points A, B, C, D, E, and F.

2. These values from the graph in Example 6.2 are tabulated below.

E_c (volts)	I_b (ma)
0	5.65
−1	4.40
−2	3.23
−3	2.03
−4	1.02
−5	0.28

3. Graphing these values, with I_b as the ordinate,

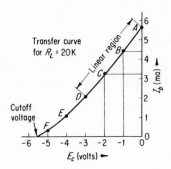

4. The transfer curve gives the plate current for any applied grid voltage. Thus, if the grid voltage is −1 volt, 4.4 ma will flow in the plate circuit. Or, if the grid voltage is reduced to −2 volts, the plate current is reduced to 3.23 ma. If the grid voltage is made negative enough, plate current will stop. This occurs here for a grid voltage of about −5.5 volts and is called the *cutoff point* for the tube. It is not a constant for a given tube, but depends slightly upon the applied dc voltage E_{bb}. Therefore, for E_{bb} = 350, the cutoff voltage is −5.5 volts in this example. The transfer curve is fairly linear between the points A and D.

To show how the amplification of a tube can be obtained from the transfer curve, a simulated signal can be produced by varying E_c sinusoidally with time. This variation can be superimposed on the transfer curve and the corresponding change in plate voltage plotted. This is shown in Fig. 6.12. The grid voltage is made to vary from −1 to 0, to −1, to −2, back to −1 volts sinusoidally. This corresponds to the points A, B, C, D, and E on the curve labeled "signal voltage." There is a resultant change in plate current. This change in plate current is constructed by transferring the points A, B, C, D, and E representing grid voltage into the corresponding points A', B', C', D', and E', representing the

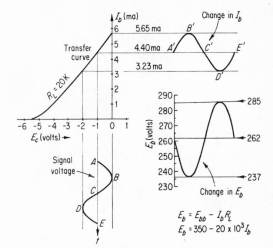

Fig. 6.12 Graphical solution for the triode amplifier utilizing the transfer curve.

change in I_b. The transfer curve has been used in this graphical construction. To obtain the resultant change in plate voltage, or output voltage, Eq. (6.14) is rewritten and solved:

$$E_b = E_{bb} - I_b R_L \qquad (6.16)$$

In terms of the example,

$$E_b = 350 - 20 \times 10^3 \, I_b \qquad (6.17)$$

From Eq. (6.17), the curve in Fig. 6.12, labeled "Change in E_b," can be constructed.

Comparing the change in E_b against the change in E_c, the following observations can be made:

1. There is voltage amplification, since a change in E_c equal to 2 volts produces a change in E_b of about 48 volts. Thus

$$A_v = \frac{\Delta E_b}{\Delta E_c} = \frac{48}{-2} = -24$$

2. The negative sign in A_v shows that the change in E_b is opposite to the change in E_c. This is also seen graphically. The output voltage is 180 deg out of phase with the input voltage.

3. The output voltage is a faithful reproduction of the input voltage because the linear portion of the transfer curve was used.

The transfer curve can also be used to predict nonlinear results. For example, suppose that the signal voltage is made to vary between the limits -4 to -6 volts. The corresponding output voltage can be found by finding the change in plate current, and then finding from this the change in output voltage by use of Eq. (6.17). This is illustrated in Fig. 6.13. Transferring the points A, B, C, D, and E of the signal voltage produces the distorted

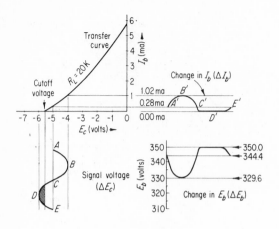

Fig. 6.13 Distortion explained with the transfer curve.

current wave-form points A', B', C', D', and E'. Using Eq. (6.17), the curve for the output voltage ΔE_b is obtained. Because a portion of the signal voltage is more negative than the cutoff voltage of the tube, the output voltage is distorted, since the tube is cut off for almost half a period. This type of distortion is called *clipping*.

6.8 Effect of R_L on A_v

The plate load resistor has a definite effect on the voltage gain A_v. In general, the larger the R_L, the greater the A_v. To find out exactly how A_v depends upon R_L, a set of load lines will be constructed for R_L values from 10 to 200 K. Then A_v will be found graphically for each load line. The load lines are shown in Fig. 6.14.

The supply voltage E_{bb} has been chosen as 350 volts, and the values of R_L as 10, 50, 100, and 200 K. To obtain A_v from the load line, two grid voltages are chosen for each load line. This constitutes ΔE_c. The corresponding change in plate voltage is found on the abscissa. The change in grid voltage is chosen as 2 volts, between $E_c = -1$ and $E_c = -3$ volts, for each load line. For the $R_L = 10$-K load line, points A and B denote the change in E_c. The corresponding values of E_b are 295 and 325 volts. Thus, for $R_L = 10$ K, the voltage gain becomes

$$A_v = \frac{\Delta E_b}{\Delta E_c} = \frac{325 - 295}{(-3) - (-1)} = -15$$

Similarly, for the load line $R_L = 20$ K, points C and D give

$$\Delta E_b = 309 - 291 = 48 \text{ volts}$$

$$\Delta E_c = (-3) - (-1) = -2 \text{ volts}$$

Therefore $A_v = -24$. For $R_L = 50$ K, using points E and F,

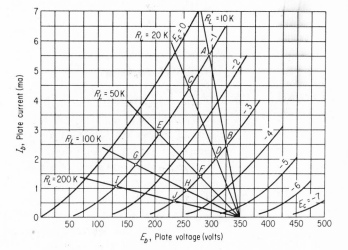

Fig. 6.14 Load lines for various values of R_L.

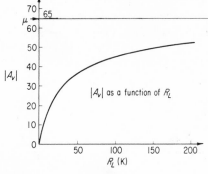

Fig. 6.15 Voltage gain as a function of R_L.

$$\Delta E_b = 280 - 207 = 73 \text{ volts}$$

$$\Delta E_c = (-3) - (-1) = -2 \text{ volts}$$

$$A_v = -36.5$$

For $R_L = 100$ K, using points G and H,

$$\Delta E_b = 236 - 132 - 104 \text{ volts}$$

$$\Delta E_c = (-3) - (-1) = -2 \text{ volts}$$

$$A_v = -52$$

Plotting $\mid A_v \mid$ versus R_L produces the graph in Fig. 6.15. The graphical result shows that A_v approaches μ asymptotically. This graph implies that μ is the maximum voltage gain which a tube can produce in a circuit. If the load resistor R_L is made too large, the load line will fall into the nonlinear region that lies below the 200-K load line, and distortion will be produced. The effect of R_L on A_v, then, is that as R_L is increased, A_v increases until R_L is made so large that the load line falls into the nonlinear region of low plate current.

6.9 RC Coupling

The simple circuit shown in Fig. 6.11 is not used in practice. It is sufficient to explain one stage of amplification, but in practice, several stages of amplification usually are joined to produce a complete system. If one desires a voltage gain greater than 100, then more than one stage of amplification must be used in cascade. To join these stages of amplification requires circuitry more complex than that shown in Fig. 6.11.

A practical circuit is shown in Fig. 6.16. The generator, labeled e_{in}, represents the signal to be amplified. Capacitor C_1, inserted between e_{in} and the control grid, seves two purposes. First, it blocks any dc voltages in e_{in} from being impressed between cathode and grid. Second, the reactance of C_1 is made low so that the ac part of the signal is not attenuated and appears in the grid circuit to be amplified. The resistor R_g is used as the load resistor across

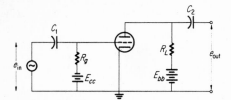

Fig. 6.16 (*Above*) A triode voltage amplifier.

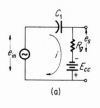

(a)

(b)

Fig. 6.17 (*Right*) The effect of R_g is to produce a signal in series with the dc bias.

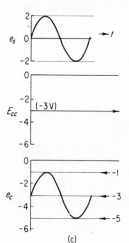

(c)

which the ac part of e_{in} will appear. This action is shown in Fig. 6.17(a). The voltage across R_g is given by

$$e_s = e_{in}\left(\frac{R_g}{R_g - jX_{C_1}}\right) \tag{6.18}$$

This should be recognized as a high-pass filter. For the case $X_{C_1} \ll R_g$,

$$e_s \cong e_{in} \tag{6.19}$$

and this voltage is developed across R_g. There are two ways of making $X_{C_1} \ll R_g$. First, C_1 can be made as large as possible. For practical purposes this value usually does not exceed 1.0 μf. Second, R_g can be made very large. There is a practical limit here also. If R_g is made too large ($R_g > 1$ meg), then the electron flow in the grid circuit is reduced to such a low value that electrons will accumulate on the grid, charging it negatively. Within a short amount of time (seconds), the control grid can become so negative that it could very well reduce the plate current to an extremely low value (near cutoff). This would cause the tube to operate in the nonlinear portion of the transfer curve. The manufacturer usually determines the maximum value of R_g that can be safely used in a tube circuit and furnishes this value in tube manuals. It is usually of the order of 1 meg.

The circuit of C_1, R_g, and E_{cc} can be represented as an ac generator e_s in series with E_{cc}, according to Eq. (6.19). This equivalent circuit is shown in Fig. 6.17(b). What the tube sees is diagrammed in Fig. 6.17(c). The sum of e_s and E_{cc} is illustrated for the case where the peak-to-peak value of e_s is 4 volts ($+2$ volts to -2 volts) and the bias voltage E_{cc} is -3 volts. Summing these up, point by point, produces the voltage e_c.

Thus the arrangement of C_1, R_g, and E_{cc} changes an ac voltage e_{in} into a pulsating negative voltage e_c. The choice of the negative bias voltage is determined by many factors. Two of them are the peak-to-peak voltage of the signal and the portion of the transfer

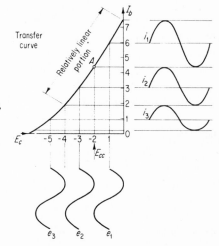

Fig. 6.18 Output signal as a function of bias.

curve to be used. For linear operation, that is, the output current proportional to the input voltage, the linear portion of the transfer curve should be used. Figure 6.18 shows that the selection of E_{cc}

is not too critical if a small signal is to be amplified linearly. For example, input voltages e_1 and e_2 will produce identical output voltages because the changes in i_1 and i_2 are identical. The signal e_3, biased as shown, will not be on the linear portion of the transfer curve and will thus produce a distorted output i_3. In general, if the signal expectations are not known, E_{cc} is chosen so that the tube operating point is in the center of the linear portion of the transfer curve, point A. Then a signal introduced will vary on both sides of point A, producing a linear output as long as the signal voltage does not exceed the limits of the linear portion of the transfer curve.

The reason for C_2 is exactly the same as that for C_1: to keep the large positive dc voltage at the plate from appearing between grid and cathode of the next stage of amplification, and to pass the ac part of the plate voltage (the amplified signal voltage).

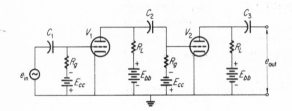

Fig. 6.19 A two-stage voltage amplifier.

Two stages of voltage amplification utilizing R_g and C for coupling are shown in Fig. 6.19. This type of coupling between stages is called *RC coupling*. The total gain is given by

$$A_v \text{ (total)} = A_{v_1} A_{v_2}$$

where A_{v_1} is the gain of the first stage, and A_{v_2} is the gain of the second stage.

6.10   Biasing Methods

There are three common methods used to obtain the proper dc bias for the grid circuit. These are fixed bias, cathode bias, and grid-leak bias. The most common is cathode bias. The circuits are illustrated in Fig. 6.20. In fixed bias, a battery is inserted in the grid circuit to establish the necessary negative dc voltage between grid and cathode. The battery in this circuit draws no current and is limited by shelf life. It may corrode and will need occasional replacing.

In cathode bias, the battery is replaced by a parallel combination of R_k and C_k. The voltage drop across this parallel combination is such that the grid is negative with respect to the cathode, as required. To ensure that the voltage drop across R_k will be a constant dc value, C_k is chosen so that $X_{C_k} \ll R_k$. This means that the ac voltage drop across the R_k and C_k combination will be close to zero, since Z (parallel) will be nearly zero (X_{C_k} very small). The dc component of the plate current, however, will produce the desired voltage drop across R_k. The value of R_k is determined by the value of E_c and I_b at the operating point:

$$R_k = \frac{|E_c|}{I_b} \quad \text{ohms} \quad (6.20)$$

where E_c and I_b are obtained from the load line.

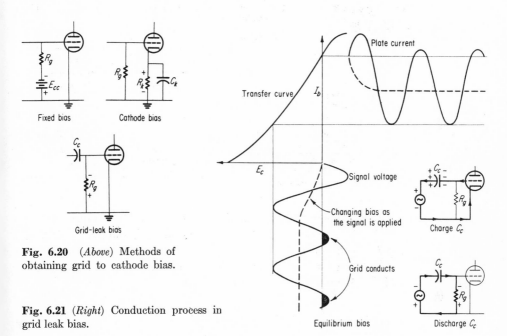

Fixed bias

Cathode bias

Grid-leak bias

Fig. 6.20 (*Above*) Methods of obtaining grid to cathode bias.

Fig. 6.21 (*Right*) Conduction process in grid leak bias.

Plate current

Transfer curve

I_b

E_c

Signal voltage

Changing bias as the signal is applied

Grid conducts

Equilibrium bias

Charge C_c

Discharge C_c

In grid-leak bias, the grid resistor is purposely made very large (of the order of 10 meg) so that the small grid current that flows will produce a voltage drop of several volts across R_g, since the dc electron flow in the grid circuit will be negative with respect to the cathode. The diagram in Fig. 6.21 serves to explain how grid-leak bias works. With no ac signal on the grid, the tube operates near zero bias. As soon as a signal is introduced, Fig. 6.21, the grid will become positive with respect to the cathode on the positive cycles of the signal voltage, and during the positive cycle the grid-cathode circuit will act like a diode. The resultant grid current charges up C_c and flows through R_g. During the negative half-cycle, capacitor C_c discharges through R_g, keeping the voltage drop across R_g essentially constant. Thus, within a few cycles, the voltage across R_g reaches an equilibrium value, with grid conduction occurring only during a fraction of the positive cycle. If the values of R_g and C_C are chosen correctly, blocking will not occur and the voltage across R_g will be a function of the amplitude of the applied signal, keeping the signal voltage essentially on the negative side of the transfer curve and thus on the linear portion of the transfer curve.

Example 6.4 Using the 50-K load line of Fig. 6.14, select a value of R_k such that E_c is in the middle of the linear portion of the transfer curve.

1. The transfer curve for 50-K load line in Fig. 6.14 is first constructed.

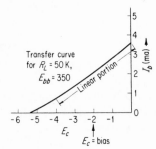

E_c	I_b
0	3.6
−1	2.9
−2	2.1
−3	1.9
−4	0.7
−5	0.2

The linear portion of the transfer curve falls in the range $E_c = 0$ to $E_c = -4$ volts. Thus the dc bias voltage E_c should be chosen as $E_c = -2$ volts.

2. For $E_c = -2$ volts, $I_b \cong 2.1$ ma. Thus R_k is determined by Eq. (6.20):

$$R_k = \frac{2}{2.1 \times 10^{-3}} = 950 \text{ ohms}$$

3. The reason that this value of R_k is correct is the fact that $R_k \ll R_L$ and will not affect the slope of the load line appreciably. If R_k were of the order of R_L, then this method, using Eq. (6.20), would yield incorrect results and another method of graphical construction would have to be used.

6.11 Class of Amplification

Amplifiers are classified generally by the dc bias on the grid with respect to the signal voltage used. The three general classes are illustrated in Fig. 6.22.

In the class A amplifier, plate current flows during the entire signal cycle. In class B amplification, plate current flows for only one half-cycle of signal voltage because E_c is at cutoff. This is

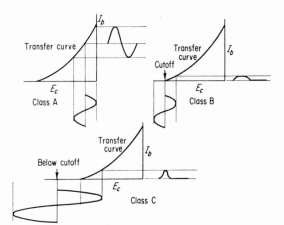

Fig. 6.22 Classes of amplification.

equivalent to half-wave rectification of the signal voltage. In class C amplifiers, the dc grid bias is well below cutoff, so that no plate current will flow until a relatively large signal is impressed on the grid, and then pulses of current will appear in the plate circuit. For class C operation, the plate current flows for less than a half-cycle of signal voltage. Classes B and C produce marked distortion. For the present, the voltage amplifiers being discussed are class A amplifiers, with E_c chosen so that the linear portion of the transfer curve is being used.

6.12 Pentodes

By insetting more grids between the control grid and the plate, a device can be made that will have a very large ac plate resistance. This means that the amplification factor will increase correspondingly, since

$$\mu = g_m r_p$$

This device is called a *pentode*. It has two more grids than the triode. The symbol for a pentode is shown in Fig. 6.23(a). The biasing voltages are shown in Fig. 6.23(b). The screen grid is held at a positive potential (of the order of 100 volts); the suppressor grid is usually connected to the cathode and is therefore at the cathode potential.

In this tube the screen grid attracts the electrons from the cathode. However, because the screen grid is literally a screen, the electrons go through the find themselves in the attractive field of the plate. The electrons strike the plate with such kinetic energy that secondary emission is produced at the plate. Many of these electrons would be lost to the screen grid if the suppressor

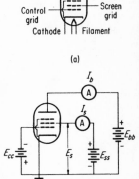

Fig. 6.23 The pentode.

grid were not present. However, the suppressor grid, being at cathode potential, repels the emitted electrons back to the plate where they are reabsorbed and form the plate current. The screen grid voltage has a greater effect on the plate current than has the plate voltage because the screen grid is closer to the cathode. But most of the electrons accelerated by the screen grid continue on to the plate circuit. Thus the plate current is a function of the screen grid voltage and does not depend too much on the plate voltage. Since the screen voltage is kept constant, the plate current changes very little with change in plate voltage. The plate characteristics show this.

A typical set of pentode plate characteristics is shown in Fig. 6.24. These are obtained in exactly the same way that the plate

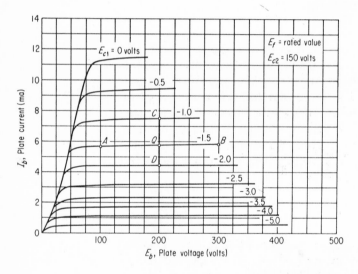

Fig. 6.24 Typical plate characteristics for a pentode.

characteristics for a triode are obtained. The screen grid voltage is held constant. In this example, the screen grid voltage was held constant at $+150$ volts. A cursory examination of Fig. 6.24 shows that the $E_c = $ const curves are nearly horizontal, indicating that I_b does not depend too much on E_b. Also, the $E_c = $ const curves tend to crowd together as E_c becomes more negative. In fact the spacing between the E_c curves is continuously changing, implying that the pentode characteristics are inherently nonlinear.

Since the slopes of the $E_c = $ constant curves are very shallow, r_p must be very high.

Example 6.5 Using the plate characteristics of Fig. 6.24, find μ, g_m, and r_p for the point $V_c = -1.5$ volts and $V_b = +200$ volts.

1. For g_m, the construction line CD is used (Fig. 6.24).

$$g_m = \left(\frac{\Delta I_b}{\Delta E_c}\right)_{E_b} \cong \frac{(7.4 - 4.4) \times 10^{-3}}{(-1) - (-2)} = 3000 \; \mu mho$$

2. For r_p, the construction line AB is used (Fig. 6.24):

$$r_p = \left(\frac{\Delta E_b}{\Delta I_b}\right)_{E_c} \cong \frac{300 - 100}{(5.8 - 5.6) \times 10^{-3}} = 2 \; meg$$

3. μ cannot be obtained graphically with any precision, and therefore is usually calculated for a pentode:

$$\mu = g_m r_p$$

$$= 3 \times 10^{-3} \times 2 \times 10^6$$

$$= 6000$$

The results of Example 6.5 show that both μ and r_p are large for a pentode as compared with those for a triode.

The transfer curve for a pentode depends markedly upon the choice of R_L. In Fig. 6.25, the load lines for $R_L = 10, 20, 50, 100$, and 200 K are shown. The corresponding transfer curves are

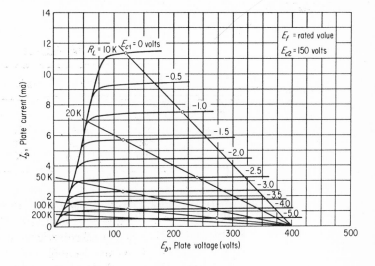

Fig. 6.25 Load lines for various values of R_L.

shown in Fig. 6.26. The plate supply voltage, E_{bb}, is 400 volts. The only transfer curve useful over an extended range of E_c is the one labeled "$R_L = 10$ K." This curve comes from the load line that crosses the characteristics above the knee of the $E_c =$ const curves. The linear portion for $R_L = 10$ K is approximately from $E_c = 0$ to $E_c = -1$ volt. For the transfer curve $R_L = 20$ K, the linear portion extends from $E_c = -1.5$ to $V_c = -2.5$ for $R_L = 50$ K (the linear portion is roughly the region $E_c = -3$ to $E_c = -4$ volts), while for $R_L = 100$ K, it extends from $E_c = -4$

to $E_c = -5$. Using these points, the voltage gain can be calculated for each plate load resistor. The resulting values are shown in Table 6.1

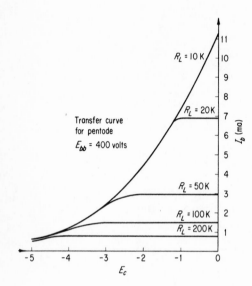

Transfer curve for pentode

$E_{bb} = 400$ volts

E_c	10 K I_b	20 K I_b	50 K I_b	100 K I_b	200 K I_b
0	11.3	6.9	2.95	1.5	0.8
−1	7.4	6.9	2.95	1.5	0.8
−2	4.4	4.4	2.95	1.5	0.8
−3	2.4	2.4	2.3	1.5	0.8
−4	1.2	1.2	1.2	1.1	0.8
−5	0.6	0.6	0.6	0.6	0.5

Fig. 6.26 The transfer curves for a pentode as a function of R_L.

TABLE 6.1

R_l, K	ΔE_c	ΔE_b	A_v
10	−1	94	−94
20	−1	124	−124
50	−1	140	−140
100	−1	145	−145

As in the triode, A_v increases with increase in R_L. However, the control grid bias is more critical, since the linear portion of the transfer curve is so small. To obtain the voltage gain of −145 volts as for $R_L = 100$ K, for example, the dc bias voltage must be $E_c = -4.5$ volts, and the signal voltage cannot exceed 1 volt peak to peak; otherwise, distortion will result.

Example 6.6   To show the production of distortion in a pentode circuit, using the characteristics of Fig. 6.25 and the 50-K load line, bias the grid at −2 volts and impress a 4-volt peak-to-peak sinusoidal voltage. Plot the resultant plate current.

1.   Redrawing the transfer curve for $R_L = 50$ K (page 201),

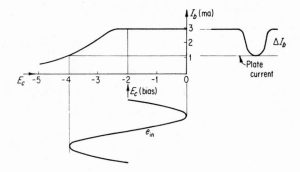

and superimposing the 4-volt peak-to-peak input voltage on $E_c = -2$ volts, it can be seen that the resulting plate current bears no resemblance to the input voltage. There is marked distortion.

The circuit for a pentode is shown in Fig. 6.27. Cathode bias is being used and the screen grid is kept at a dc potential of about $+100$ volts by means of the resistor-capacitor combination R_s and C_s. The current flowing in the screen grid circuit has an ac

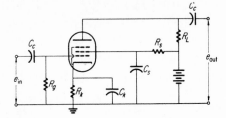

Fig. 6.27 The pentode voltage amplifier.

and a dc component. The dc component produces a voltage drop across R_s such that

$$E_{bb} - I_s R_s = E_s \qquad (6.21)$$

where E_s is the desired screen voltage. Equation (6.21) is used for determining R_s. The capacitor C_s is chosen to have a low reactance to the signal frequencies to be encountered, so that the ac voltage drop across C_s is very low. Therefore E_s is composed of a dc voltage of about $+100$ volts, and superimposed on this is a small ac ripple. Using R_s and C_s in this way enables one to use a single battery to operate the pentode circuit.

The value of R_k is found by the following equation:

$$R_k = \frac{|E_c|}{I_s + I_b} \quad \text{ohms} \qquad (6.22)$$

It should be noted that both I_s and I_b flow through R_k.

Problems

6.1 The 6AV6 is a high-μ triode. Determine μ, g_m, and r_p graphically for the 6AV6 at $E_c = -1.0$ volt and $E_b = 160$ volts.

6.2 The 6AU6 is a high-gain pentode. Determine μ, g_m, and r_p graphically for the 6AU6 at $E_c = -1.0$ volt, $E_b = 200$ volts.

6.3 The 6S4 is a power triode. Determine μ, g_m, and r_p for this tube at $E_c = -4$ volts and $E_b = 250$ volts.

6.4 The 6CA5 is a power pentode. Determine μ, g_m, and r_p for this tube at $E_c = -3.0$ volts and $E_b = 150$ volts.

6.5 From the plate characteristics for the 12AU7, for $E_b = 150$ volts, what is the current when $E_c = 0$, -2, -4 volts? Draw the transfer curve at $E_b = 150$ volts for $R_L = 0$. From the slope of this curve, determine g_m at $E_c = -2$.

6.6 Draw the transfer curve for the 6SJ7 pentode for $E_b = 200$ volts. Determine (a) g_m at $E_c = -1.5$ volts; and (b) the cutoff voltage. Assume $R_L = 0$.

6.7 Draw the transfer curve for the 6SK7 pentode for $E_b = 200$ volts ($R_L = 0$). Determine (a) g_m for $E_c = -1.5$, -3.0, -6.0, -9.0 volts; (b) the cutoff voltage.

6.8 For the 6SK7 pentode, draw a load line for $R_L = 20$ K and $E_{bb} = 400$ volts. From this determine the voltage gain as a function of E_c. Show the result in the form of a graph.

6.9 Determine R_k for a 6SQ7 triode, where $R_L = 100$ K, $E_{bb} = 300$ volts and $E_c = -1.0$ volts.

6.10 Using a 6SL7 triode, determine the voltage gain for this tube, in the linear portion of its transfer curve, for the following R_L values: $R_L = 50$, 100, 250, and 500 K. Use $E_{bb} = 350$ volts.

6.11 For the 6AU6 pentode in the circuit shown, determine R_k, R_s, and I_b. Assume that $E_s = 150$ volts and $E_c = -1.5$ volts.

6.11

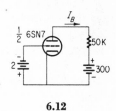

6.12

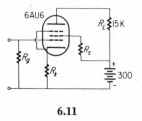

6.13

6.12 Find I_b for the circuit above.

6.13 Determine the Q point for the circuit shown (left).

6.14 Determine E_b and I_b for the circuit shown at right.

6.15 One half of a 6SN7 operates with $E_b = 200$ volts and a bias voltage $E_c = -4$ volts. The load resistance is 50 K. Determine the plate supply voltage E_{bb}, the cathode resistor R_k, and the amplification in the vicinity of $E_c = -4$ volts.

6.16 Draw a load line for one half of a 12AX7 under the following conditions: $R_L = 100$ K, $E_{bb} = 400$ volts. From the load line, draw the transfer curve and determine the plate current wave form for an input voltage that is sinusoidal and 2 volts peak to peak. Graph two cases: (a) $E_c = -1$ volt, (b) $E_c = -4$ volts.

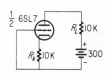

6.14

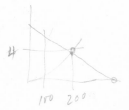

7.1 AC Voltage-Equivalent Circuit

7

Voltage Amplifiers — Analytic

When the operation of a vacuum tube is restricted to the linear portion of its transfer curve, an analytical approach can be used which will predict the operation of a given amplifier circuit. The analytical approach can also take into account the reactive components in a circuit, such as capacitors and inductors. Thus it can predict the operation of a given circuit as a function of frequency. The approach consists of replacing the tube with an equivalent voltage generator and resistance, replacing the circuitry associated with the tube by an ac equivalent circuit, and then performing calculations on this equivalent circuit.

Before this can be done, a few terms should be defined. In Fig. 7.1, typical grid and plate voltages and currents are shown when a signal is applied to the grid. The composite voltage between control grid and cathode is labeled e_c, Fig. 7.1(a). This voltage is made up of a dc component E_c and an ac component e_g, the signal.

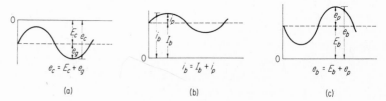

$$e_c = E_c + e_g \qquad\qquad i_b = I_b + i_p \qquad\qquad e_b = E_b + e_p$$

(a)   (b)   (c)

Fig. 7.1 Waveforms in a voltage amplifier.

The dc voltage E_c is the bias voltage between grid and cathode, usually provided by cathode bias. The resulting plate current i_b is shown in Fig. 7.1(b). This current can be thought of as the sum of a dc current I_b and an ac component i_p. The plate voltage is shown in Fig. 7.1(c). The pulsating voltage e_b is made up of a dc component E_b and an ac component e_p, the amplified signal voltage.

With these definitions in mind, Eq. (6.9) can be rewritten as

$$\Delta I_b = \frac{1}{r_p}\,\Delta E_b + g_m\,\Delta E_c$$

For small changes in E_c, E_b, and I_b, the following approximations are made; let

$$\Delta I_b \cong i_p$$
$$\Delta E_b \cong e_p \qquad\qquad (7.1)$$
$$\Delta E_c \cong e_g$$

Substituting the results of Eq. (7.1) into Eq. (6.9),

$$i_p = \frac{1}{r_p}\,e_p + g_m e_g \qquad\qquad (7.2)$$

Solving for e_p,

$$e_p = i_p r_p - g_m r_p e_g \qquad (7.3)$$

Since $g_m r_p = \mu$,

$$e_p = i_p r_p - \mu e_g \qquad (7.4)$$

Now e_p is the ac voltage between plate and cathode of the tube. Equation (7.4) shows that it is equivalent to a voltage generator, with an output equal to $-\mu e_g$, in series with a resistor r_p through which the current i_p is flowing. This is illustrated in Fig. 7.2(a).

Fig. 7.2 AC voltage equivalent circuit for a triode.

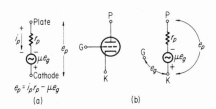

This means that for ac voltages, the equivalent circuit for a tube, from plate to cathode, is as shown in Fig. 7.2(a). In the grid circuit the ac voltage e_g, the signal voltage to be amplified, is impressed between grid and cathode. Since no grid current flows as a result of this voltage, the equivalent circuit is an open circuit between grid and cathode. Therefore the grid circuit can be converted to an ac equivalent circuit by representing it as an open circuit with a voltage e_g between grid and cathode. The two equivalent circuits completely describe the tube for ac calculations. This composite ac equivalent circuit for a tube is shown in Fig. 7.2(b). Because the tube has been replaced by an ac generator and internal resistance r_p, this is called an ac voltage-equivalent circuit and can be recognized as the ac equivalence of Thévenin's theorem, where $-\mu e_g = V_0$ and $r_p = R_0$.

In illustrating the input and output voltage wave forms for a pentode, the results would be the same as those shown for a triode in Fig. 7.1, that is, there would be ac components for the grid voltage, plate voltage, and plate current only. Therefore Eq. (6.9) also predicts the ac behavior of a pentode. Thus the ac voltage-equivalent circuit shown for a triode, in Fig. 7.2(b), is also the ac equivalent circuit for a pentode. The only difference lies in the respective values for μ and r_p for a triode and pentode. For ac calculation, a pentode can also be replaced by a voltage generator and internal resistance, as shown in Fig. 7.3.

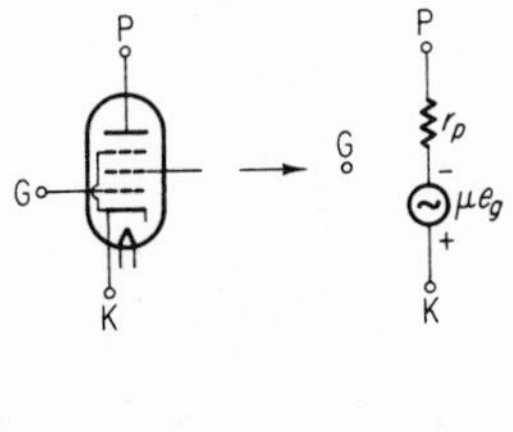

Fig. 7.3 AC voltage equivalent circuit for a pentode.

7.2 Voltage Gain as a Function of R_L

The most common type of voltage amplifier has the input voltage applied between the grid and ground and the output voltage taken off between the plate and ground. This is shown in Fig. 7.4(a). It is a simplified diagram, but is sufficient for this discussion. The voltage generator e_s is assumed not to have a dc component and can be connected as shown.

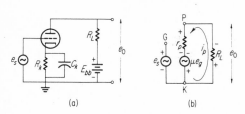

Fig. 7.4 The reduction of a circuit into an ac equivalent circuit.

(a)   (b)

In order to make ac calculations on this circuit, it must be converted into an ac equivalent circuit. To do this, first replace the tube by its ac voltage-equivalent circuit and then redraw the circuitry, short-circuiting all components having zero ac voltage drops. The resultant ac equivalent circuit for the grounded cathode circuit is shown in Fig. 7.4(b). It is from this circuit that the name "grounded cathode" is derived. As can be seen by comparing the two circuits of Fig. 7.4, the power supply E_{bb} has been replaced by a short circuit, and the parallel combination R_k and C_k has been replaced by a short circuit.

Batteries or dc sources are replaced by short circuits because of the very low reactance of such devices to ac current. The parallel combination R_k and C_k has been short-circuited because C_k is purposely made large so that X_{C_k} will be small.

In Fig. 7.4(b), the ac generators e_s and μe_g have $(+)$ and $(-)$ signs to indicate the relative polarities between generators. This shows that the output voltage will be 180 deg out of phase with the input voltage, as has already been shown to be the case in the graphical solutions. To solve this circuit for voltage gain A_v, the circuital equations for Fig. 7.4(b) are written down.

In the grid circuit,

$$e_g = e_s \tag{7.5}$$

In the plate circuit,

$$-\mu e_g + i_p(R_L + r_p) = 0 \tag{7.6}$$

$$-i_p R_L = e_0 \tag{7.7}$$

and the definition for A_v is

$$A_v = \frac{e_0}{e_s} \tag{7.8}$$

Eliminating e_g between Eqs. (7.5) and (7.6),

$$-\mu e_s + i_p(r_p + R_L) = 0 \qquad (7.9)$$

Solving Eqs. (7.7) and (7.9) for e_s,

$$e_s = \frac{\begin{vmatrix} 0 & r_p + R_L \\ e_0 & -R_L \end{vmatrix}}{\begin{vmatrix} -\mu & r_p + R_L \\ 0 & -R_L \end{vmatrix}} = -\frac{e_0(r_p + R_L)}{\mu R_L} \qquad (7.10)$$

Inserting this value of e_s in Eq. (7.8),

$$A_v = -\frac{\mu R_L}{r_p + R_L} \qquad (7.11)$$

The negative sign indicates that the input and output voltages are 180 deg out of phase.

As can be seen by inspection, if $R_L = 0$, then $A_v = 0$; if $R_L \to \infty$, then $A_v \to -\mu$. For the limit $R_L \to \infty$, the result can be more easily seen if numerator and denominator are first divided by R_L. This is just the result obtained graphically for A_v as a function of R_L, Fig. 6.15. So, the equivalent circuit is valuable in that it can "predict" experimental results for a circuit.

Example 7.1 In the circuit shown, calculate the voltage gain A_v

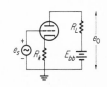

1. First the circuit must be changed to an ac equivalent circuit. The tube is replaced by its ac voltage-equivalent circuit and the circuit external to the tube is redrawn, short-circuiting all components having negligible ac voltage drops. This procedure produces an ac equivalent circuit.

As can be seen, R_k remains in this circuit because it was not by-passed by a capacitor C_k.

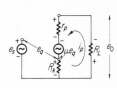

2. The circuital equations become as follows:

(a) In the grid circuit,

$$e_g = -i_p R_k + e_s$$

(b) In the plate circuit,

$$-\mu e_g + i_p(r_p + R_L + R_k) = 0$$

$$e_0 = -i_p R_L$$

and the voltage gain is given by

$$A_v = \frac{e_0}{e_s}$$

3. Rewriting the first three equations for solution by determinants and expressing the unknowns e_g, e_s, and i_p in terms of e_0,

$$e_g + i_p R_k - e_s \qquad = 0$$
$$\mu e_g - i_p(r_p + R_L + R_k) = 0$$
$$-i_p R_L \qquad = e_0$$

Solving for e_s,

$$e_s = \frac{\begin{vmatrix} 1 & R_k & 0 \\ \mu & -(r_p + R_L + R_k) & 0 \\ 0 & -R_L & e_0 \end{vmatrix}}{\begin{vmatrix} 1 & R_k & -1 \\ \mu & -(r_p + R_L + R_k) & 0 \\ 0 & -R_L & 0 \end{vmatrix}}$$

$$= -\frac{e_0 \left[r_p + R_L + (\mu + 1) R_k \right]}{\mu R_L}$$

4. Solving for A_v,

$$A_v = -\frac{\mu R_L}{r_p + R_L + (\mu + 1) R_k}$$

5. Comments: As can be seen, with R_k unbypassed (no C_k), the gain is reduced, as compared to the case where C_k is in the circuit, Eq. (7.11).

Example 7.2 In the circuit of Fig. 7.4, find the value of R_L that will give a gain $| A_v |$ of 20. The ac parameters for the tube are $r_p = 60$ K, $g_m = 2000 \ \mu$mho.

1. Using Eq. (7.11),

$$A_v = -\frac{\mu R_L}{r_p + R_L}$$

and solving for R_L,

$$R_L = -r_p \left(\frac{A_v}{A_v + \mu} \right)$$

2. $\mu = g_m r_p$; thus $\mu = 120$.
3. Solving for R_L, numerically (remember that $A_v = -20$),

$$R_L = 60 \text{ K} \left(\frac{20}{120 - 20} \right) = 12 \text{ K}$$

Example 7.3   A 5-henry inductor is used as the load for a voltage amplifier. Calculate the voltage gain as a function of frequency. Let $r_p = 30$ K and $g_m = 2000$ μmho.

1.   The circuit diagram is shown at right (top).
2.   The ac equivalent circuit becomes as shown at right (bottom).

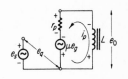

3.   The circuit equations are

$$e_g = e_s$$

$$-\mu e_g + i_p(r_p + jX_L) = 0$$

$$e_0 = -i_p(jX_L)$$

$$A_v = \frac{e_0}{e_s}$$

4.   Solving for the ratio e_0/e_s,

$$A_v = -\frac{\mu jX_L}{r_p + jX_L}$$

Dividing numerator and denominator by jX_L

$$A_v = -\frac{\mu}{1 - j(r_p/X_L)}$$

$$= -\frac{\mu}{1 - j(r_p/2\pi fL)}$$

This can be recognized as the form for a high-pass filter, where f_{co} is given by $f_{co} = r_p/2\pi L$. A_v can then be put in the form

$$A_v = -\frac{\mu}{1 - j(f_{co}/f)}$$

and

$$|A_v| = \frac{\mu}{\sqrt{1 + (f_{co}/f)^2}}$$

5.   Using the values given for the circuit,

$$\mu = g_m r_p = 60$$

$$f_{co} = \frac{r_p}{2\pi L} = 955 \text{ cps}$$

then

$$|A_v| = \frac{60}{\sqrt{1 + (955/f)^2}}$$

6. The curve of A_v versus frequency can be plotted.

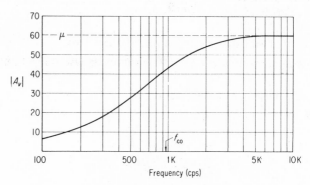

7. As can be seen, the gain approaches μ as the frequency increases, and drops off to zero at low frequencies.

7.3 Interelectrode Capacitance

In order to make more meaningful calculations on an ac equivalent circuit, certain capacitive reactances must be taken into account. Any objects having a difference in potential constitute a capacitance. Such a condition exists in vacuum tubes. The cathode, grid, and plate are made up of concentric cylinders, and thus there are capacitive effects among the three elements. These capacitances

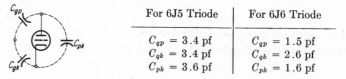

	For 6J5 Triode	For 6J6 Triode
	$C_{qp} = 3.4$ pf	$C_{qp} = 1.5$ pf
	$C_{qk} = 3.4$ pf	$C_{qk} = 2.6$ pf
	$C_{pk} = 3.6$ pf	$C_{pk} = 1.6$ pf

Fig. 7.5 Interelectrode capacitance.

are usually small, so that their effects are not felt until high frequencies are being amplified. At these high frequencies, the capacitive reactances between elements can be so low as to effect seriously the operation of the amplifier. Figure 7.5 shows schematically the interelectrode capacitances that are important in a triode and the values for typical triodes.

7.4 Miller Capacitance

It is possible to simplify the diagram shown in Fig. 7.5 by replacing C_{gp}, the capacitance between grid and plate, by an equivalent capacitance between grid and cathode. This will make the equivalent circuit simpler also. This equivalent capacitance for C_{gp} is called the *Miller capacitance*. To obtain it, Fig. 7.4(b) is redrawn to include the interelectrode capacitances, as shown in Fig. 7.6(a). To simplify the calculation, the plate-to-cathode voltage is represented as a generator of voltage $A_v e_s$, as shown in Fig. 7.6(b).

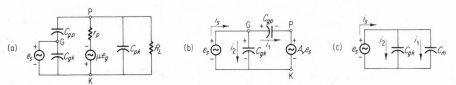

Fig. 7.6 Reduction of the ac equivalent circuit in order to obtain the input impedance.

Applying Kirchoff's laws to the circuit of Fig. 7.6(b),

$$i_s = i_1 + i_2 \tag{7.12}$$

$$e_s - i_1(-jX_{C_{gp}}) - A_v e_s = 0 \tag{7.13}$$

$$e_s - i_2(-jX_{C_{gk}}) = 0 \tag{7.14}$$

Solving Eq. (7.13) for i_1,

$$i_1 = \frac{e_s(1 - A_v)}{-jX_{C_{gp}}} = \frac{e_s[2\pi f C_{gp}(1 - A_v)]}{-j} = \frac{e_s}{-jX_{C_m}} \tag{7.15}$$

where

$$X_{C_m} = \frac{1}{2\pi f C_{gp}(1 - A_v)} \tag{7.16}$$

and

$$C_m = C_{gp}(1 - A_v) \tag{7.17}$$

Solving Eq. (7.14) for i_2,

$$i_2 = \frac{e_s}{-jX_{C_{gk}}} \tag{7.18}$$

Putting Eqs. (7.15) and (7.18) into Eq. (7.12),

$$i_s = \frac{e_s}{-jX_{C_m}} + \frac{e_s}{-jX_{C_{gk}}} = \frac{e_s}{-j}\left(\frac{1}{X_{C_m}} + \frac{1}{X_{C_{gk}}}\right) \tag{7.19}$$

or

$$i_s = \frac{e_s}{-j} (2\pi f C_m + 2\pi f C_{gk})$$

$$= \frac{e_s}{-j} 2\pi f (C_m + C_{gk}) = \frac{e_s}{-jX_{C_{in}}} \tag{7.20}$$

where

$$X_{C_{in}} = \frac{1}{2\pi f (C_m + C_{gk})} \qquad (7.21)$$

Equation (7.21) shows that the capacitance seen by the generator e_s is due to two capacitors in parallel, C_m and C_{gk}, Fig. 7.6(c). This is also shown in Eq. (7.20), where the total current is capacitive, that is, the denominator is a pure $-j$ term. Thus interelectrode capacitance C_{gp} can be replaced by capacitor C_m between grid and cathode.

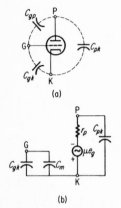

This is shown diagrammatically in Fig. 7.7. The parallel combination of C_{gk} and C_m is put between grid and cathode and C_{pk} is between plate and cathode. The circuit in Fig. 7.7(b) is the ac voltage-equivalent circuit for a tube when the interelectrode capacitance is taken into account. The Miller capacitance C_m is dependent upon the gain of the stage, since $C_m = C_{gp}(1 - A_v)$, Eq. (7.17). The parallel combination of C_m and C_{gk} will be called C_{in}, the input capacitance of a given amplifier stage.

Fig. 7.7   Miller capacitance.

7.5   AC Equivalent Circuit as a Function of Frequency

The complete ac equivalent circuit for a stage of amplification will now be solved. In this case, interelectrode capacitances, Miller capacitances, wiring capacitances, and coupling capacitances will be accounted for. Two RC-coupled voltage amplifiers are shown in Fig. 7.8(a), and their corresponding ac equivalent circuits are shown in Fig. 7.8(b). The capacitance $C_{in} = C_{gp}(1 - A_v) + C_{gk}$, and C_w represents the wiring capacitance that exists between components and the chassis, which is used as the common ground conductor. Closer scrutiny of Fig. 7.8(b) discloses that the ac equivalent circuit can be broken up into units that repeat with additional stages. Each unit is identical in appearance, although individual values may vary.

The unit being discussed is shown in Fig. 7.9. This unit appears twice in Fig. 7.8(b). If the voltage gain for one of these units is calculated, then the gain for a number of stages would simply be the product of the respective gains, that is, A_v (total) $= A_{v_1} A_{v_2}$. Therefore, whether one stage or several stages are used, the general calculation is made on a single stage such as that shown in Fig. 7.9. The voltage gain for the unit shown in Fig. 7.9 is given as the output voltage divided by the input voltage:

$$A_v = \frac{e_0}{e_g} \qquad (7.22)$$

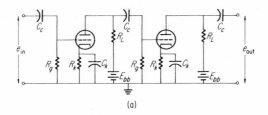

(a)

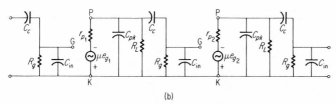

(b)

Fig. 7.8 (a) Two-stage voltage amplifier schematic. (b) AC equivalent circuit of the two-stage amplifier.

In order to obtain this ratio, the ac network in Fig. 7.9 must be solved. This is a rather complex network where e_0 is dependent upon frequency. The method usually used for solving this network as a function of frequency breaks the problem up into three parts, each much simpler than the original problem. The three parts are formed by breaking the frequency range up into three ranges: low, intermediate, and high. The capacitances C_{pk}, C_w, and C_{in} are of the order of pf. At low frequencies the reactances of these capacitors are so high that they have no appreciable effect upon the impedance of the parallel combination of R_L and R_g. Also, the coupling capacitor reactance is large at lower frequencies. Since C_c is in series with other circuit components, its reactance is important at low frequencies. At very high frequencies, the shunt capacitors C_{pk}, C_w, and C_{in} have low reactances. Being parallel with R_L and R_g, these reactances will now effect the overall impedance. The coupling capacitor C_c, however, will have a negligible effect because it is in series with other components.

The three frequency ranges into which the original problem is split are defined by the effect that the reactances will have on the circuit.

The low-frequency range is defined by the following conditions:

$$X_{C_{pk}} \gg R_L$$

$$X_{C_w} \gg R_L$$

$$X_{C_{in}} \gg R_g \qquad (7.23)$$

$$X_{C_c} \geq R_g$$

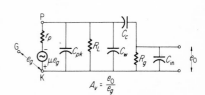

Fig. 7.9 One stage of amplification.

The intermediate frequency range is defined by the following conditions:

$$X_{(C_{pk}+C_w+C_{in})} \gg R_L$$
$$X_{C_c} < R_g \tag{7.24}$$

The high-frequency range is defined by the following conditions:

$$X_{(C_{pk}+C_w+C_{in})} < R_L$$
$$X_{C_c} \ll R_g \tag{7.25}$$

Each condition allows us to draw a different version of the circuit shown in Fig. 7.9. For the conditions expressed in Eq. (7.23), the shunt capacitors C_{pk}, C_w, and C_{in} can be omitted, but the coupling capacitor must be included. This produces the circuit shown in Fig. 7.10(a).

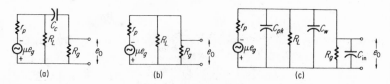

(a)   (b)   (c)

Fig. 7.10 (a) Low-frequency equivalent circuit. (b) Mid-frequency equivalent circuit. (c) High-frequency equivalent circuit.

For the conditions expressed in Eq. (7.24), the shunt capacitors can still be omitted, but now the coupling capacitor has such a low reactance that it can be replaced by a short circuit. This produces the circuit shown in Fig. 7.10(b). For the high-frequency conditions expressed in Eq. (7.25), the shunt capacitances must now be included, while the reactance of C_c is even lower. Thus the ac equivalent circuit becomes that shown in Fig. 7.10(c).

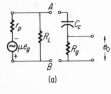

(a)

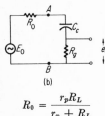

(b)

$$R_0 = \frac{r_p R_L}{r_p + R_L}$$

$$E_0 = -g_m e\, R_0$$

Fig. 7.11 Reduction of low-frequency circuit by Thévenin's theorem.

7.6 Low-Frequency Range

In order to find the voltage gain A_v for the circuit in Fig. 7.10(a), the voltage e_0 must be found as a function of frequency, and then the ratio e_0/e_g must be taken. In order to simplify the solution process, the circuit of Fig. 7.10(a) can be changed into a simple series circuit by Thévenin's theorem. The load will be taken to be the capacitor C_c and resistor R_g, Fig. 7.11(a). The remaining circuit is then converted into an equivalent generator voltage E_0 and series resistor R_0; see Fig. 7.11(b). To find R_0, the generator is shorted out in Fig. 7.11 (a) and the resistance between points A and B is found. The resistors r_p and R_L are seen to be in parallel:

$$R_{AB} = R_0 = \frac{r_p R_L}{r_p + R_L} \tag{7.26}$$

This parallel combination will be called R_0. The voltage E_0 is just E_{AB} in Fig. 7.11 (a). This becomes

$$E_{AB} = E_0 = I_{total} R_L = \left(\frac{-\mu e_g}{r_p + R_L}\right)(R_L) \tag{7.27}$$

Multiplying numerator and denominator by r_p and regrouping terms,

$$E_0 = -\left(\frac{\mu}{r_p}\right)(e_g)\left(\frac{R_L r_p}{R_L + r_p}\right)$$

$$= -g_m e_g R_0 \tag{7.28}$$

The output voltage can now be written down by inspection from Fig. 7.11 (b):

$$e_0 = E_0 \left(\frac{R_g}{R_g + R_0 - jX_{C_c}}\right) \tag{7.29}$$

In terms of Eqs. (7.26) and (7.28),

$$e_0 = -\frac{g_m e_g R_0 R_g}{R_g + R_0 - jX_{C_c}} \tag{7.30}$$

Factoring out $R_g + R_0$ in the denominator,

$$e_0 = -(g_m e_g)\left(\frac{R_0 R_g}{R_0 + R_g}\right)\left(\frac{1}{1 - j[X_{C_c}/(R_g + R_0)]}\right) \tag{7.31}$$

Let

$$R_{||} = \frac{R_0 R_g}{R_0 + R_g} \tag{7.32}$$

Using the definition in Eq. (7.32), Eq. (7.31) can be put into the form

$$e_0 = -\frac{g_m e_g R_{||}}{1 - j[X_{C_c}/(R_0 + R_g)]} \tag{7.33}$$

The voltage gain $A_v = e_0/e_g$ is thus

$$A_v = -\frac{g_m R_{||}}{1 - j[X_{C_c}/(R_0 + R_g)]} \tag{7.34}$$

Because of the $-j$ term in the denominator, there is a phase shift between e_0 and e_g, and because of the X_{C_c} term, A_v is a function of frequency.

Rewriting Eq. (7.34) in terms of absolute value and phase angle,

$$| A_v | = \frac{g_m R_{||}}{\sqrt{1 + [X_{C_c}/(R_0 + R_g)]^2}} \qquad (7.35)$$

$$\theta = \arctan\left(\frac{X_{C_c}}{R_0 + R_g}\right) + 180° \qquad (7.36)$$

The 180 deg in θ comes from the minus sign in Eq. (7.34). Eqs. (7.35) and (7.36) can be rewritten in terms of the half-power point. Equation (7.35) should be recognized as the form for the voltage output across a high-pass filter. In this case, the half-power point is found by equating the denominator to $\sqrt{2}$, Eq. (3.38).

$$\sqrt{1 + \left(\frac{X_{C_c}}{R_0 + R_g}\right)^2} = \sqrt{2} \qquad \text{for half-power point} \qquad (7.37)$$

$$f_1 = \frac{1}{2\pi(R_0 + R_g)C_c} \qquad (7.38)$$

where f_1 is the low-frequency half-power point. In terms of this definition, Eqs. (7.35) and (7.36) can be put into the forms

$$| A_v | = \frac{g_m R_{||}}{\sqrt{1 + (f_1/f)^2}} \qquad (7.39)$$

$$\theta = \arctan\left(\frac{f_1}{f}\right) + 180° \qquad (7.40)$$

These equations can be plotted as a function of frequency, Fig. 7.12. Both $| A_v |$ and θ are plotted on the same graph. The abscissa is a log scale, so as to accommodate a large range of frequencies, and the ordinate is linear. As can be seen, the voltage gain asymptotically approaches the value $g_m R_{||}$ as f increases. The phase angle between e_0 and e_g changes from 270 deg at very low frequencies to 180 deg as f increases.

The reason that the gain decreases for small f is that the reactance of C_c is large compared with R_g, thus dropping more of the voltage across C_c rather than R_g, Fig. 7.11(b). The capacitor C_c was inserted in the circuit to separate the large positive dc voltage on the plate of one circuit from the small negative dc voltage on the grid of the next circuit. However, it turns out, as shown in this section, that this same capacitor is now instrumental in decreasing the gain of an amplifier at low frequencies.

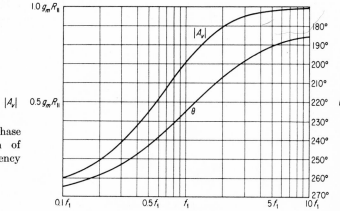

Fig. 7.12 Voltage gain and phase characteristics as a function of frequency for the low-frequency circuit.

In order to calculate the maximum value of the C_c to be put into a circuit, the maximum value of R_g is determined from a tube manual. If the information is not given, use the value $R_g = 1$ meg as a rule of thumb. Then, given a value of the lower half-power point f_1, C_c can be calculated from Eq. (7.38).

Example 7.4 A particular amplifier is to have a lower half-power point of 1 cps. The circuit is shown. Determine C_c.

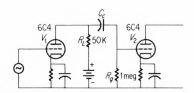

1. In order to calculate C_c, Eq. (7.38) must be used:

$$C_c = \frac{1}{2\pi f_1 (R_0 + R_g)}$$

As can be seen, $(R_0 + R_g)$ must also be determined. This is done by using Eq. (7.26).

$$R_0 + R_g = \frac{r_p R_L}{r_p + R_L} + R_g$$

The values of R_L and r_p are those for V_1.

2. The average ac plate resistance for a 6C4 can be obtained from a tube manual:

$$r_p = 7700 \text{ ohms}$$

3. $\quad R_0 + R_g = \dfrac{50 \times 10^3 \times 7.7 \times 10^3}{50 \times 10^3 + 7.7 \times 10^3} + 1 \times 10^6$

$$\cong 1.0067 \times 10^6 \cong 1 \times 10^6$$

4. C_c can now be calculated:

$$C_c = \frac{1}{2\pi \times 1 \times 1 \times 10^6}$$

$$= 0.16 \;\mu\text{f}$$

7.7 Intermediate Frequency Range

The circuit of Fig. 7.10(b), can be converted into a simple series circuit by Thévenin's theorem. From this new circuit, e_0 can be written down by inspection. Using R_g as the load, the circuit can be put in terms of E_0 and R_0, Fig. 7.13(b). To determine R_0, the generator in Fig. 7.13(a) is shorted out and R_{AB} found. This is r_p and R_L in parallel:

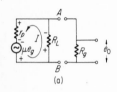

(a)

(b)

$$R_{AB} = R_0 = \frac{r_p R_L}{r_p + R_L} \tag{7.41}$$

To find E_0, E_{AB} is determined in Fig. 7.13(a). This becomes

$$E_{AB} = E_0 = -IR_L = -\left(\frac{\mu e_g}{r_p + R_L}\right) R_L \tag{7.42}$$

$R_0 = \dfrac{r_p R_L}{r_p + R_L}$

$E_0 = g_m e_g R_0$

Fig. 7.13 Reduction of the mid-frequency circuit by Thévenin's theorem.

Multiplying numerator and denominator by r_p and regrouping terms,

$$E_0 = -\left(\frac{\mu}{r_p}\right) e_g \left(\frac{r_p R_L}{r_p + R_L}\right)$$

$$= -g_m e_g R_0 \tag{7.43}$$

Using Fig. 7.13(b), the output voltage can be written as

$$e_0 = -E_0 \left(\frac{R_g}{R_g + R_0}\right) \tag{7.44}$$

Using Eq. (7.43) in Eq. (7.44),

$$e_0 = -\frac{g_m e_g R_0 R_g}{R_g + R_0}$$

$$= -g_m e_g R_{||} \tag{7.45}$$

The voltage gain becomes

$$A_v = \frac{e_0}{e_g} = -g_m R_{||} \qquad (7.46)$$

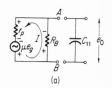

(a)

Thus, at the intermediate frequencies, the voltage gain is not a function of frequency and is in fact constant. The phase angle between e_0 and e_g is also constant and equal to 180 deg.

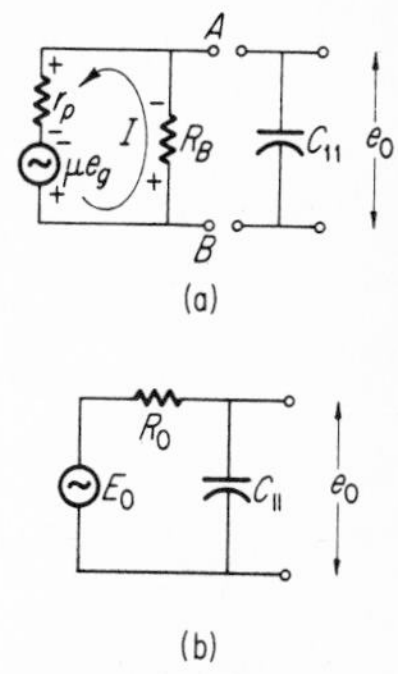

(b)

7.8 High-Frequency Range

$$R_0 = R_{||}$$
$$E_0 = -g_m e_g R_{||}$$

The circuit that must be solved in order to obtain the voltage gain as a function of f, when f is high, is shown in Fig. 7.10(c). This can be simplified before calculations are made. The resistors R_L and R_g are in parallel and can be combined. Also, the capacitors C_{pk}, C_w, and C_{in} are in parallel and can be combined. The new circuit is shown in Fig. 7.14(a).

Fig. 7.14 Reduction of the high-frequency circuit by Thévenin's theorem.

The resistor R_B is given by

$$R_B = \frac{R_L R_g}{R_L + R_g} \qquad (7.47)$$

and

$$C_{||} = C_{pk} + C_w + C_{in} \qquad (7.48)$$

The capacitor is taken as the load, and the remainder of the circuit, Fig. 7.14(a), is converted into a generator E_0 and series resistor R_0, Fig. 7.14(b). To obtain R_0, the generator is shorted in Fig. 7.14(a) and the resultant resistance R_{AB} is found:

$$R_{AB} = R_0 = \frac{r_p R_B}{r_p + R_B} = R_{||} \qquad (7.49)$$

where $R_{||}$ is the parallel combination of r_p, R_g, and R_L, as in the previous low and intermediate cases.

To obtain E_0, the potential between points A and B is found in Fig. 7.15(a), that is, E_{AB}.

$$E_{AB} = E_0 = -I R_B = -\left(\frac{\mu e_g}{r_p + R_B}\right) R_B$$

Multiplying numerator and denominator by r_p,

$$E_0 = -\left(\frac{\mu}{r_p}\right) e_g \left(\frac{r_p R_B}{r_p + R_B}\right)$$

$$= -g_m e_g R_{||} \qquad (7.50)$$

The circuit in Fig. 7.14(b) can be recognized as a low-pass filter. The output voltage is given by

$$e_0 = E_0 \frac{-jX_{C_{11}}}{R_0 - jX_{C_{11}}} \tag{7.51}$$

Replacing E_0 and R_0 by the values given in Eqs. (7.49) and (7.50),

$$e_0 = -\frac{g_m e_g R_{11}(-jX_{C_{11}})}{R_{11} - jX_{C_{11}}} \tag{7.52}$$

The voltage gain A_v is then given by

$$A_v = \frac{e_0}{e_g} = -g_m R_{11} \left(\frac{1}{1 + j(R_{11}/X_{C_{11}})} \right) \tag{7.53}$$

The absolute value of A_v and the phase angle between e_0 and e_g can be given as

$$|A_v| = \frac{g_m R_{11}}{\sqrt{1 + (R_{11}/X_{C_{11}})^2}} \tag{7.54}$$

$$\theta = \arctan\left(-\frac{R_{11}}{X_{C_{11}}}\right) + 180° \tag{7.55}$$

These equations can be simplified by defining the half-power point and then rewriting the equations in terms of the half-power point.

For the half-power point,

$$\sqrt{1 + \left(\frac{R_{11}}{X_{C_{11}}}\right)^2} = \sqrt{2} \tag{7.56}$$

$$f_2 = \frac{1}{2\pi R_{11} C_{11}} \tag{7.57}$$

where f_2 is the upper or high-frequency half-power point. In terms of Eq. (7.57), Eqs. (7.54) and (7.55) become

$$|A_v| = \frac{g_m R_{11}}{\sqrt{1 + (f/f_2)^2}} \tag{7.58}$$

$$\theta = \arctan\left(-\frac{f}{f_2}\right) + 180° \tag{7.59}$$

The results can be plotted to show how $|A_v|$ and θ vary with frequency, Fig. 7.15.

At frequencies low with respect to f_2, the upper half-power

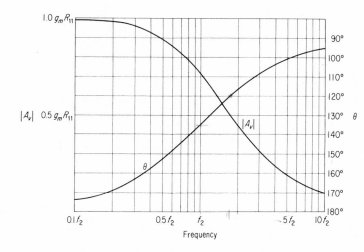

Fig. 7.15 Voltage gain and phase characteristics for the high-frequency circuit.

point, the output voltage lags the input voltage by 180°. Also, the gain is given by $g_m R_{11}$. For high frequencies, the lag angle decreases to 90 deg and the gain approaches zero.

7.9 Band Width of RC-Coupled Voltage Amplifier

Three separate solutions obtained may now be joined to show the overall band width of an RC-coupled voltage amplifier. In Fig. 7.16, the low-frequency, intermediate-frequency, and high-fre-

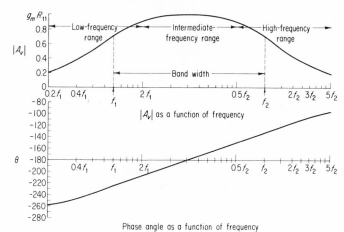

Fig. 7.16 Overall band width and phase characteristics for an RC-coupled voltage amplifier.

quency solutions are joined to show the complete solution. The band width is given by BW $= f_2 - f_1$.

As can be seen, the gain drops off at the high- and low-frequency ends. The dropoff at the low-frequency end is due to the increasing reactance of C_c, and the dropoff at the high-frequency end is due to the decreasing reactance of $C_{||}$. In the intermediate range, both voltage gain and phase shift are constant.

The results of the previous sections have been arranged in Table 7.1 for easy reference.

TABLE 7.1

Low-frequency range	$f_1 = \dfrac{1}{2\pi(R_0 + R_g)C_c}$	$|A_v| = \dfrac{g_m R_{||}}{\sqrt{1 + (f_1/f)^2}}$
Intermediate-frequency range		$|A_v| = g_m R_{||}$
High-frequency range	$f_2 = \dfrac{1}{2\pi R_{||}C_{||}}$	$|A_v| = \dfrac{g_m R_{||}}{\sqrt{1 + (f/f_2)^2}}$

Example 7.5 For the circuit shown, the band width is found experimentally to be as given in Fig. 7.17. Find the ac parameter of the tube V_1, and the shunt capacitance $C_{||}$.

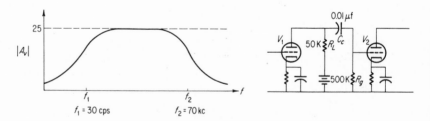

Fig. 7.17

1. Since f_1, f_2, and the intermediate-frequency gain can be read from the graph, the following equations can be written down:

$$|A_v| = g_m R_{||} = \frac{g_m r_p R_B}{r_p + R_B}$$

$$f_1 = \frac{1}{2\pi C_c(R_0 + R_g)}$$

$$f_2 = \frac{1}{2\pi C_{||}R_{||}}$$

where R_B stands for the parallel combination of R_L and R_g.

2. R_B can be determined from the circuit values

$$R_B = \frac{R_g R_L}{R_g + R_L} = \frac{500 \times 10^3 \times 50 \times 10^3}{550 \times 10^3} = 45.5 \text{ K}$$

3. R_0 can be determined from the circuit values. Solving for R_0 in the equation for f_1,

$$R_0 = \frac{1}{2\pi f_1 C_c} - R_g$$

$$= \frac{1}{2\pi \times 30 \times 0.01 \times 10^{-6}} - 1 \times 10^6$$

$$\cong 30 \text{ K}$$

4. Once R_0 is known, r_p can be determined.

$$R_0 = \frac{r_p R_L}{r_p + R_L}$$

Solving for r_p,

$$r_p = \frac{R_L R_0}{R_L - R_0} = \frac{50 \text{ K} \times 30 \text{ K}}{20 \text{ K}} = 75 \text{ K}$$

5. $R_{||}$ can now be calculated:

$$R_{||} = \frac{r_p R_B}{r_p + R_B} = \frac{75 \text{ K} \times 45.5 \text{ K}}{120.5 \text{ K}} = 28.3 \text{ K}$$

6. Substituting into the equation for $|A_v|$ and solving for g_m,

$$g_m = \frac{|A_v|}{R_{||}} = \frac{25}{28.3 \text{ K}} = 833 \; \mu\text{mho}$$

7. The equation for f_2 can also be solved for $C_{||}$.

$$C_{||} = \frac{1}{2\pi f_2 R_{11}} = \frac{1}{2\pi \times 70 \times 10^3 \times 28.3 \times 10^3}$$

$$= 80.3 \text{ pf}$$

8. Answers:

$$r_p = 75 \text{ K}$$

$$g_m = 883 \; \mu\text{mho}$$

$$\mu = g_m r_p = 66.2$$

$$C_{||} = 80.3 \text{ pf}$$

Example 7.6 If R_L for the amplifier of Example 7.5 is reduced to 10 K, calculate the new band width and intermediate gain.

1. Use the values given in part 8 of Example 7.5.
2. The new $R_{\parallel}$ will be given by the parallel combination of $r_p = 75$ K, $R_L = 10$ K, and $R_g = 500$ K.

$$R_{\parallel} = \frac{R_g R_L r_p}{R_g R_L + R_g r_p + R_L r_p}$$

$$\cong 8.68 \text{ K}$$

3.
$$|A_v| = g_m R_{\parallel}$$

$$= 7.65$$

4.
$$f_1 = \frac{1}{2\pi (R_0 + R_g) C_c}$$

where

$$R_0 = \frac{r_p R_L}{r_p + R_L}$$

$$= 8.83 \text{ K}$$

$$R_0 + R_g \cong 509 \text{ K}$$

Thus

$$f_1 = \frac{1}{2\pi \times 509 \times 10^3 \times 0.01 \times 10^{-6}}$$

$$= 31.2 \text{ cps}$$

5.
$$f_2 = \frac{1}{2\pi R_{\parallel} C_{\parallel}}$$

$$= \frac{1}{2\pi \times 8.68 \times 10^3 \times 80.3 \times 10^{-12}}$$

$$= 228 \text{ kc}$$

6. In Table 7.2 these results are compared with the values in Example 7.5.

TABLE 7.2

R_L K	$|A_v|$	f_1, cps	f_2, kc	BW, kc
10	9.72	31.2	228	228
50	25	30	70	70

Thus, if the value of R_L is reduced, there is a reduction in overall gain, but the band width goes up. In fact there is a definite relationship between A_v and BW.

(a)   For $R_L = 10$ K,

$$| A_v | \times \text{BW} = 1750$$

(b)   For $R_L = 50$ K,

$$| A_v | \times \text{BW} = 1750$$

In other words, for a given tube and $C_{||}$, the (gain $\times$ band width) product is a constant.

(c)   The (gain $\times$ band width) product can be put into the following form:

$$| A_v | \times \text{BW} = g_m R_{||}(f_2 - f_1) \cong g_m R_{||} f_2$$

$$= g_m R_{||} \frac{1}{2\pi R_{||} C_{||}}$$

$$| A_v | \times \text{BW} = \frac{g_m}{2\pi C_{||}}$$

Example 7.7   Assume that a two-stage RC-coupled amplifier has identical stages. Calculate the overall band width if the band width of a single stage is as given by the curve.

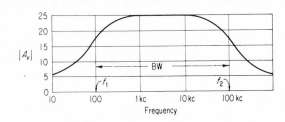

1.   Graphical Solution for Two Identical Stages.   This problem can be done by the graphical approach. Since the amplifications are the same for each stage,

$$A_v \text{ (total)} = A_{v_1} \times A_{v_2} = A_v{}^2$$

Thus the graphical approach consists of multiplying the curve by itself and determining the new band width and intermediate gain (see diagram on next page).

2.   The overall gain in the intermediate-frequency range has increased to $| A_v | = 625$, but the band width has decreased. The

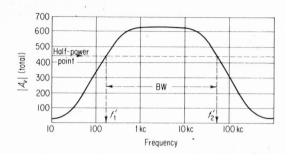

Frequency

lower half-power point f_1 has increased and the upper half-power point f_2 has decreased.

Thus, for two (or more) stages of amplification, the overall band width is less than the band width for a single stage. Conversely, to achieve a specified overall band width, the individual stages of an amplifier must have band widths in excess of the desired overall band width.

Analytically the equation for voltage gain is given by

$$A_v \text{ (total) } = A_{v_1} A_{v_2} \cdots A_{vN}$$

where

N = number of stages and

A_{v_1} = voltage gain of a single stage as a function of frequency.

This equation would have to be solved for the new half-power points.

3. **Analytical Solution for Two Identical Stages.** To solve the problem analytically rather than graphically (as was done), the problem is again split up into three frequency ranges.

(a) For the low-frequency range,

$$A_v \text{ (total) } = \left[\frac{g_m R_{||}}{\sqrt{1 + (f_1/f)^2}}\right]^2 = \frac{(g_m R_{||})^2}{1 + (f_1/f)^2}$$

For the lower half-power point,

$$1 + \left(\frac{f_1}{f'_1}\right)^2 = \sqrt{2}$$

$$f'_1 = \frac{f_1}{\sqrt{\sqrt{2} - 1}}$$

$$= 1.55 f_1$$

where f'_1 is the overall, lower half-power point and f_1 is the half-power point for a single stage. As can be seen, the overall half-power point is higher.

(b) For the high-frequency range,

$$A_v \text{ (total)} = \left[\frac{g_m R_{||}}{\sqrt{1 + (f/f_2)^2}}\right]^2 = \frac{(g_m R_{||})^2}{1 + (f/f_2)^2}$$

For the half-power point,

$$1 + \left(\frac{f'_2}{f_2}\right)^2 = \sqrt{2}$$

$1 + (X)^2 = 1.414$

$1 + (X)^{1/4} = 1.414$

f_{comp} ⟶ $f'_2 = \sqrt{\sqrt{2} - 1}\, f_2$

$$= 0.643 f_2$$

where f'_2 is the overall, upper half-power point and f_2 is the half-power point for a single stage. The overall half-power point is lower than that for the single stage.

The overall BW, that is, $f'_2 - f'_1$ is thus less than that for a single stage.

7.10   Input and Output Impedance

An entire stage of amplification can be described as a four-terminal network. This is useful because it can predict what effect an amplifier will have in loading a circuit to which it is connected. Conversely, the loading effect on an amplifier can be predicted when various loads are applied to the amplifier. For example, consider the ac generator in Fig. 7.18(a). The voltage e_{AB} depends

Fig. 7.18 Effect of R_g on output voltage.

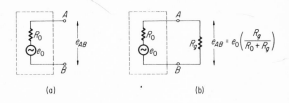

(a)   (b)

entirely upon the load that is placed across the generator. The voltage e_0 is the emf of the generator, and R_0 is the output impedance of the device. If a load R_g is connected across the generator, Fig. 7.18(b), then the output voltage is given by

$$e_{AB} = e_0 \left(\frac{R_g}{R_0 + R_g}\right) \tag{7.60}$$

For maximum e_{AB}, R_g must be made as large as possible, for if $R_g \gg R_0$, Eq. (7.60) can be approximated by e_0:

$$e_{AB} = e_0 \left(\frac{R_g}{R_0 + R_g} \right) \cong e_0 \qquad (7.61)$$

Thus, as $R_g \to \infty$, $e_{AB} \to e_0$. This is why, for example, the grid resistor R_g is made as large as possible so that the voltage developed across R_g will be as large as possible. In cases where power must be transferred from generator to load, other considerations are important, (see Sec. 1.15). If the power developed in R_g must be a maximum, R_g must equal R_0. In either case, a knowledge of R_0 is important and useful.

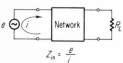

$$Z_{in} = \frac{e}{i}$$

Fig. 7.19 Circuit configuration for calculating input impedance.

INPUT IMPEDANCE

To determine the input impedance of a network, the following procedure is used:

1. Connect a voltage generator e to the input terminals of the network.
2. Convert the network into an ac equivalent network.
3. Determine i, the current flowing in the branch containing the generator.
4. $Z_{in} = e/i$ (if there are no reactances $Z_{in} = R_{in}$).

This is illustrated in Fig. 7.19.

Example 7.8 In the triode amplifier shown, below left, determine the input resistance R_{in}.

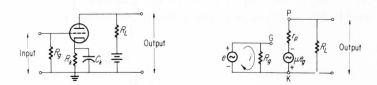

1. Using the steps outlined above, the circuit to be calculated becomes as shown.

As can be seen, the current i is determined only by R_g; therefore the solution is quite simple:

$$i = \frac{e}{R_g}$$

and

$$Z_{in} = R_{in} = \frac{e}{i} = R_g$$

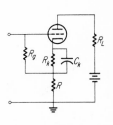

Example 7.9 Determine the input resistance of the accompanying amplifier in the intermediate-frequency range.

1. Drawing the ac voltage-equivalent circuit, adding the generator e, and loop currents results in the lower diagram.

2. The equations become

$$i(R_g + R) + i_p R = e$$

$$iR + i_p(R + R_L + r_p) - \mu e_g = 0$$

$$iR_g - e_g = 0$$

$$R_{in} = e/i$$

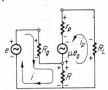

$R_L = 100$ K

$R_g = 0.5/$meg

$R = 100, 10,$ and 1 K

$r_p = 60$ K

$\mu = 90$

3. Solving the first three equations for i,

$$i = \frac{\begin{vmatrix} e & R & 0 \\ 0 & R + R_L + r_p & -\mu \\ 0 & 0 & -1 \end{vmatrix}}{\begin{vmatrix} R_g + R & R & 0 \\ R & R + R_L + r_p & -\mu \\ R_g & 0 & -1 \end{vmatrix}}$$

$$= \frac{e(R + R_L + r_p)}{(R_g + R)(R + R_L + r_p) + R(\mu R_g - R)}$$

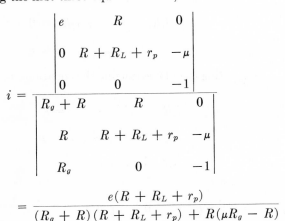

4. Solving for the ratio $R_{in} = e/i$,

$$R_{in} = R_g + R + \frac{\mu R R_g - R^2}{R_L + R + r_p}$$

Comparing this to the ac equivalent circuit, the input resistance is made up of R_g and R in series plus an interaction term due to the fact that both i and i_p flow through R.

5. A plot of R_{in} versus R is meaningful and helps indicate the effect of the interaction term.
The horizontal axis is a logarithmic scale. This allows a large range in R to be shown. The curve for R_{in} increases for large R because the interaction term rapidly increases with R. The important consequence is that the effective input resistance can be several times the grid resistor R_g, depending upon the choice of R.

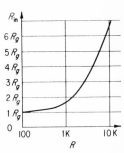

6. Conclusion. A circuit of this type, where both grid and plate ac current pass through a common cathode resistor R, is useful in raising R_{in} to a value greater than R_g.

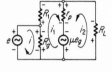

Example 7.10 Determine R_{in} for the circuit shown. Use the same circuit values as those given in Example 7.9. The capacitor C is used as a dc blocking capacitor so that the large, positive dc voltage on the plate does not appear on the control grid. Its reactance to the ac signal is made negligible

1. Drawing the ac voltage-equivalent circuit, adding the generator e, and indicating loop currents, and the accompanying illustration results.

2. The resulting equations are

$$e - iR_g \qquad\qquad + i_1R_g = 0$$

$$\mu e_g - i_2(R_L + r_p) - i_1r_p = 0$$

$$e - i_1R_1 \qquad\qquad + i_2R_L = 0$$

$$e_g = e$$

3. Eliminating e_g and rearranging the equations for determinant use,

$$iR_g - i_1R_g \qquad\qquad = e$$

$$i_1r_p + i_2(R_L + r_p) = \mu e$$

$$i_1R_1 - i_2R_L \qquad\qquad = e$$

and solving for i,

$$i = \frac{\begin{vmatrix} e & -R_g & 0 \\ \mu e & r_p & (R_L + r_p) \\ e & R_1 & -R_L \end{vmatrix}}{\begin{vmatrix} R_g & -R_g & 0 \\ 0 & r_p & (R_L + r_p) \\ 0 & R_1 & -R_L \end{vmatrix}}$$

$$= \frac{e\{[r_pR_L + R_1(R_L + r_p)] + R_g[R_L(\mu + 1) + r_p]\}}{R_g(r_gR_L + R_1R_L + R_1r_p)}$$

4. Solving for $R_{in} = e/i$, the resulting equation can be put in the form

$$R_{in} = R_g \frac{1}{1 + R_g\{[(\mu + 1)R_L + r_p]/[r_pR_L + R_1R_L + R_1r_p]\}}$$

Therefore R_{in} will always be less than R_g.

5. To show what effect R_1 has on R_{in}, R_{in} is shown as a function of R_1 for several values of R_1.

6. Conclusions. In this circuit, part of the signal at the plate is presented on the control grid along with the original signal. Since the signal from the plate is 180 deg out of phase with the input signal, the net signal to be amplified will be reduced. As a consequence, it can be said qualitatively that the voltage gain is reduced for this amplifier. Also, the input resistance will be lower than the resistor R_g. Depending upon the choice of R_1, the input resistance can be made to take on any value in the range $0 < R_{in} < R_g$. A circuit like this is useful when a low input resistance is desired.

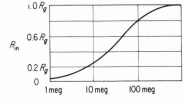

OUTPUT IMPEDANCE

The output impedance of a circuit is obtained by using the following procedure (see Fig. 7.20):

1. Draw the ac voltage-equivalent circuit.
2. Apply a voltage generator e to the output terminals.
3. Short-circuit the input generator emf.
4. Obtain the current flow i through the applied generator branch.
5. The output impedance is given by $Z_{out} = e/i$ (if there are no reactance terms, then $Z_{out} = R_{out}$, a pure resistance).

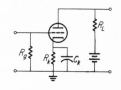

Fig. 7.20 Circuit configuration for calculation of the output impedance.

Step 3 is necessary so that the current i found in step 4 is due only to the applied generator e.

Example 7.11 Find R_{out} for the amplifier shown in the intermediate-frequency range.

1. Applying steps 1, 2, and 3, results in the accompanying illustration.

2. As can be seen by inspection, $e_g = 0$; therefore the generator $\mu e_g = 0$. The circuit can be redrawn as shown.

3. There is no need to calculate this circuit. It can be seen by inspection that R_{out} is given by the parallel combination of r_p and R_L.

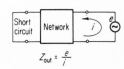

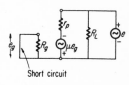

$$R_{out} = \frac{r_p R_L}{r_p + R_L}$$

Example 7.12 Find R_{out} in the intermediate-frequency range for the amplifier shown on the next page. For numerical results, use the circuital values of Example 7.9.

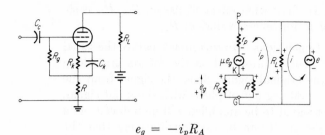

$$e_g = -i_p R_A$$

where

$$R_A = \frac{R_g R}{R_g + R}$$

$$\mu e_g + e_g - i_p(R_L + r_p) + i_1 R_L = 0$$

$$e - i R_L + i_p R_L = 0$$

Eliminating e_g and ordering the equations for determinant use,

$$-i_p[(\mu + 1)R_A + R_L + r_p] + i R_L = 0$$

$$i_p R_L \qquad\qquad - i R_L = -e$$

Solving for i,

$$i = \frac{\begin{vmatrix} -[(\mu + 1)R_A + R_L + r_p] & 0 \\ R_L & -1 \end{vmatrix} \; e \atop \; R_L}{\begin{vmatrix} -[(\mu + 1)R_A + R_L + r_p] & 1 \\ R_L & -1 \end{vmatrix} \; R_L}$$

$$= \frac{e[(\mu + 1)R_A + R_L + r_p]}{R_L\{[(\mu + 1)R_A + R_L + r_p] - R_L\}}$$

$$R_{\text{out}} = \frac{e}{i} = R_L\left[1 - \frac{R_L}{(\mu + 1)R_A + R_L + r_p}\right]$$

This equation reduces to the result obtained in Example 7.11 if $R = 0$. As can be seen from the result, R_{out} is less than R_L. A plot of R_{out} as a function of R is shown. For comparison, the same circuit values used in Example 7.9 are used here.

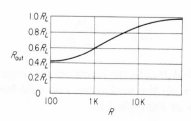

The output resistance R_{out} approaches $0.375R_L$ for small R.

Example 7.13 Find the output impedance for the accompanying circuit in the intermediate-frequency range.

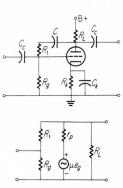

1. Since the intermediate-frequency range is being used, the capacitors C, C_c, and C_k can be replaced by short circuits in the ac voltage-equivalent circuit.

2. The ac equivalent circuit becomes as shown.

3. Shorting the input terminals and applying a generator e to the output terminals.

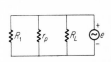

As can be seen in step 2, if the input terminals are shorted, R_g is short-circuited. This also means that $e_g = 0$. Therefore the circuit across the generator consists simply of R_1, r_p, and R_L in parallel.

4.
$$R_{\text{out}} = \frac{R_1 R_L r_p}{R_1 R_L + R_1 r_p + R_L r_p}$$

5. Thus the output impedance is less than that for a conventional amplifier. The results for input and output impedance calculations are summarized in Table 7.3.

TABLE 7.3

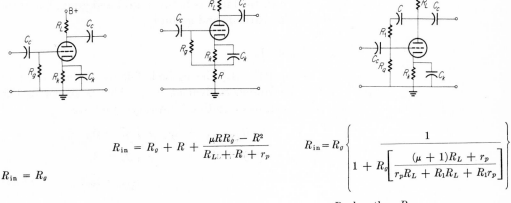

$R_{\text{in}} = R_g$	$R_{\text{in}} = R_g + R + \dfrac{\mu R R_g' - R^2}{R_L + R + r_p}$	$R_{\text{in}} = R_g\left\{\dfrac{1}{1 + R_g\left[\dfrac{(\mu + 1)R_L + r_p}{r_p R_L + R_1 R_L + R_1 r_p}\right]}\right\}$
	R_{in} greater than R_g	R_{in} less than R_g
$R_{\text{out}} = \dfrac{r_p R_L}{r_p + R_L} = R_{11}$	$R_{\text{out}} = R_L\left[1 - \dfrac{R_L'}{(\mu + 1)R_A + R_L + r_p}\right]$ where $R_A' = \dfrac{R_g R}{R_g + R}$	$R_{\text{out}} = \dfrac{R_L R_1 r_p}{R_L R_1 + R_L r_p + R_1 r_p}$
	R_{out} less than R_{11}	R_{out} less than R_{11}

7.11 Cathode Follower

As an example of the use of current feedback to effect a change in input and output impedance, as well as gain, a circuit called the *cathode follower* will be analyzed. The important characteristics can be pointed out by confining the calculations to the mid-frequency range.

The circuit to be evaluated is shown in Fig. 7.21(a). The R_k and C_k combination furnishes the dc bias so as to operate the

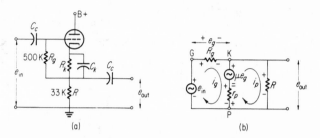

Fig. 7.21 The cathode follower and its ac equivalent circuit.

tube on the proper portion of the transfer curve. R_g furnishes the dc path from cathode to control grid so that the bias can be established between cathode and grid. Instead of placing a load resistor R_L between plate and $B+$, the load is placed between the R_k and C_k combination and ground. Therefore, as far as the ac signal is concerned, the input is between grid and ground, and the output is between cathode and ground.

Calculation of A_v

In order to calculate A_v, the cathode-follower circuit must be put into the ac voltage-equivalent form. This is shown in Fig. 7.21(b). The equations resulting from this circuit are

$$e_{\text{in}} - i_g(R_g + r_p) - \mu e_g + i_p r_p = 0$$
$$e_{\text{in}} - i_g R_g \qquad\qquad - i_p R = 0$$
$$e_{\text{out}} = i_p R \qquad (7.62)$$
$$e_{\text{out}} = A_v e_{\text{in}}$$
$$e_g = i_g R_g$$

Putting the first two equations into determinate form and eliminating e_g, with the aid of the last equation,

$$i_g[(\mu + 1)R_g + r_p] - i_p r_p = e_{\text{in}}$$
$$i_g R_g \qquad\qquad + i_p R = e_{\text{in}} \qquad (7.63)$$

Solving for i_p,

$$i_p = \frac{\begin{vmatrix} (\mu + 1)R_g + r_p & 1 \\ & & e_{\text{in}} \\ R_g & 1 \end{vmatrix}}{\begin{vmatrix} (\mu + 1)R_g + r_p & -r_p \\ R_g & R \end{vmatrix}}$$

$$= \frac{e_{\text{in}}[(\mu + 1)R_g + r_p] - R_g}{R[(\mu + 1)R_g + r_p] + r_p R_g} \tag{7.64}$$

Since $A_v = i_p R / e_{\text{in}}$, using Eq. (7.64),

$$A_v = \frac{R\{[(\mu + 1)R_g + r_p] - R_g\}}{R[(\mu + 1)R_g + r_p] + r_p R_g} \tag{7.65}$$

This can be put in the form

$$A_v = \frac{1 - \{R_g / [(\mu + 1)R_g + r_p]\}}{1 + \{r_p R_g / R[(\mu + 1)R_g + r_p]\}} \tag{7.66}$$

At this point the equation can be substantially simplified if recourse is made to numerical examples. The tube used will be a pentode having a transconductance of the order of 5000 μmho. For a pentode, the amplification factor will be of the order of 10^3. To make the simplification more meaningful, the characteristics for a 6AK5 tube will be used so that numerical estimates of the various terms can be made. For a 6AK5, $r_p = 500$ K, $g_m = 5000$ μmho, and $\mu = 2500$. Therefore, in Eq. (7.66), the quantity $(\mu + 1)R_g \gg r_p$, and the equation can be simplified into

$$A_v \cong \frac{\mu / (\mu + 1)}{1 + [r_p / R(\mu + 1)]} \tag{7.67}$$

Since R will be of the order of r_p, the ratio $r_p / R(\mu + 1) \ll 1$ and can be neglected.

A_v further reduces to

$$A_v \cong \frac{\mu}{\mu + 1} \tag{7.68}$$

Thus, if the correct tube is chosen for the cathode follower (large g_m and μ), the equation for A_v is independent of the circuit parameters, to a good approximation. Putting the value of $\mu = 2500$ into Eq. (2.69),

$$A_v = \frac{2500}{2501} = 0.9996 \tag{7.69}$$

Gain for this cathode follower is unity. Since A_v was positive, there is no phase reversal between input and output voltage.

CALCULATION OF R_{in}

The circuit of Fig. 7.21 (b) is the required circuit for obtaining R_{in}. The e_{in} generator will represent the generator placed across the input circuit to determine R_{in}. Thus the equations in Eq. (7.63) can be solved for i_g, to obtain $R_{in} = e_{in}/i_g$:

$$i_g = \frac{\begin{vmatrix} 1 & -r_p \\ & & e_{in} \\ 1 & R \end{vmatrix}}{\begin{vmatrix} (\mu + 1)R_g + r_p & -r_p \\ R_g & R \end{vmatrix}}$$

$$= \frac{e_{in}(R + r_p)}{R[(\mu + 1)R_g + r_p] + r_p R_g} \tag{7.70}$$

and

$$R_{in} = \frac{e_{in}}{i_g} = \frac{R[(\mu + 1)R_g + r_p] + r_p R_g}{R + r_p} \tag{7.71}$$

This can be put into the following form, so that approximations can be made:

$$R_{in} = \frac{(\mu + 1)R_g + r_p + (r_p/R)R_g}{1 + (r_p/R)} \tag{7.72}$$

In the numerator,

$$(\mu + 1)R_g \gg r_p \qquad \text{and} \qquad (\mu + 1)R_g \gg \left(\frac{r_p}{R}\right)R_g$$

so that the equation for R_{in} becomes

$$R_{in} \cong \frac{(\mu + 1)R_g}{1 + (r_p/R)} \tag{7.73}$$

which is good to within 1 percent of the actual answer given by Eq. (7.71). Putting values into Eq. (7.73),

$$R_{in} = 156 R_g$$

Thus the input resistance for the cathode follower is very high.

CALCULATION OF R_{out}

The circuit of Fig. 7.21 (b) must be modified in order to calculate

R_{out}. First, e_{in} must be shorted out and a generator e applied to the output terminals. The circuit thus formed is shown in Fig. 7.22(a). This circuit can be simplified somewhat by noting that R_g and R are in parallel. This combination will be called R_G.

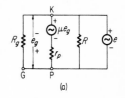

$$R_G = \frac{R_g R}{R_g + R} \tag{7.74}$$

This produces the circuit in Fig. 7.22(b). From this circuit, the following equation can be written:

$$e_g = -e$$

$$\mu e_g - i_g(R_G + r_p) - iR_G = 0 \tag{7.75}$$

$$e - i_g R_G \qquad - iR_G = 0$$

Fig. 7.22 Reduction of cathode follower circuit for output impedance calculations.

Putting the equation into determinant form and eliminating e_g,

$$i_g(R_G + r_p) + iR_G = -\mu e$$
$$i_g R_G \qquad + iR_G = e \tag{7.76}$$

Solving for i,

$$i = \frac{\begin{vmatrix} R_G + r_p & -\mu \\ R_G & 1 \end{vmatrix} e}{\begin{vmatrix} R_G + r_p & 1 \\ R_G & 1 \end{vmatrix}} = \frac{e[(\mu + 1)R_G + r_p]}{R_G r_p} \tag{7.77}$$

and

$$R_{\text{out}} = \frac{e}{i} = \frac{R_G r_p}{(\mu + 1)R_G + r_p} \tag{7.78}$$

In order to simplify Eq. (7.78), by using the proper approximations, it is put into the following form:

$$R_{\text{out}} = \frac{R_{||}}{[R_G \mu / (R_G + r_p)] + 1}$$

where

$$R_{||} = \frac{R_G r_p}{R_G + r_p}$$

By multiplying and dividing by r_p in the denominator, the equation can be written as

$$R_{\text{out}} = \frac{R_{||}}{g_m R_{||} + 1}$$

Since $g_m R_{||} \gg 1$, the equation becomes

$$R_{\text{out}} \cong \frac{1}{g_m} \qquad (7.79)$$

For the numerical example, $R_{\text{out}} = 200$ ohms and is essentially independent of the circuit parameters R and R_g. Thus, for a cathode follower, R_{out} is very small. The results are tabulated in Table 7.4.

TABLE 7.4

Cathode Follower

$A_v = \dfrac{\mu}{\mu + 1}$	$R_{\text{in}} = \dfrac{(\mu + 1)R_g}{1 + (r_p/R)}$	$R_{\text{out}} = \dfrac{1}{g_m}$
(equal to unity)	(very high)	(very low)

Input Capacitance for a Cathode Follower

The circuit for a cathode follower also decreases the input capacitance significantly, with respect to a conventional amplifier. A simplified approach to this problem is as follows: In drawing the ac equivalent circuit, including the interelectrode capacitances, the output circuit is represented by a generator of magnitude $e_{\text{out}} = e_{\text{in}}[\mu/(\mu + 1)]$. The circuit is then drawn as shown in Fig. 7.23(a).

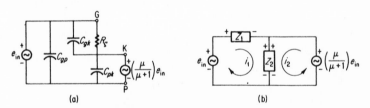

(a)   (b)

Fig. 7.23 Cathode follower circuit reduction for input impedance calculations.

The capacitance between grid and plate, C_{gp}, is seen to be in parallel with the input generator. Thus the input capacitance is

$$C_{\text{in}} = C_{gp} + C_x \qquad (7.80)$$

where C_x is yet to be determined. To determine C_x, the circuit is redrawn as shown in Fig. 7.23(b), omitting C_{gp} because we know that it is in parallel with the generator. Z_1 is the parallel combination of R_g and C_{gk}, and Z_2 is the capacitance C_{pk}. From Fig. 7.23(b), the equation about the outside loop is

$$e_{\text{in}} - i_1 Z_1 - \left(\frac{\mu}{\mu + 1}\right) e_{\text{in}} = 0 \qquad (7.81)$$

Solving for $Z_{in} = e_{in}/i$,

$$Z_{in} = (\mu + 1)Z_1$$

$$\frac{1}{Z_{in}} = \frac{1}{\mu + 1}\left(\frac{1}{R_g} + \frac{1}{-jX_{C_{gk}}}\right)$$

$$= \frac{1}{(\mu + 1)R_g} + \frac{2\pi f}{-j}\left(\frac{C_{gk}}{\mu + 1}\right)$$

$$= \frac{1}{R_{in}} + \frac{1}{-jX_{C_x}}$$

where

$$C_x = \frac{C_{gk}}{\mu + 1} \tag{7.82}$$

and

$$R_{in} = (\mu + 1)R_g \tag{7.83}$$

This simplified approach gives a value for R_{in} that differs from Eq. (7.73) by a factor of about $1/10$. Thus the value for C_x is not exact, but as can be seen, C_x is so small that a factor of $1/10$ would not be appreciable here. A numerical example will suffice to point this out. Combining Eq. (7.80) and Eq. (7.82),

$$C_{in} = C_{gp} + \frac{C_{gk}}{\mu + 1} \tag{7.84}$$

For a pentode such as the 6AK5, $C_{gp} = 0.02$ pf, and $C_{gk} = 4$ pf. Thus, numerically,

$$C_{gp} \gg \frac{C_{gk}}{\mu + 1}$$

and the equation for C_{in}, to a good approximation, becomes

$$C_{in} \cong C_{gp} \tag{7.85}$$

The input capacitance for a cathode follower is thus very small.

FREQUENCY CHARACTERISTICS

To determine f_1, Eq. (7.38) must be modified:

$$f_1 = \frac{1}{2\pi(R_0 + R_g)C_c}$$

R_0 is the parallel combination of R_L and r_p of the previous stage, while R_g is taken as the input impedance of the cathode follower.

In the low-frequency region, this reduces to the input resistance, Eq. (7.73). Usually R_{in} for a cathode follower is much greater than R_0, and R_0 can be neglected. The lower half-power point becomes

$$f_1 = \frac{1}{2\pi\{(\mu + 1)R_g/[1 + (r_p/R)]\}C_c}$$

Also, since R_{out} is so low, the upper half-power point in the output circuit will be much higher for a cathode follower than for a conventional amplifier, for a given $C_{||}$:

$$f_2 = \frac{1}{2\pi(1/g_m)C_{||}}$$

The cathode follower has a very wide band width, of the order of megacycles per second.

USES OF THE CATHODE FOLLOWER

The input and output properties of a cathode follower make it very useful, either as an isolating device or as a device insensitive to shunt capacitance.

The input properties are essentially (1) very high input resistance, and (2) small input capacitance. This combination means that Z_{in} is very high, and therefore a cathode follower will not appreciably load any circuit to which it is connected. This, coupled with the fact that the output is equal to the input (that is, $A_v = 1$), makes the cathode follower an excellent isolating device. For example, an oscilloscope is a voltage-measuring device wherein the leads from the first amplifier in the oscilloscope must be connected to the circuit under test. If this first amplifier is a conventional one, the impedance placed in parallel with the circuit under test will consist of about 500 K and 100 pf in parallel. This can seriously affect the circuit under test (loading the circuit). If this first circuit is a cathode follower, however, then the input impedance can be of the order of 200 meg and 4 pf in parallel. Such a high input impedance will produce negligible loading for most applications, Fig. 7.24. The output of the cathode follower is then connected to conventional voltage amplifiers within the oscilloscope. In this application, the cathode follower has isolated the circuit under test from the amplifiers in the oscilloscope.

The output properties of a cathode follower are (1) very low R_{out}, and (2) small C_{out}. The output resistance is equal to $1/g_m$ and $C_{out} = C_{pk}$. For a typical cathode follower, the numerical values are of the order of $R_{out} = 100$ ohms and $C_{out} = 3$ pf. The low value for R_{out} means that a relatively large shunt capacitance can be placed across the output terminals of a cathode

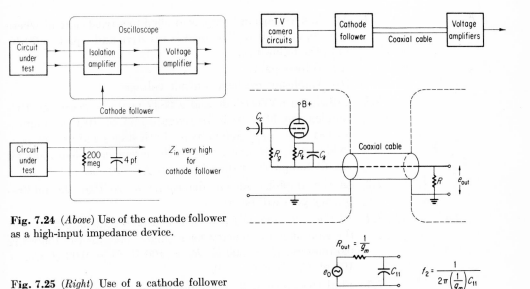

Fig. 7.24 (*Above*) Use of the cathode follower as a high-input impedance device.

Fig. 7.25 (*Right*) Use of a cathode follower as a low-output impedance device.

follower and still have a high upper half-power frequency f_2. Consider the following example of using the low output resistance of the cathode follower.

In a television studio, the TV cameras are movable; their complex signals must be sent to voltage amplifiers for further amplification before the signals are strong enough to be transmitted. The complex signals must be transmitted from camera to studio amplifiers by means of a coaxial cable. Usual band widths required in television work are of the order of 4 Mc or higher; that is, f_2 is about 4 Mc or higher. A coaxial cable consists of a two-wire transmission line composed of a hollow conducting cylinder and a concentric inner conductor, Fig. 7.25. A cable such as this has considerable interconductor capacitance, and it is not unusual for the capacitance for a length of cable to be about 100 to 200 pf. If this is connected to the output of a conventional amplifier, f_2 will be very low, owing to the high shunt capacitance of the coaxial cable. If a cathode follower is used, however, the very low value of R_{out} in conjunction with this shunting capacitance still produces a high value of f_2. The cathode follower and coaxial cable circuit forms a wide band width and unit-voltage gain circuit capable of transferring a complex signal several hundred feet, from TV camera to studio amplifier. Thus, in applications where a large band width is to be preserved in the face of large values of shunting capacitance, a cathode follower may provide a very useful answer.

Problems

7.1 An input voltage of 6 vpp, is applied to a triode circuit where $\mu = 20$, $r_p = 10$ K. The load is resistive and equal to 20 K. Find the following:
(a) Voltage gain.
(b) Amplitude of the ac output voltage.

7.2 Utilizing a 6AV6 triode and a 6SJ7 pentode, design an RC-coupled amplifier with an overall gain of 1000 at midfrequency. Use $E_{bb} = 350$ volts for both stages and use the pentode as the first stage. Sketch the circuit and determine all circuit component values to ensure $f_1 = 20$ cps.

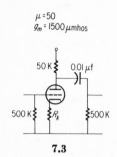

μ=50
g_m = 1500 μmhos

7.3

7.3 In the amplifier shown, determine R_k so that the midfrequency gain will be 10.

7.4 In Fig. 7.8, assume V_1 is a 6AV6 triode. Calculate f_1, f_2, and the gain at midfrequency for a single stage for the following parameters: $R_g = 500$ K, $R_L = 100$ K, $C_c = 0.02$ μf, $C_{in} = 100$ pf, $C_{Pk} = 0$, $C_w = 0$.

7.5 Find the upper half-power frequency for the accompanying circuit, where $R_k = 1$ K, $C_k = 0.01$ μf, $r_p = 10$ K, and $g_m = 1500$ μmho.

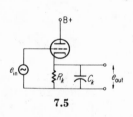

7.5

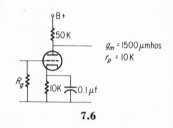

7.6

7.6 Determine the value of f_1 for this circuit (*Hint*: Take into account the effect of C_k.)

7.7 A pentode is operated as a voltage amplifier. The parameters for the tube are $\mu \doteq 1000$, $r_p = 250$ K, $R_L = 50$ K.

$$C_{gp} = 2.4 \text{ pf}, \qquad C_{gk} = 3.2 \text{ pf}, \qquad C_{pk} = 1.6 \text{ pf}$$

(a) Calculate the input capacitance.
(b) Calculate the input impedance of the tube.

7.8 In the circuit shown in Example 7.4, using the following parameters for V_1,

$R_L = 75$ K $\qquad\qquad C_c = 0.02$ μf

$r_p = 1$ meg $\qquad\qquad C_{gk} = 5.5$ pf

$g_m = 2000$ μ mho $\qquad C_{pk} = 5.0$ pf

$R_g = 500$ K

and taking C_{in} of the next stage as 50 pf, calculate (a) mid-frequency voltage gain, (b) f_1, and (c) f_2.

7.9 The frequency-response curve for a particular amplifier is found to be as shown. Determine (a) C_{11}, (b) C_c, (c) r_p, and (d) μ.

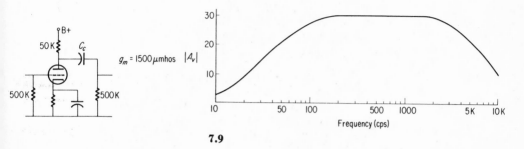

7.9

7.10 The upper half-power cutoff frequency for a three-stage amplifier utilizing identical stages is to be 20 kc. Calculate f_2 for each stage.

7.11 Determine (a) the input resistance and (b) the output resistance for the following amplifier in the midfrequency range:

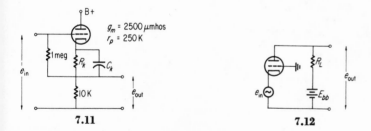

7.11   **7.12**

7.12 Determine the voltage gain for the accompanying circuit:

7.13 Determine the voltage gain for the circuit shown. Assume that the two tubes are identical. R_1 and R_2 serve to bias V_2 properly and C_2 keeps the grid of V_2 at ac ground.

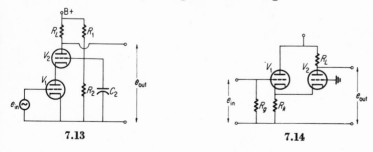

7.13   **7.14**

7.14 Assume that both tubes are identical. Calculate the voltage gain for midfrequencies.

7.15 Determine the voltage gain at resonance and the band width of the amplifier shown.

7.15   **7.16**

7.16 Replace terminals A and B by a Thévenin's voltage-equivalent circuit. Give all values in terms of e_{in}, g_m, r_p, and R_k.

7.17 Find e_{out} in terms of e_1, e_2, and the tube and circuit parameters. Assume that the tubes are identical.

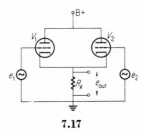

7.17

8.1 Need for Power Amplifiers

In order to point out the need for power amplifiers, a discussion in terms of a familiar system will be useful. Such a system is the phono-amplifier. A block diagram is shown in Fig. 8.1(a). The output of a typical phono crystal is of the order of millivolts. This variation in voltage must somehow drive a loudspeaker so that the voltage variations can be transformed into pressure variations in air. However, a loudspeaker, by its construction, see Fig. 8.1(c), requires a current variation of the order of amperes in its voice coil in order to move the cone sufficiently to produce audible sound.

It is the function of the intermediate amplifiers, Fig. 8.1(a), to convert the small change in voltage into a large change in current. A voltage amplifier, as discussed in Chapters 6 and 7, is necessary to amplify the voltage to about $\Delta e = 10$ volts. Then this voltage variation is used to drive a power amplifier, which in turn drives the loudspeaker. The purpose of the power amplifier

Power Amplifiers

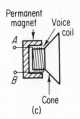

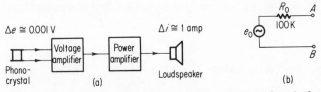

Fig. 8.1 (a) Block diagram of a phono amplifier. (b) Crystal equivalent circuit. (c) Loudspeaker construction.

is not only to produce the large current variations required to operate the loudspeaker, but also to transfer the maximum amount of ac power to the loudspeaker so as to make the system as efficient as possible. In order to transfer maximum power to a load (the loudspeaker), internal impedance of the power amplifier must match the impedance of the load. A typical loudspeaker has an impedance of 4 to 8 ohms. Since the internal resistance r_p is of the order of 5000 ohms for a power amplifier, there will be a grave mismatch as far as power transfer is concerned. See Fig. 8.2(a).

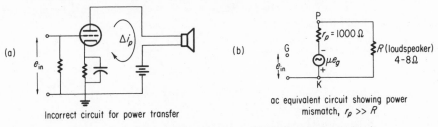

Fig. 8.2 (a) Incorrect circuit for power transfer. (b) AC equivalent circuit showing power mismatch.

For this reason a power amplifier must use an impedance-matching device. One of the properties of this impedance-matching

245

device is to make the resistance of the loudspeaker, R_{LS}, look like a much greater value R_L' in the plate circuit of the power amplifier. See Fig. 8.2(c) and (d). Also, the impedance-matching

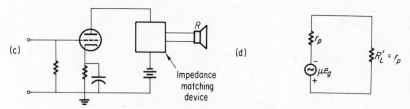

Fig. 8.2 (c) Need for impedance-matching device. (d) For correct match, $R'_L = r_p$.

device must take the plate current variation in the plate circuit, $\Delta i = 100$ ma, and change it into a current variation in the loudspeaker voice coil, of the order of amperes. A transformer has the necessary properties.

8.2 Output Transformer

In this discussion the simplest approach will be used in order to make the important concepts stand out. For this reason a perfect transformer will be discussed. Consider a generator connected to a resistance by means of a transformer. The power developed in the primary circuit is equal to the power developed in the secondary circuit.

$$\text{Power in} = \text{power out}$$
$$E_1 I_1 = E_2 I_2 \tag{8.1}$$

where the voltage and current are effective values. Also, the induction law states that the voltage induced in a coil due to a changing magnetic field is proportional to the number of turns in the coil. The change in flux is the same for primary and secondary; thus

$$\frac{N_1}{N_2} = \frac{E_1}{E_2} \tag{8.2}$$

where N_1 and N_2 are the number of turns in the primary and secondary, respectively.

Dividing both sides of Eq. (8.1) by $I_1{}^2$,

$$\frac{E_1}{I_1} = E_2 \frac{I_2}{I_1}$$

and multiplying the right-hand side by I_2/I_2,

$$\frac{E_1}{I_1} = \frac{E_2}{I_2}\left(\frac{I_2}{I_1}\right)^2 \tag{8.3}$$

From Eqs. (8.2) and Eq. (8.1),

$$\frac{N_1}{N_2} = \frac{E_1}{E_2} = \frac{I_2}{I_1} \tag{8.4}$$

and putting Eq. (8.4) in Eq. (8.3),

$$\frac{E_1}{I_1} = \frac{E_2}{I_2}\left(\frac{N_1}{N_2}\right)^2 \tag{8.5}$$

Now considering Eq. (8.5) in conjunction with Fig. (8.3), the ratio of E_2/I_2 will yield the value of R_2, since R_2 determines the current that will flow in the secondary circuit for a given applied voltage E_2. In like manner, the ratio of E_1/I_1 can be called R_1, the effective resistance in the primary circuit. Equation (8.5) can then be written as

$$R_1 = R_2\left(\frac{N_1}{N_2}\right)^2 \tag{8.6}$$

where N_1/N_2 is the turns ratio between primary and secondary, R_2 is the resistance in the secondary circuit, and R_1 is the effective resistance in the primary circuit. The effective resistance R_1 is the resistance that the generator in Fig. 8.3 "sees." In other words, the transformer makes R_2 in the secondary look like R_1 in the primary.

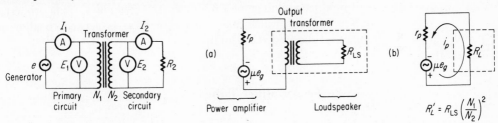

Fig. 8.3 Transformer parameters.

Fig. 8.4 The use of an output transformer for impedance matching.

This property of a transformer is illustrated in Fig. 8.4. The ac voltage-equivalent circuit for a tube is shown with a transformer connecting R_{LS} to the vacuum-tube circuit. As far as current flow in the primary circuit is concerned, it is the same as if R'_L, given by Eq. (8.6), were in the plate circuit; that is, an ac voltage drop is developed across the primary of the transformer, which can be given as $i_p R'_L$.

To make R'_L equal r_p so as to obtain maximum power dissipation in the load, the turns ratio (N_1/N_2) is made larger than 1:

$$R'_L = R_{LS} \left(\frac{N_1}{N_2}\right)^2 \tag{8.7}$$

where R_{LS} is the resistance of the loudspeaker and R'_L is the effective resistance in the primary circuit. For maximum power transfer, in Fig. 8.4(b), $R'_L = r_p$, and Eq. (8.7) becomes

$$r_p = R_{LS} \left(\frac{N_1}{N_2}\right)^2$$

$$\frac{N_1}{N_2} = \sqrt{\frac{r_p}{R_{LS}}} \tag{8.8}$$

A numerical example will suffice to give relative values for N_1 and N_2. A typical value for r_p is 8000 ohms, and 8 ohms for R_{LS}. Thus $N_1/N_2 = \frac{32}{1}$, approximately. This also shows that the output transformer is a stepdown transformer. Using Eq. (8.4),

$$I_2 = I_1 \left(\frac{N_1}{N_2}\right)$$

A change in I_1 produces a change in I_2, given by

$$\Delta I_2 = \Delta I_1 \left(\frac{N_1}{N_2}\right) \tag{8.9}$$

Thus a current change in the secondary circuit is greater than the change in the primary by the factor N_1/N_2. For the example used above, that is, $N_1/N_2 = \frac{32}{1}$, a change of 50 ma in the primary will produce a change of 1.6 amp in the secondary. This property of a stepdown transformer is exactly what is needed to allow a vacuum tube to drive a loudspeaker.

8.3 Power Amplifier Tubes and Circuits

Power amplifier tubes are constructed to control relatively large currents in the plate circuit. As a result the ac plate resistance of a triode power tube is low, and the amplification factor will be relatively low, as shown in Table 8.1. For this reason a high-μ power tube called a *beam power* tube, essentially a pentode, has been developed. Because it is a pentode, its plate resistance will be high and therefore μ will be high. See Table 8.1. The beam power tube is shown diagrammatically in Fig. 8.5.

The major difference between this pentode and the voltage amplifier pentode is the way that suppression of secondary emission is accomplished. Instead of a suppressor grid, the beam power tube has beam-forming plates, connected electrically to the cathode. Therefore the beam plates repel the electrons and the electron flow from cathode to plate is confined to a fairly dense stream of electrons. Electrons striking the plate produce secondary electrons, which are immediately returned to the plate because of the large negative charge of the approaching beam of electrons. Thus the beam itself acts like the suppressor grid of a conventional pentode.

TABLE 8.1   Power Amplifier Parameters

Tube	r_p	g_m	μ	Power Output, Watts
2A3 (Triode)	800	5250	4.2	3.5
6C4 (Triode)	7,700	2200	17.0	1.0
6AQ5 (Beam Pentode)	58,000	3700	214	2.0
6V6 (Beam Pentode)	80,000	3750	300	5.0

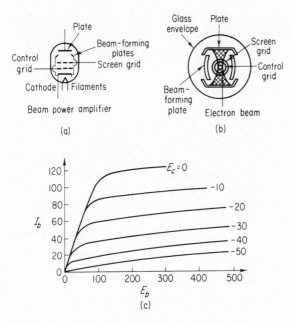

Fig. 8.5 (a) Beam power amplifier symbol. (b) Beam power tube cross-section. (c) Plate characteristics.

The plate characteristics for a beam power pentode look like those for a conventional pentode. It should be noted in Fig. 8.5(c) that the ordinate has larger numbers than those for the conventional pentode. Also, the increments in E_c are larger. These two characteristics denote a power tube.

The circuit for a power amplifier utilizing a triode is shown in Fig. 8.6(a), and the circuit for a power amplifier utilizing a beam pentode is shown in Fig. 8.6(b). Schematically, the only difference between a power amplifier and a voltage amplifier is the replacement of R_L by a power transformer.

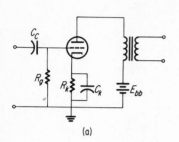 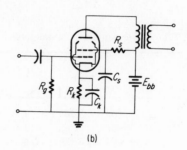

(a)   (b)

Fig. 8.6 (a) Triode power amplifier circuit. (b) Beam power amplifier circuit.

PLATE VOLTAGE—NO SIGNAL

The voltage drops in the plate circuit of a power amplifier are due to dc components only, if there is no signal present. Thus, in the circuit of Fig. 8.6(a),

$$E_{bb} = E_{R_k} + E_b + E_L \qquad (8.10)$$

where

E_{R_k} = dc voltage drop across R_k

E_b = dc voltage drop across the tube

E_L = dc voltage drop across the transformer primary.

As a first-order approximation this can be simplified by considering the order of magnitude of each term. E_{R_k} and E_L are of the order of 10 volts each. Thus $E_b \gg E_{R_k}$ and E_b and, to a good approximation,

$$E_{bb} \cong E_b \qquad (8.11)$$

PLATE VOLTAGE—WITH SIGNAL

With a signal present on the grid of the power amplifier, there will be dc and ac voltages present on the plate of the tube. The complex voltage between plate and ground will be e_b, and is made up of two components, Fig. 8.7(a):

$$e_b = E_{bb} + e_L \qquad (8.12)$$

where

e_b = instantaneous voltage between plate and ground (neglecting E_{R_k})

E_{bb} = dc voltage across the power supply

e_L = ac voltage across the transformer primary (neglecting E_L)

The voltage across the inductance of the primary is given by

$$e_L = -L \frac{di_p}{dt} \qquad (8.13)$$

where i_p is the plate current and is given by

$$i_p = I_b + I_p \sin \omega t \qquad (8.14)$$

with a dc component I_b (the average value) and an ac component (the signal). Putting Eq. (8.14) in Eq. (8.13), one obtains

$$e_L = -LI_p\omega \cos \omega t \qquad (8.15)$$
$$= -X_L I_p \cos \omega t$$

Calling the quantity $X_L I_p$ the *amplitude* of the ac voltage, $X_L I_p = E_p$, and putting this result in Eq. (8.12),

$$e_b = E_{bb} - E_p \cos \omega t \qquad (8.16)$$

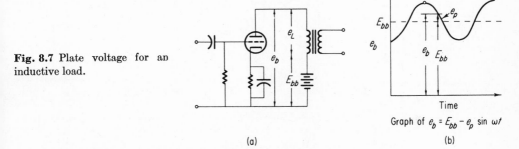

Fig. 8.7 Plate voltage for an inductive load.

Graph of $e_b = E_{bb} - e_p \sin \omega t$

(a)

(b)

This result is shown graphically in Fig. 8.7(b). As can be seen, at time A the voltage between plate and ground is greater than the supply voltage E_{bb}. This is due to the inductive voltage produced across the primary of the transformer.

8.4 Nonlinear Operation

Up to now it has been assumed that the signal present on the control grid of a tube does not exceed the linear portion of the transfer curve. In a power amplifier, the signal on the control grid must have a large amplitude so that the voltage and current variations in the plate circuit will be large. This ensures that the output power of stage will be sufficient to operate a loudspeaker.

In using a large amplitude signal on the control grid, a large portion of the transfer curve is used, and nonlinear operation results; that is, the output voltage waveform is distorted. This will be illustrated graphically for both a triode and pentode power amplifier.

In Fig. 8.8(a), the plate characteristics for a triode power amplifier are shown. Note the ordinate. The current scale goes as high as 100 ma. Also, the voltage increments for E_c are much higher than those for a voltage amplifier. Here the increments are in

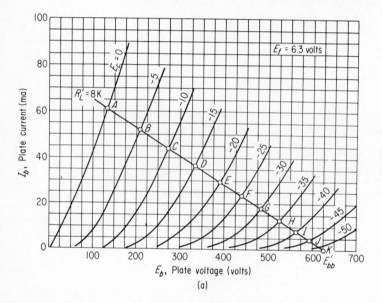

(a)

Fig. 8.8 (a) Load line for a triode power amplifier. (b) Graphical solution utilizing the transfer curve.

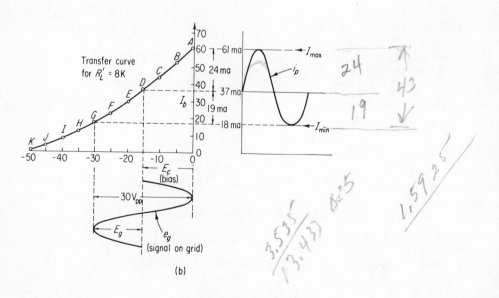

(b)

steps of 5 volts. A load line for $R'_L = 8000$ ohms is superimposed on the graph through the point $E_b = 300$ volts, $I_b = 40$ ma. The transfer curve for this load line is shown in Fig. 8.8(b). The graph shows the distortion that is produced when a large signal is impressed on the control grid. A 30-volt peak-to-peak (pp) signal is drawn in. The resultant current i_p is shown. As can be seen, the first half-cycle varies from 37 to 61 ma, that is, $\Delta i_p = 24$ ma. The second half-cycle varies from 37 to 18 ma, that is, $\Delta i_p = 19$ ma. There is a tendency for the second half-cycle to be flattened due as a result of operation on the nonlinear portion of the transfer curve.

Figure 8.9(a) shows the plate characteristics for a pentode power amplifier. The load line is placed above the knee of the curves so that the linearity of the transfer curve will be maximum. This will enable one to place a large amplitude signal on the control grid. The load line is, for an R'_L of 3000 ohms, through the point $E_b = 200$, $I_b = 60$ ma. The transfer curve for this load line is shown in Fig. 8.9(b). A 30-volt pp signal is impressed on the grid of this tube also. The resultant plate-current variation is seen to be highly distorted because the transfer curve of this tube is more nonlinear over a 30-volt range in E_c.

This distortion in the plate current can be predicted analytically also. For the simplest approach, it is assumed that the transfer curve can be approximated by a parabola. The relationship between i_p and e_c can then be written in general as

$$i_p = A + Be_c + Ce_c^2 \tag{8.17}$$

where A, B, and C are undetermined coefficients. Equation (8.17) should be recognized as the general equation for a parabola. Let e_c be represented as

$$e_c = E_c + E_g \sin \omega t \tag{8.18}$$

where E_c is the dc bias and $E_g \sin \omega t$ is the signal impressed on the grid. Inserting Eq. (8.18) into Eq. (8.17) and expanding the squared term, one obtains

$$i_p = A + BE_c + BE_g \sin \omega t + CE_c^2 + 2CE_cE_g \sin \omega t + CE_g^2 \sin^2 \omega t \tag{8.19}$$

Replacing the $\sin^2 \omega t$ term by $\sin^2 \omega t = \frac{1}{2} - \frac{1}{2} \cos 2\omega t$ and grouping terms, Eq. (8.19) becomes

$$i_p = A + BE_c + CE_c^2 + \tfrac{1}{2}CE_g^2 + (BE_g + 2CE_cE_g) \sin \omega t$$
$$- \tfrac{1}{2} CE_g^2 \cos 2\omega t \tag{8.20}$$

As can be seen, not only is the fundamental present but also a

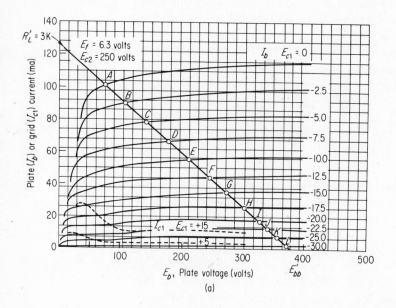

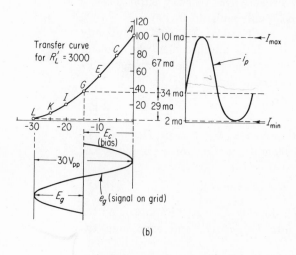

Fig. 8.9 (a) Load line for a pentode power amplifier. (b) Graphical solution utilizing the transfer curve.

second harmonic. If the signal is not present, $E_g = 0$ and Eq. (8.20) becomes

$$i_p = A + BE_c + CE_c{}^2 = I_b \qquad (8.21)$$

which has a dc value. Thus the term $A + BE_c + CE_c{}^2$ is equal to the dc value of the plate current with no signal present. In Fig. 8.8, for example, this would correspond to the point D on the transfer curve, and the value $I_b = 37$ ma. Therefore the dc term $CE_g{}^2/2$ is introduced because of the distortion that occurs when a signal is present. Also, the term $(BE + 2CE_cE_g)$ is the amplitude of the fundamental and will be called I_1. The term $CE_g{}^2/2$ is the amplitude of the second harmonic and will be called I_2. In terms of these definitions, Eq. (8.20) can be written as

$$i_p = I_b + I_2 + I_1 \sin \omega t - I_2 \cos 2\omega t \qquad (8.22)$$

In effect, Eq. (8.22) is the analytical form for the graph of the distorted current wave form seen in Fig. 8.8(b) or Fig. 8.9(b); that is, if the terms in Eq. (8.22) are graphed, the distorted current wave forms described will result. This connection between analytical and graphical results is shown in Fig. 8.10(b). Because the characteristic distortion of the second half-cycle is caused by the introduction of the second harmonic, this type of distortion is called *second-harmonic distortion*. A parabolic-shaped transfer curve, therefore, introduces second-harmonic distortion. It is possible to find the values of I_1 and I_2 in terms of the terminal points on the load line, that is, I_{max} and I_{min}. At point A in Fig. 8.10(b), the value of i_p is equal to I_{max} and is given by

$$I_{max} = I_b + 2I_2 + I_1 \qquad (8.23)$$

Similarly, at point B in Fig. 8.10(b), I_{min} is given by

$$I_{min} = I_b + 2I_2 - I_1 \qquad (8.24)$$

It should be noted that Eq. (8.23) can be obtained from Eq. (8.22) by letting $\omega t = \pi/2$, and Eq. (8.24) can be obtained from Eq. (8.22) by letting $\omega t = \frac{3}{2}\pi$. Adding Eqs. (8.23) and (8.24) and solving for I_2,

$$I_2 = \frac{I_{max} + I_{min} - 2I_b}{4} \qquad (8.25)$$

Subtracting Eq. (8.24) from Eq. (8.23) and solving for I_1,

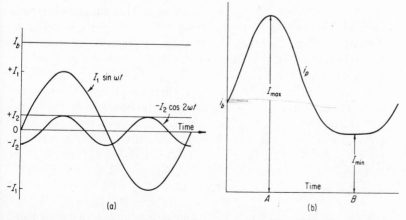

Fig. 8.10 The construction of a curve having second-harmonic distortion.

$$I_1 = \frac{I_{max} - I_{min}}{2} \tag{8.26}$$

Thus the amplitudes of the fundamental and second harmonic can be obtained graphically from a load-line graph, since the points I_b, I_{max}, and I_{min} are easily read off.

Example 8.1 Using the characteristics in Fig. 8.8 and the load line shown, determine I_1 and I_2 for a 30-volt pp signal on the control grid.

1. Since the signal on the grid is 30 volts pp, the value of E_c is half this value; $E_c = -15$ volts. This ensures that e_c does not go positive. From this value of E_c, I_b can be determined. From the graph in Fig. 8.8,

$$I_b = 37 \text{ ma} \qquad \text{at } E_c = -15 \text{ volts}$$

2. The maximum value of I occurs at $E_c = 0$ and the minimum value at $E_c = -30$. Thus, from the graph,

$$I_{max} = 61 \text{ ma} \qquad \text{at } E_c = 0$$

$$I_{min} = 18 \text{ ma} \qquad \text{at } E_c = -30$$

3. Substituting these values into Eqs. (8.25) and (8.26),

$$I_1 = \frac{I_{max} - I_{min}}{2} = \frac{61 - 18}{2} = 21.5 \text{ ma}$$

$$I_2 = \frac{I_{max} + I_{min} - 2I_b}{4} = \frac{61 + 18 - 74}{4} = 1.25 \text{ ma}$$

Equation (8.22) shows that when a signal is present, and there is second harmonic distortion present, the dc value of the current increases by the amount I_2; that is, the change in I_{dc} is just equal to the amplitude of the second harmonic. This suggests a simple method for keeping a check on an amplifier circuit in order to determine when distortion is occurring, and even to determine the relative amount of distortion. Figure 8.11 shows how the method works.

A dc milliammeter in the plate circuit will read the average value of plate current. As long as the signal operates on the linear portion of the transfer curve, the ammeter will read the same value for I_b even though different amplitude signals may be present, Fig. 8.11(b). However, as soon as the signal amplitude is increased to the point where second-harmonic distortion becomes appreciable, the dc or average value of i_p will increase by an amount equal to I_2. Thus a dc ammeter in the plate circuit can tell one when distortion is occurring and can give the amplitude

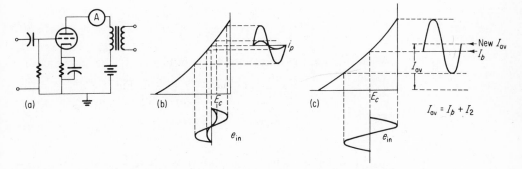

Fig. 8.11 The change in average current due to second-harmonic distortion.

of the second harmonic. In this discussion it has been assumed that the bias E_c has not been changed, since this also affects the value of I_b.

SECOND-HARMONIC DISTORTION

In order to make the distortion meaningful, a quantity termed *percent* of second-harmonic distortion is defined as

$$\text{Percent 2nd HD} = \frac{I_2}{I_1} \times 100 \qquad (8.27)$$

On the basis of listening tests, manufacturers have decided to adopt a rule of thumb that the percent of second-harmonic distortion in an amplifier will not exceed 5 percent. It is thought that if this distortion figure is exceeded, the effect is noticeable and not tolerated. In actual practice, it is easy to keep the distortion well below this value.

AC POWER

The average power delivered to the load (the loudspeaker) is given by

$$P_{ac} = E_{eff}I_{eff} \qquad (8.28)$$

where E_{eff} is the effective value of the plate voltage variation and I_{eff} is the effective value of the plate current variation; it will be assumed that I_1 is kept below the 5 percent level, as defined in Eq. (8.27). Therefore the power contained in the second harmonic can be neglected. Equation (8.28) becomes

$$P_{ac} = \frac{E_1}{\sqrt{2}} \cdot \frac{I_1}{\sqrt{2}} \qquad (8.29)$$

The equation for E_1 will be of the same form as that for I_1

and can be written down as

$$E_1 = \frac{E_{\max} - E_{\min}}{2} \tag{8.30}$$

where $E_{\max}$ is the maximum value of e_b and $E_{\min}$ is the minimum value of e_b, as obtained from the load line. In terms of Eqs. (8.26) and (8.30), the average power becomes

$$P_{ac} = \frac{(E_{\max} - E_{\min})(I_{\max} - I_{\min})}{8} \text{ watts} \tag{8.31}$$

This equation tells how much power is dissipated in R'_L. But this is the same power that is dissipated in the loudspeaker, since the transformer is treated as perfect.

A figure of merit is sometimes used to show how efficient a stage of power amplification is when converting the available dc power into ac power. It is termed the *efficiency* of a power amplifier and is given as a percentage:

$$\text{Percent eff} = \frac{P_{ac}}{P_{dc}} \times 100 \tag{8.32}$$

where the term P_{dc} refers to the dc power supplied by the power supply. This is just equal to the product of the supply voltage E_{bb} and the average current $I_b + I_2$:

$$P_{dc} = E_{bb}(I_b + I_2) \cong E_{bb}I_b$$

Example 8.2 For the power amplifier load line of Fig. 8.8, determine the percent 2nd HD, P_{ac} and percent eff for an input voltage of 30 volts pp.

1. From Example 8.1 it has been determined that

$$I_b = 37 \text{ ma}$$

$$I_{\max} = 61 \text{ ma}$$

$$I_{\min} = 18 \text{ ma}$$

From these values it was determined that

$$I_1 = 21.5 \text{ ma}$$

$$I_2 = 1.25 \text{ ma}$$

Thus

$$\text{Percent 2nd HD} = \frac{1.25}{21.5} \times 100 = 5.8 \text{ percent}$$

2. To obtain the average power, $E_{\max}$ and $E_{\min}$ must be deter-

mined. Using the load line of Fig. 8.8 and dropping perpendiculars from points A and G to the E_b axis, it is found that

$$E_{max} = 482 \text{ volts}$$

$$E_{min} = 135 \text{ volts}$$

Thus

$$E_1 = \frac{E_{max} - E_{min}}{2} = 173.5 \text{ volts}$$

From Eq. (8.29), P_{ac} becomes

$$P_{ac} = \frac{E_1 I_1}{2} = 1.87 \text{ watts}$$

3. The percent efficiency for this power amplifier is given by

$$\text{Percent eff} = \frac{P_{ac}}{P_{dc}} \times 100$$

To obtain P_{dc}, the values of E_{bb} and I_b must be determined. Using the approximation of Eq. (8.11), E_{bb} and I_b are to be evaluated at the Q point, or point D in Fig. 8.8:

$$E_{bb} \cong E_b = 332 \text{ volts}$$

$$I_b = 37 \text{ ma}$$

and

$$P_{dc} = 332 \times 37 \times 10^{-3} = 12.3 \text{ watts}$$

Thus the efficiency becomes

$$\text{Percent eff} = 15.2 \%$$

8.5 Graphical Solution

In designing a power amplifier, one wants the maximum power output possible and the smallest second-harmonic distortion. The two conditions are not compatible, and thus a compromise must be made. In this section, a method of successive approximations will be used to find the maximum power output possible, consistent with keeping the second-harmonic distortion below 5 percent. The problem will be worked out in more detail than is really necessary so as to be inclusive and bring out the salient points. In essence, the method will consist of drawing several load lines, calculating percent 2nd HD and P_{ac} for each load line, and then determining the optimum value for R'_L.

Before the load lines can be drawn, the maximum power dissipation curve for the tube must be constructed. This was not necessary for voltage amplifiers because the maximum power ratings for the tube are not exceeded in normal use and therefore the topic was not introduced. In power amplifier circuits, however, it is relatively easy for the power dissipation of the tube to be exceeded. For this reason the maximum power-dissipation curve for a power tube is superimposed on the plate characteristics as shown in Fig. 8.12. The maximum allowable power is 15 watts for this tube, and the curve AB is a locus of points where $E_b I_b = 15$. This is an equation for a hyperbola. All load lines must fall on or below this curve. It should be noted that the figure shows points on the characteristics above the maximum power curve AB. This is due to the fact that the maximum power rating of the tube can be exceeded instantaneously, as long as the average power is at or below 15 watts. Cases like this come up when tubes are operated in pulse circuits, described in Chapter 10.

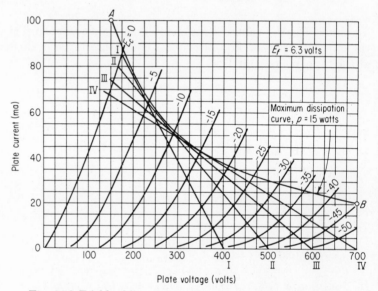

Fig. 8.12 Trial load line constructions in power-amplifier design.

Several trial load lines are now drawn in. Each is drawn tangent to the maximum power curve so as to help ensure maximum ac power conditions. Four load lines are shown in Fig. 8.12. The values of E'_{bb} were arbitrarily chosen as 400, 500, 600, and 700 volts. In load line I, it can be seen that if ΔE_c is chosen as 40 volts, there will be a large amount of distortion, since the plate characteristics suggest that the tube will cut off at about $E_c = -35$ volts. If ΔE_c is made less, say, $\Delta E_c = 30$ volts, then the percent second-harmonic distortion becomes about 15 percent. Using the same load line I, and $\Delta E_c = 20$ volts, the percent second-harmonic distortion becomes about 9 percent. For $\Delta E_c = 10$ volts, percent of second HD becomes about 6 percent. It is evident that as the change in E_c is made smaller, the distortion decreases. However, the ac power developed will also decrease. To see how ac power depends upon the load line and ΔE_c, the curves in Fig. 8.13 have been obtained for the load lines that are shown in Fig. 8.12.

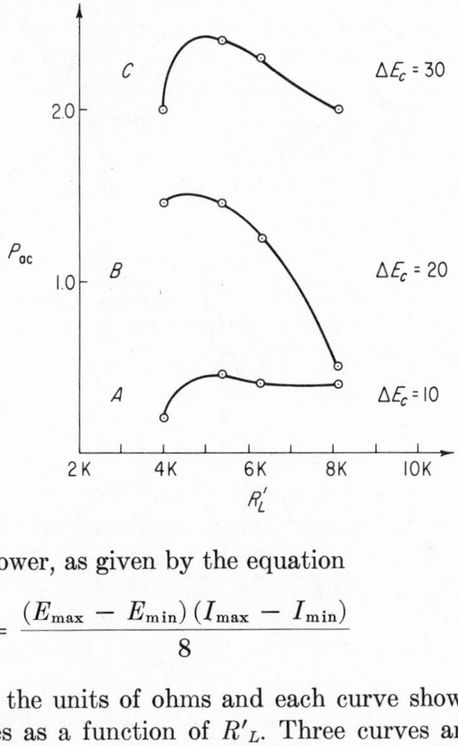

Fig. 8.13 AC power as a function of signal voltage and load line.

The vertical axis is ac power, as given by the equation

$$P_{\text{ac}} = \frac{E_1 I_1}{2} = \frac{(E_{\max} - E_{\min})(I_{\max} - I_{\min})}{8}$$

The horizontal axis has the units of ohms and each curve shows how the ac power varies as a function of R'_L. Three curves are shown: $\Delta E_c = 30$ volts, $\Delta E_c = 20$ volts, and $\Delta E_c = 10$ volts. The smaller ΔE_c, the smaller the ac power. Also, each curve peaks at about $R'_L = 5000$ ohms. In curve A (that is, $\Delta E_c = 10$ volts), the 5 percent harmonic-distortion point occurs at about $R'_L = 8000$ ohms. The other points have larger distortions. For curve B (that is, $\Delta E_c = 20$ volts), the 5 percent distortion point is near $R'_L = 6.3$ K. And for curve C ($\Delta E_c = 30$ volts), the 5 percent distortion point occurs near $R'_L = 8000$ ohms. Curves for $\Delta E_c = 40$ volts and $\Delta E_c = 50$ volts are not shown, since all points have second-harmonic distortions greater than 5 percent. Thus curve C would be a good compromise at $R'_L = 8000$ ohms because the ac power is greatest for the three curves.

The graphical construction has shown that the largest grid voltage swing, compatible with the distortion condition percent second HD = 5 per cent and the condition that ac power be a maximum, is $\Delta E_c = 30$ volts. Therefore the bias voltage for this power amplifier would be $E_c = -15$ volts. Knowing the impedance of the loudspeaker that will be connected to the power amplifier will then enable one to find the turns ratio needed for the output transformer. Assuming for this example that the impedance of the loudspeaker is 8 ohms, the turns ratio, as given by Eq. (8.7), becomes

$$\frac{N_1}{N_2} = \sqrt{\frac{R'_L}{R_{LS}}} = \sqrt{\frac{8000}{8}} = \frac{32}{1}$$

Furthermore, R_k can be evaluated, since the Q point will be situated where the $E_c = -15$ volt curve intersects load line IV; that is $E_b = 324$ volts, $I_b = 43$ ma as in Fig. 8.12. R_k can be calculated from

$$R_k = \frac{|E_c|}{I_b} \cong 350 \text{ ohms} \qquad \text{(to 2 significant figures)}$$

The power dissipated is given by

$$I_b^2 R_k = 0.64 \text{ watt}$$

Thus a 1-watt, 350-ohm resistor would be used for R_k. Needless to say, if the signal on the grid were less than 15 volts in amplitude, the percent second-harmonic distortion would be less than 5 percent. The circuit diagram for the power amplifier finally adopted is shown in Fig. 8.14. R_g will be of the order of 500 K and C_c and C_k are determined by the choice of f_1. The value for E_{bb} is given by the value of E_b at the Q point, Eq. (8.11).

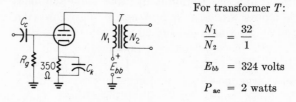

For transformer T:

$$\frac{N_1}{N_2} = \frac{32}{1}$$

$$E_{bb} = 324 \text{ volts}$$

$$P_{ac} = 2 \text{ watts}$$

Fig. 8.14 Typical circuit values for a triode power amplifier.

In summary: To find the correct value for R'_L when only the tube characteristics are given for a triode power amplifier:

1. A maximum power dissipation curve is drawn.
2. Trial load lines are drawn.
3. Trial E_c values are selected, starting with the largest.

4. Percent second HD is calculated for each case, using Eq. (8.27).
5. Average ac power is calculated, using Eq. (8.31).
6. As soon as a percent distortion ≤ 5 percent is reached, the value of R'_L is determined.
7. The grid bias is established as $E_c = -\frac{1}{2}\Delta E_c$.
8. The value of R_k is determined.
9. The value of E_{bb} is determined.

Example 8.3 For the characteristics shown for a power triode, find the ac power, percent second HD, R'_L, the Q point, and E_{bb} for $\Delta E_c = 30$ volts. Repeat for $\Delta E_c = 20$ volts, for the load line given.

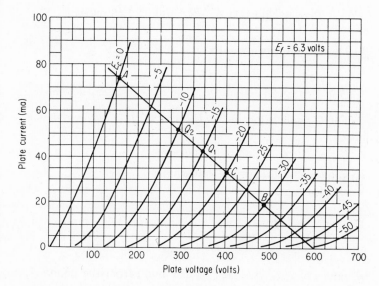

1. For the case $\Delta E_c = 30$ volts, points A, Q_1, and B mark the points used to evaluate I_{max}, I_b, and I_{min}, respectively. The values for E_{max} and E_{min} can also be obtained:

$$I_{max} = 74 \text{ ma} \qquad E_{max} = 485 \text{ volts}$$

$$I_b = 42 \text{ ma}$$

$$I_{min} = 18 \text{ ma} \qquad E_{min} = 155 \text{ volts}$$

2. From these values, R'_L can be determined. Using the load-line equation,

$$R'_L = \frac{E'_{bb} - E_b}{I_b}$$

$$E'_{bb} = 600 \text{ volts}$$

and selecting $E_b = E_{\min} = 155$ volts, $I_b = I_{\max} = 74$ ma,

$$R'_L = \frac{600 - 155}{0.074} = 6020 \text{ ohms}$$

3. For percent second HD, I_1 and I_2 must be found:

$$I_1 = \frac{I_{\max} - I_{\min}}{2} = \frac{74 - 18}{2} = 28 \text{ ma}$$

$$I_2 = \frac{I_{\max} + I_{\min} - 2I_b}{4} = \frac{74 + 18 - 84}{4} = 2 \text{ ma}$$

Thus

$$\text{Percent 2nd HD} = \frac{I_2}{I_1} \times 100 = 7.15 \text{ percent}$$

4. The ac power is given by

$$P_{ac} = \frac{(I_{\max} - I_{\min})(E_{\max} - E_{\min})}{8} = 2.3 \text{ watts}$$

5. The Q point is located at $E_c = -15$, $I_b = 42$ ma; that is, at point Q_1.

6. $E_{bb} = E_b$ (at Q point) $= 345$ volts.

7. For the case $\Delta E_c = 20$ volts,

$$I_{\max} = 74 \text{ ma} \qquad E_{\max} = 400 \text{ volts}$$

$$I_b = 52 \text{ ma}$$

$$I_{\min} = 33 \text{ ma} \qquad E_{\min} = 155 \text{ volts}$$

The values of $I_{\max}$, I_b, and $I_{\min}$ are found from the points A, Q_2, and C, respectively.

8. Calculating I_1 and I_2,

$$I_1 = \frac{I_{\max} - I_{\min}}{2} = 20.5 \text{ ma}$$

$$I_2 = \frac{I_{\max} + I_{\min} - 2I_b}{4} = 0.75 \text{ ma}$$

9. $$\text{Percent 2nd HD} = \frac{I_2}{I_1} \times 100 = 3.66 \text{ percent}$$

10. $$P_{ac} = \frac{(I_{\max} - I_{\min})(E_{\max} - E_{\min})}{8} = 1.26 \text{ watts}$$

11. The Q point is located at $E_c = -10$ volts, $I_b = 52$ ma; that is, at point Q_2.

12. $E_{bb} = 290$ volts (E_b at the Q point).

13. R'_L does not change, since it depends upon the slope of the load line.

Example 8.4 For the plate characteristics given, a power pentode, find the optimum load line and ΔE_c for second HD $\leq 5\%$, and P_{ac} = max. Assume that P_{max} for the tube is 20 watts. Draw the circuit for the power amplifier.

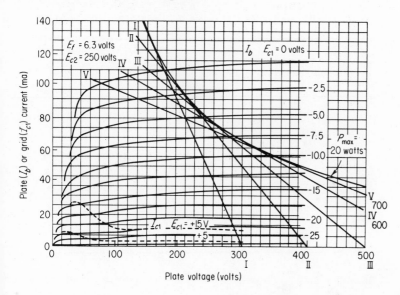

1. The locus of points for $I_b E_b = 20$ are drawn in first, as shown in the diagram.

2. Arbitrarily selecting values of E'_{bb} of 300, 400 and 500, 600, and 700 volts, load lines are drawn tangent to the maximum power curve, as shown. These are curves I, II, III, IV, and V.

3. For load line I:

ΔE_c	20	15
I_{max}	109	109
I_b	55	67.5
I_{min}	18	32.5
E_{max}	280	261
E_{min}	177	177
% 2nd HD	9.3	4.3
P_{ac}, watts	2.3	1.6

4. For load line II:

ΔE_c	20	15	10
I_{max}	109.5	109.5	109.5
I_b	56	68	81.5
I_{min}	18	35	56
E_{max}	364	328	286
E_{min}	179	179	179
% 2nd HD	5.7	5.7	2.3
P_{ac}, watts	4.2	2.8	1.4

5. For load line III:

ΔE_c	15	12.5
I_{max}	108	108
I_b	68	75
I_{min}	36.5	46
E_{max}	388	355
E_{min}	157	157
% 2nd HD	5.9	3.2
P_{ac}, watts	4.1	3.1

6. For load line IV:

ΔE_c	12.5
I_{max}	106
I_b	75
I_{min}	46
E_{max}	391
E_{min}	120
% 2nd HD	1.7
P_{ac}, watts	4.1

7. For load line V, the transfer curve as constructed on the next page shows that the distortion will occur at both ends of the grid swing. A flattening of both peaks results. A clue that something like this happened is given by the result for I_2 when calculated for this case.

ΔE_c	12.5
I_{max}	101
I_b	75
I_{min}	46

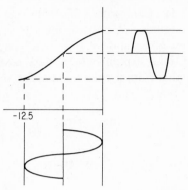

$$I_2 = \frac{I_{max} + I_{min} - 2I_b}{4} = -0.75 \text{ ma}$$

The negative value for I_2 implies that other types of distortion are occurring.

8. Inspecting each case, it can be seen that P_{ac} increases as the slope of the load line becomes less vertical. The final choice would be the one obtained in Sec. 6, that is, $\Delta E_c = 12.5$ volts. For this case, the ac power is greatest, 4.1 watts, and the second-harmonic distortion is rather low, 1.7 percent.

9. Thus, for power pentodes, the selection of a load line is rather simple.

(a) Draw in the maximum power-dissipation curve.

(b) Draw in a load line just above the knee of the $E_c = 0$ curve, tangent to the maximum power curve.

(c) Test for I_2 to see that I_2 is not negative.

10. The Q point for curve IV is determined as $E_c = -6.25$ volts, $I_b = 75$ ma; that is, where the $E_c = -6.25$ curve intersects load line IV. E_b at this point is determined as 258 volts.

$$Q \text{ point} \begin{cases} E_c = -6.25 \text{ volts} \\ E_b = 258 \text{ volts} \\ I_b = 75 \text{ ma} \end{cases} \text{ load line IV}$$

11. The circuit to be used is shown.

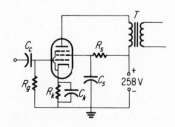

12. Comments. To complete the dc calculations, that is, to determine R_k and R_s, the screen grid current is necessary. This is not given and must be obtained by either setting up the dc conditions experimentally and measuring I_s, or by obtaining the value from a tube manual for the particular tube used. Then R_k would be given by

$$R_k = \frac{|E_c|}{I_b + I_s}$$

and R_s would be given by

$$R_s = \frac{E_{bb} - E_s}{I_s}$$

8.6 Analytical Approach

In the graphical approach, not only R'_L was varied, but ΔE_c was changed also. From these data, a family of curves was drawn, Fig. 8.13. Then the optimum conditions were determined. Analytically, the problem must be simplified so much that one wonders whether the results are meaningful. The results to be calculated here, when compared with experiment, will show good agreement, verifying the need for this approach.

The plate characteristics will be assumed to be completely linear, Fig. 8.15, since the nonlinear problem would be nearly intractable. The line OA and the others parallel to it represent

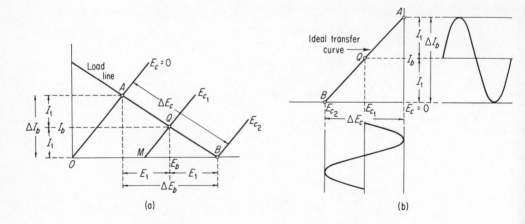

Fig. 8.15 (a) Ideal plate characteristics. (b) Ideal transfer curve.

the E_c = constant curves. The line AQB represents a load line. E_1 and I_1 are the amplitudes of the plate voltage and plate current variations, respectively. Since the characteristics are assumed linear, ΔE_c covers the entire transfer curve from A to B. The first problem is to find the condition on R'_L for maximum P_{ac}. Thus P_{ac} must be written as a function of R'_L.

In triangle QMB, the reciprocal of the slope of line QB is R'_L. It is given by

$$R'_L = \frac{E_1}{I_1} \qquad (8.33)$$

Along line OA, that is, $E_c = 0$, r_p can be calculated as

$$r_p = \left(\frac{\Delta E_b}{\Delta I_b}\right)_{E_c} = \frac{E_{\min}}{I_{\max}} = \frac{E_b - E_1}{2I_1} \qquad (8.34)$$

Eliminating E_1 from these two equations and solving for I_1,

$$I_1 = \frac{E_b}{R'_L + 2r_p} \qquad (8.35)$$

The ac power is given by

$$P_{ac} = \tfrac{1}{2}I_1{}^2 R'_L \qquad (8.36)$$

Substituting the value for I_1 as given by Eq. (8.35),

$$P_{ac} = \frac{1}{2}\frac{E_b{}^2 R'_L}{(R'_L + 2r_p)^2} \qquad (8.37)$$

To maximize P_{ac} as a function of R'_L, Eq. (8.37) is differentiated with respect to R'_L and set equal to zero:

$$\frac{dP_{ac}}{dR'_L} = 0 \qquad (8.38)$$

$$\frac{1}{2}\frac{E_b{}^2}{(R'_L + 2r_p)^2} - \frac{E_b{}^2 R'_L}{(R'_L + 2r_p)^3} = 0$$

Getting a lowest common denominator,

$$\frac{R'_L - 2r_p}{2(R'_L + 2r_p)^3} = 0$$

This can only be zero if the numerator is zero.

$$R'_L = 2r_p \qquad (8.39)$$

Thus the proper value of R'_L for maximum ac power transfer is obtained when R'_L is made twice as large as the plate resistance of the tube. Under these maximum power conditions, the ac power can be written in terms of E_b and r_p. Using the value of R'_L found in Eq. (8.39), I_1 can be written as

$$I_1 = \frac{E_b}{4r_p} \tag{8.40}$$

and the ac power, Eq. (8.36), is written as

$$P_{ac} = \frac{E_b{}^2}{16r_p{}^2} \quad \text{(max)} \tag{8.41}$$

The next problem is to calculate the theoretical efficiency of such a linear model and assume that this will be the limiting value of the efficiency of a real, nonlinear tube circuit. The efficiency can be calculated once P_{ac} and P_{dc} are known:

$$P_{dc} = E_b I_b = E_b I_1 = \frac{E_b{}^2}{4r_p} \tag{8.42}$$

Taking the ratio of Eq. (8.41) to Eq. (8.42) and expressing it in percentage form,

$$\text{Percent eff} = \frac{P_{ac}}{P_{dc}} \times 100 = 25 \text{ percent} \tag{8.43}$$

Thus a single-tube power amplifier, operating in class A, has a maximum theoretical efficiency of 25 percent.

A note on the value of R'_L found in Eq. (8.39) should be interjected at this time. In Fig. 8.13, a study of the family of curves showing P_{ac} versus R'_L shows that the peak value of ac power for all three curves occurs at a value of R'_L of about 5000 ohms. The ac plate resistance of the tube used is determined graphically to be about 2.4 K, that is, $r_p = 2.4$ K. Thus there is experimental confirmation of the theoretically devised result, $R'_L = 2r_p$, for maximum ac power transfer.

Unfortunately the theoretical result $R'_L = 2r_p$ for maximum ac power is not the sole requirement, as Sec. 8.5 made clear. Thus $R'_L = 2r_p$ should be used only as a starting point in determining the optimum value of R'_L for a triode power amplifier. As for a pentode power amplifier, a load-line construction of $2r_p$ superimposed on the plate characteristics will show how useless this is. In a pentode, the necessary requirement is that the load line fall above the knee of the $E_c = 0$ curve. This will usually give a value of R'_L much less than r_p. Therefore a pentode cannot be used efficiently in a class A power amplifier.

8.7 Push-Pull Operation

In order to increase efficiency and linearity of the transfer curve, a circuit called a *push-pull amplifier* is used. It consists of two tubes operating near cutoff and in such a manner that when one is increasing its conduction, the other is decreasing its conduction. First of all, it can be seen that if a tube is operating in class B, its dc power requirements drop, enabling the efficiency to go up. This occurs because the tube is conducting for only one half-cycle and therefore its average current is reduced considerably. Secondly, the push-pull amplifier parameters can be so arranged that the resulting transfer curve becomes linear for a range of ΔE_c about four times greater than that for a single tube operating in class A. Thus the ac power is more than doubled, and the percent second-harmonic distortion is practically eliminated.

The operation of a push-pull amplifier operating in class B will be explained first, and then the necessary small changes will be made to produce a practical amplifier having excellent linearity. In Fig. 8.16 the schematic for a push-pull amplifier is shown in

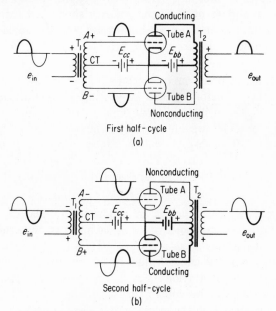

Fig. 8.16 Conduction process in a push-pull power amplifier circuit.

simplified form. A sinusoidal signal impressed on the primary of the input transformer T_1 is followed with wave forms for each half-cycle at various points in the circuit. The potential E_{cc} biases the tubes to cutoff so that with no signal impressed on the control grid, neither tube conducts. The voltage E_{bb} is the conventional plate-supply voltage. Both transformers, T_1 and T_2, have center taps. Transformer T_1 has a center-tapped secondary that serves as a phase splitter. The output transformer, T_2, serves as a summing device.

Consider Fig. 8.16(a) during the first half-cycle of the input wave form. Assume that point A is positive with respect to the cathodes, the control grid of tube A is above cutoff and the control grid of tube B is below cutoff. Thus only tube A conducts for the first half-cycle. This produces a current flow in the primary of the output transformer T_2 as shown, with the first half-cycle of the signal appearing as a voltage across the secondary of T_2.

During the second half-cycle, Fig. 8.16(b), point A of the input transformer will be negative with respect to point B. With respect to the cathodes, point B will be above cutoff for tube B and point A will be below cutoff for tube A. Thus, for the second half-cycle, tube B is conducting while tube A is nonconducting. The current flow through the primary of T_2 is opposite in direction to that of tube A during the first half-cycle. Therefore the voltage appearing across the secondary of T_2 is the polarity shown. As can be seen, each tube conducts on alternate half-cycles, and the transformer T_2 combines these two currents to produce the resultant output voltage.

Unfortunately, if the tubes are biased at cutoff as illustrated in Fig. 8.16, severe distortion will be produced. This distortion can be predicted by using the load-line representation. A simple graphical process allows one to predict the resultant transfer curve for the complete push-pull system. Since current due to tube A is opposite in direction to the current due to tube B in the primary of T_2, the two currents can be represented as positive and negative currents on a common coordinate system.

The graphical construction is produced by placing together the transfer curves for both tubes in such a way that the one for tube B is upside down, as shown in Fig. 8.17. The two transfer curves are matched at the E_c value. In this example, $E_c = 35$ volts. To illustrate the distortion that is produced, a signal voltage E_c is impressed on the control grids of the two tubes and the resultant current ΔI_b in the primary of T_2 is obtained. This current was constructed by projecting corresponding points of ΔE_c from the resultant transfer curve. For example, at point A in ΔE_c, both tubes are cut off and the point projects to point A in I_b, that is, to zero current. Point B in ΔE_c projects to point B in ΔI_b, etc.

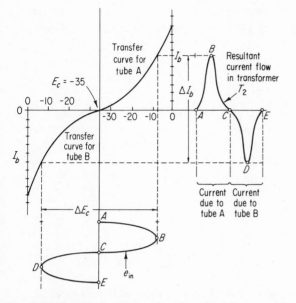

Fig. 8.17 Graphical construction for obtaining overall transfer curve.

Fig. 8.18 Third harmonic distortion.

The distortion shown in Fig. 8.17 is called *third-harmonic distortion* because when the third harmonic is added to the fundamental, with the proper phase, a wave form having this distortion is produced. This is illustrated in Fig. 8.18. In this diagram, the heavy line e_t represents the sum of the fundamental and the third harmonic. To overcome this effect, an attempt is made to make the overall transfer curve more linear. This can be done if the two tubes are not biased at cutoff but at some value above cutoff. The exact value of bias must be determined graphically by moving the individual transfer curves so that different values of E_c coincide, and graphing the resultant transfer curves until a linear resultant transfer curve is obtained. The process is illustrated in Fig. 8.19.

In Fig. 8.19(a), the two transfer curves are lined up at the cutoff point, $E_c = -35$ volts, and the resultant transfer curve coincides with the individual transfer curves. For Fig. 8.19(b), the transfer curves are lined up at $E_c = -21$ volts, the optimum value for this example, and the resultant transfer curve can be seen to be quite linear over its entire range. For Fig. 8.19(c), the transfer curves are lined up at $E_c = -10$ volts, and the resultant transfer curve can again be seen to be nonlinear. Graphically, then, it has been determined that for the tubes possessing the indicated transfer curves, a bias of $E_c = -21$ volts will produce a resultant transfer curve that is linear over its entire range. This means that a large-magnitude voltage can be placed on the control grids, producing large plate-current and plate-voltage variations.

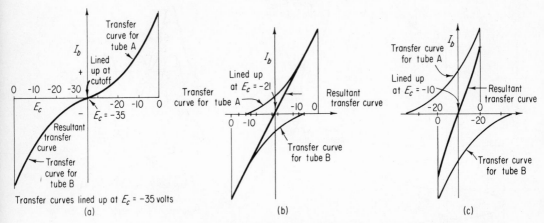

Fig. 8.19 Effect of bias voltage on the overall transfer curve.

Thus, for a push-pull amplifier, the following graphical procedure should be followed:

1. Triode Tubes
(a) Construct maximum power-dissipation curve.
(b) Determine r_p for the tube.
(c) Draw load line tangent to maximum power curve so that $R'_L = 2r_p$, for maximum ac power.
(d) Construct transfer curves.
(e) Determine proper bias voltage E_c by method of reversing one transfer curve and graphically constructing resultant transfer curve.
(f) From the resultant transfer curve, ΔE_c and corresponding ΔI_b can be determined.
(g) From the plate characteristics and load line, ΔE_b and E_{bb} can be determined.

2. Pentode Tubes
(a) Draw a load line that terminates above the knee of the $E_c = 0$ curve and ends on the E_b axis at the highest permissible plate voltage.
(b) Draw transfer curve for this load line and follow the procedure outlined above for finding proper grid bias E_c and resultant transfer curve.

The grid bias is usually obtained by utilizing cathode bias, where the cathodes are connected together and a common R_k is used. Circuit diagrams for push-pull amplifiers using triode and pentode tubes are shown in Fig. 8.20. In each diagram, the cathode resistor and bypass capacitor are common to both tubes. The resistor R is a variable resistor of about 10 ohms whose movable tap is connected to R_k. Its function is to balance the two tubes.

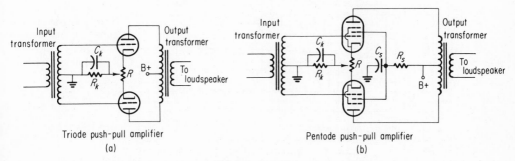

Fig. 8.20 Two push-pull amplifier circuits.

No two tubes of the same type are really exactly alike, so by varying the movable tap while observing the signal voltage on the plates of the two tubes, R can be adjusted so that the outputs of the two tubes are equal. In the pentode circuit, the screen RC filter circuit is common to both screen grids.

To illustrate the great improvement that a push-pull amplifier system makes over a single-tube system, the following example uses the same tube used to illustrate Example 8.4.

Example 8.5 Using a pentode power tube having the characteristics shown, design a push-pull system and calculate the ac output power.

1. The load line is drawn in as shown. The value of R'_L, as obtained from the graph, is $R'_L = 3.34$ K.

2. From the intersection of the load line and plate characteristics, a transfer curve is obtained.

As can be seen, the transfer curve is nonlinear. The cutoff point is at $E_c = -32.5$ volts.

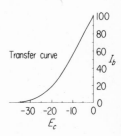

3. Using the technique of inverting the transfer curve for one tube and graphically determining the value of E_c that will produce a linear, resultant transfer curve, it is found that a value of $E_c = -21$ volts produces the accompanying result.

The resultant transfer curve is extremely linear and a vast

improvement over the individual transfer curves.

4. The maximum peak-to-peak voltage on the control grid is 42 volts and the bias must be $E_c = -21$ volts. I_b for each tube for this value of E_c is about 16 ma.

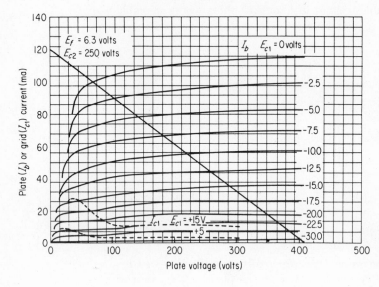

The pentode power-tube characteristics that are called for in Example 8.5.

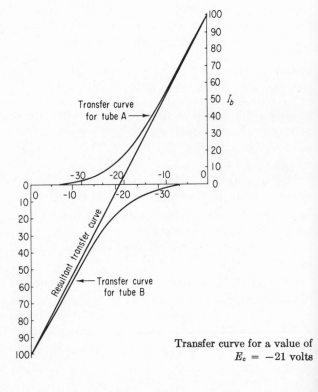

Transfer curve for a value of $E_c = -21$ volts

5. The second-harmonic distortion is negligible for this transfer curve.

6. The ac power is obtained by noting that $I_1 = 100$ ma. Since R'_L can be determined from the load line as $R'_L = 3.34$ K, the ac power becomes

$$P_{ac} = \tfrac{1}{2}I_1^2 R'_L = 16.7 \text{ watts}$$

Compared to 4.1 watts, the maximum power available for a single tube, the gain in power utilizing two tubes is well worth the more complicated circuitry.

7. In order to calculate R_s and R_k, more must be known about the tubes than can be obtained from the plate characteristics. The following information is typical of data obtained from a tube manual.

PUSH-PULL AMPLIFIER

Typical Operation (values for each tube):
Plate supply voltage (maximum)..........................375 volts
Grid No. 2 supply voltage................................250 volts
Zero-signal grid No. 2 current............................ 8 ma

Maximum Circuit Values:
Grid No. 1 circuit resistance
   for fixed bias...0.1 meg max
   for cathode bias.......................................0.5 meg max

8. From the data given above, R_k can be determined:

$$R_k = \frac{|E_c|}{I_b + I_s}$$

Since both tubes have a common cathode resistor, the values for I_b and I_s are double that for a single tube:

$$R_k = 440 \text{ ohms}$$

(to two significant figures).

9. To determine R_s, the value of I_s for a single tube is doubled, since the screen grids are connected together. The screen grid is at a potential of 250 volts and the plate supply is 352 volts. Thus

$$R_s = \frac{E_{bb} - E_s}{I_s} = \frac{352 - 250}{2 \times 8 \times 10^{-3}} = 8.4 \text{ K}$$

(to two significant figures).

10. The circuit diagram becomes as shown.

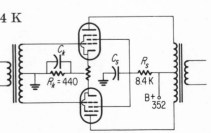

8.8 Phase Splitters

A push-pull amplifier has a specialized input requirement, that is, two signal voltages of equal amplitudes, 180 deg out of phase with each other. These input signals must be derived from a single source. Any device or circuit that converts a single signal into two equal-amplitude, -180 deg out-of-phase signals is termed a phase splitter. Figure 8.21 illustrates this requirement. The output signals e_1 and e_2 may or may not be greater in ampli- tube than the input signal e_0, depending upon the type of phase splitter used. Several of the more common types of phase splitters will be described in this section.

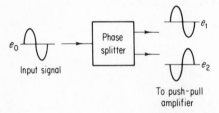

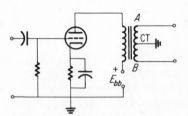

Fig. 8.21 Block diagram of phase-splitter action.

Fig. 8.22 Transformer phase splitter.

TRANSFORMER

The simplest type of phase splitter is a transformer having a center-tapped secondary. Figure 8.22 shows a triode voltage amplifier with a transformer load. The secondary of this trans- former produces the phase splitting. The center tap is grounded and the two outputs A and B will be of equal amplitude but 180 deg out of phase with each other. If the transformer is made to be a step-up transformer, there will be an effective voltage gain in the phase splitter. A transformer providing a correct frequency response in the audio range must first of all have a large core for better low-frequency response and carefully shielded windings for better high-frequency response. These requirements make a good audio transformer relatively expensive.

PARAPHASE AMPLIFIER

The simplest form of vacuum-tube phase splitter, and one having an excellent frequency response, is shown in Fig. 8.23. The action of this circuit can be traced qualitatively. During the positive half-cycle of the input voltage e_{in}, current will increase in the plate circuit. This will produce an increase in the voltage drop across the cathode resistor R (and a decrease in the plate voltage) as more voltage is dropped across the plate load resistor R.

For the negative half-cycle, the current will decrease in the plate circuit, decreasing the voltage drop across the cathode resistor R and increasing the plate voltage. Thus e_{BG} will have the same phase as the input voltage and e_{AG} will be 180 deg out of phase with the input. Since the voltages e_{BG} and e_{AG} are produced across equal resistances (R in the cathode lead and R in the plate lead)

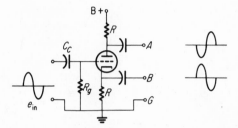

Fig. 8.23 Paraphase amplifier phase splitter.

and the plate current is common to both resistors, e_{AG} and e_{BG} will have the same amplitude. This type of phase splitter has an excellent frequency response but an overall voltage gain less than unity.

Two-Tube Phase Splitter

In this circuit both tubes act like conventional amplifiers. The first tube inverts the phase of the signal and amplifies it. The other tube amplifies a fraction of this output voltage and reinverts the phase. A schematic is shown in Fig. 8.24. Tube V_1 is a conventional stage whose output is between points A and G. The voltage

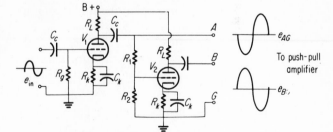

Fig. 8.24 Two-tube phase splitter.

gain is greater than unity. A fraction of voltage e_{AG} appears on the grid of V_2 through the voltage divider R_1 and R_2. The ratio of voltage division is adjusted so that the output e_{BG} of tube V_2 is of the same magnitude as e_{AG}. Also, e_{AG} and e_{BG} are 180 deg out of phase with each other.

Cathode-coupled Phase Splitter

An interesting variation on the two-tube phase splitter is that of the cathode-coupled phase splitter. This circuit utilizes less components and still has a gain greater than unity. A schematic diagram is shown in Fig. 8.25. The input voltage e_{in} is amplified by tube V_1 and inverted in phase. The output voltage from this stage is e_{AG}. The circuit for V_1 is conventional except for R_k. This resistor is shared by both V_1 and V_2 and the resistor is unbypassed.

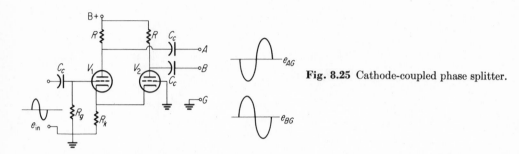

Fig. 8.25 Cathode-coupled phase splitter.

The voltage e_k produced across R_k is proportional to e_{in}. This voltage appears between grid and cathode of V_2 and serves as the signal input for V_2. The voltage e_k is in phase with e_{in}; thus, with respect to the control grid of V_2, it is 180 deg out of phase with e_{in}. The V_2 output e_{BG} is 180 deg out of phase with e_{AG}. The value of R_k is adjusted so that e_{AG} and e_{BG} have the same amplitude.

8.9   A Brief Study of Four Audio Amplifiers

Phono-Amplifier

In this single-tube audio amplifier, there are several interesting features, Fig. 8.26. First, the on-off switch (SW-1) applies power to two circuits. The first circuit heats the filament of the 25C5 tube and starts the turntable motor. The 25C5 requires a 25-volt drop across the filament to bring the filament to the proper oper-

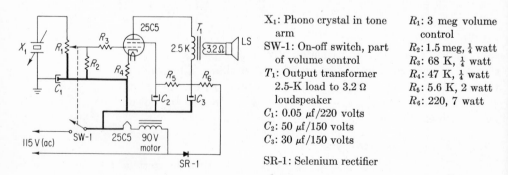

X_1: Phono crystal in tone arm
SW-1: On-off switch, part of volume control
T_1: Output transformer 2.5-K load to 3.2 Ω loudspeaker
C_1: 0.05 μf/220 volts
C_2: 50 μf/150 volts
C_3: 30 μf/150 volts

SR-1: Selenium rectifier

R_1: 3 meg volume control
R_2: 1.5 meg, ¼ watt
R_3: 68 K, ¼ watt
R_4: 47 K, ¼ watt
R_5: 5.6 K, 2 watt
R_6: 220, 7 watt

Fig. 8.26 Amplifier circuit in Westinghouse Model H73MP1 Portable Phonograph Player

ating temperature. This filament is in series with a motor winding that requires 90 volts. Thus the total voltage across the series connection of filament and motor must be 115 volts.

The second circuit is a half-wave rectifier consisting of SR-1, C_2, C_3, R_5, and R_6. The plate is connected to a higher source of voltage than the screen grid of the 25C5. In effect, the combination of R_5 and C_2 can be thought of as R_s and C_s.

In the grid circuit, R_3 is a current limiting resistor. It is possible to inject a large-amplitude signal on the grid of the 25C5 by accidentally placing one's fingers on the lead between X_1 and R_1. The signal will consist of 60 cps hum pick-up. The human body acts like an antenna and picks up the electromagnetic radiation that is ever present and due to power lines. This excess signal can drive the grid to a positive value, producing excess control grid current and damaging the tube. The series resistance R_3 keeps the current down to a safe value.

The capacitor C_1 serves to isolate the 115-volt ac line from the tone arm, which houses the crystal X_1. This is for safety purposes so that if one picks up the tone arm while in contact with a radiator or heater grating, he will not inadvertantly produce a complete circuit in the 115-volt ac circuit. The capacitor is a 0.05 μf capacitor having a reactance of about 5000 ohms at 60 cps. Thus any circuit produced is a high-resistance circuit.

Finally, R_2 in parallel with the movable tap on the volume control R_1 produces a resistance change such that the intensity of the sound has a fairly linear adjustment.

SINGLE-ENDED AUDIO AMPLIFIER

This is a utility amplifier that can be used as a phono-amplifier or to amplify the output of a microphone, Fig. 8.27. Since the output of a microphone is very low, on the order of millivolts, an extra stage of amplification is necessary. For this reason, when S_1 is in the "mike" position, an additional stage V_1 is included. In the "phono" position, the output of the phone crystal is connected to V_2. Thus, in the "mike" position, there are three stages of voltage amplification (V_1, V_2, and V_3), and in the "phono" position, there are two stages of voltage amplification (V_2 and V_3). For maximum voltage swing on the grid of the power amplifier, the output will be about 2 watts.

A brief discription of the circuit follows. The capacitors C_1 and C_4 are used to suppress high-frequency noise, such as record scratch noise, and do not affect signals in the normal audio-frequency range. R_1 is inserted to form a π type of filter in conjunction with C_1 and the interelectrode capacitance of V_1 to produce excellent filtering of noise transients from the microphone input. The combination of R_4 and C_3 is called a *decoupling network*.

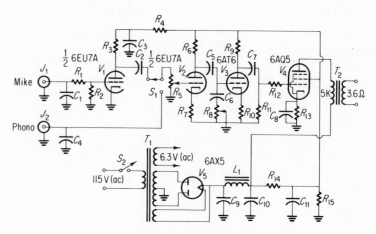

Fig. 8.27 Audio amplifier with inputs for microphone or phonograph. Maximum output of 2 watts.

R_1: 10 K, $\frac{1}{4}$ watt

R_2: 2.2 meg, $\frac{1}{4}$ watt

R_3, R_8: 100 K, $\frac{1}{4}$ watt

R_4: 47 K, $\frac{1}{4}$ watt

R_5, R_9: 500 K potentiometer

R_6, R_{15}: 220 K, $\frac{1}{2}$ watt

R_7: 3.3 K, $\frac{1}{4}$ watt

R_{10}: 2.2 K, $\frac{1}{4}$ watt

R_{11}: 470 K, $\frac{1}{4}$ watt

R_{12}: 1 K, $\frac{1}{4}$ watt

R_{13}: 250 K, 1 watt

R_{14}: 8.3 K, 1 watt

C_1, C_4: 100 pf/600 volts

C_2, C_5, C_7: 0.05/600 volts

C_3: 8 μf electrolytic/450 volts

C_6: 0.1 μf/200 volts

C_8: 50 μf/50 volts

C_9: 50μf/450 volts

C_{10}: 30 μf/450 volts

C_{11}: 20 μf/450 volts

S_1: Selector switch, SPDT

S_2: On-off switch, SPST

L_1: 5 henrys/100 ohms

T_1: Power transformer 300-0-300/60 ma

T_2: Output transformer 38/1

V_1, V_2: 6EU7A

V_3: 6AT6

V_4: 6AQ5

V_5: 6AX5

Because the power-supply connections are the same for V_1, V_2, and V_3, interactions between stages may occur through the common impedance of the filter. This could cause very low frequency oscillations called *motorboating*. The decoupling network reduces this tendency toward oscillations.

The voltage amplifier V_1 will normally be amplifying a very small signal; therefore the grid voltage can be obtained by grid leak bias, as is done here. The volume control R_5 controls either the "mike" or "phono" signal. The stages V_2 and V_3 have unbypassed cathode resistors. This decreases the gain for each stage but increases the band width for each stage. The RC combination R_8 and C_6 is used to produce a tone control. With R_8 adjusted for zero resistance, C_6 is connected from the grid of V_3 to ground. Thus the band width is decreased on the high-frequency side, producing an effective bass. With R_8 adjusted for maximum resistance, the impedance will be high for all audio frequencies, increasing the high-frequency response and producing an effective treble.

PUSH-PULL AMPLIFIER

This system consists of a pentode voltage amplifier, paraphase phase splitter, and push-pull power amplifier, Fig. 8.28. The first stage produces a voltage gain of about 530. Thus only 40 mv is necessary at the input to drive the push-pull amplifier to the full-power output of 15 watts. The circuit is conventional and needs only a cursory description.

The RC combination R_4 and C_3 in the plate circuit of V_1 is a decoupling circuit. The second stage, utilizing V_2, is a conventional paraphase amplifier. The resistor R_8 furnishes the bias for V_2. The pentode of V_1 and the triode of V_2 are incorporated in one envelope in the 6CM8. It is a nine-pin miniature tube. The power supply has a choke input filter to ensure good voltage regulation under the heavy current drain brought about by two power-amplifier tubes.

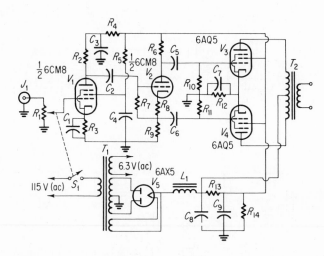

Fig. 8.28 Push-pull amplifier utilizing a paraphase amplifier for phase splitting. Maximum output is 15 watts.

S_1: On-off switch connected mechanically to volume control
T_1: Power transformer 350–0–350/130 ma
T_2: Output transformer 5 K to 3.6 ohms
R_1: 1 meg, volume control
R_2: 100 K, 1 watt
R_3: 2.2 K, $\frac{1}{4}$ watt
R_4: 10 K, $\frac{1}{4}$ watt
R_5: 22 K, $\frac{1}{4}$ watt
R_6, R_9: 47 K, $\frac{1}{2}$ watt
R_7, R_{10}, R_{11}: 470 K, $\frac{1}{4}$ watt
R_8: 1 K, $\frac{1}{4}$ watt
R_{12}: 250, 5 watts

R_{13}: 2.7 K, 2 watts
R_{14}: 270 K, $\frac{1}{2}$ watt
C_1: 2 μf/10 volts
C_2, C_5, C_6: 0.05 μf/400 volts
C_3: 8 μf/400 volts
C_4: 0.1 μf/400 volts
C_7: 50 μf/50 volts
C_8: 80 μf/400 volts
C_9: 40 μf/400 volts
V_1, V_2: 6CM8, triode pentode
V_3, V_4: 6AQ5 power tube
V_5: 6AX5 rectifier

Intercom Set

The unique feature of this circuit is the use of the loudspeakers as dynamic microphones when talking, Fig. 8.29. Permanent magnet loudspeakers can be used as microphones, since when one talks into a PM loudspeaker he moves the coil back and forth in the field of the permanent magnet, inducing a voltage in the voice coil. This voltage can then be amplified and used to drive another loudspeaker. When SW_2 is in the "Listen" position, T_1 is connected to one of the remote speakers through SW_3, and the remote speaker serves as a microphone.

The voltage is stepped up by input transformer T_1 and applied to the grid of V_1 through a volume control R_1. C_1 and R_2 bias the first stage by means of grid-leak bias. The diodes in V_1 are not needed and are simply grounded so that they cannot interfere with the circuit. R_5 is inserted to limit the grid current in V_2 if

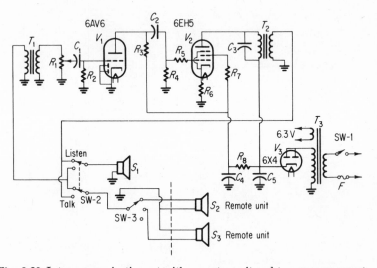

Fig. 8.29 Intercommunication set with a master unit and two or more remote units. (Courtesy of RCA.)

C_1, C_2: 0.0022 μf/200 volts
C_3: 0.005 μf/200 volts
C_4, C_5: 60 μf electrolytic/150 volts
F: Fuse, 1 amp
R_1: Volume control, potentiometer 500 K, audio taper
R_2: 6.8 meg, 0.5 watt
R_3, R_4: 470 K, 0.5 watt
R_5: 10 K, 0.5 watt
R_6, R_7: 68, 0.5 watt
R_8: 2.5 K, 1 watt

S_1, S_2, S_3: Speaker, permanent magnet, voice coil impedance 3–4 ohms
SW-1: On-off switch, SPST, attached to volume control R_1
SW-2: Talk-Listen switch, 4PDT
SW-3: Station selector switch, rotary
T_1: Input transformer, 4-ohm primary, 25-K secondary.
T_2: Output transformer, 3-K primary, 4-ohm secondary
T_3: Power transformer, 125–0/50 ma 6.3 volts/2 amp

V_2 is inadvertantly overdriven. C_3 is placed across the primary of the output transformer to lower the high-frequency response of the circuit, and in effect it will flatten the overall response because T_2 is a relatively low-inductance transformer having poor low-frequency response. The power supply is a half-wave rectifier utilizing a capacitor input filter.

Problems

8.1 A 6AQ5 power pentode is connected as a single-ended power amplifier. The load is an 8-ohm loudspeaker. Using an effective load line for $R_L = 4.2$ K and $E'_{bb} = 400$ volts, determine E_{bb}, maximum ac power, efficiency, percent second-harmonic distortion, and turns ratio for the output transformer if $E_c = -5$ volts.

8.2 A 6S4 power triode is to be used as a single-ended power amplifier. Its maximum allowable plate dissipation is 7.5 watts. Draw in the maximum power curve on its plate characteristics and determine the correct load line for maximum ac power output commensurate with the condition that the second-harmonic distortion is kept at ≤ 5 percent. What is the allowable peak-to-peak voltage on the grid? What is the voltage gain for the stage?

8.3 In the 6S4 power-amplifier plate characteristics, a load line passes through the points $I_b = 90$ ma, $E_b = 0$ volts, and $I_b = 0$ ma, $E_b = 400$ volts. Determine the Q point for the case where the second-harmonic distortion is (a) ≤ 5 percent, (b) ≤ 2 percent. Determine R_L and the output transformer turns ratio if the loudspeaker load is 16 ohms.

8.4 Determine the correct grid bias for operating two 6AQ5 tubes in push-pull if $R_L = 4.2$ K. What is the maximum ac power output and the maximum grid-voltage swing, peak to peak, allowable? Use $E'_{bb} = 400$ volts.

8.5 Draw the push-pull circuit for the tubes used in Problem 8.4. Determine the components, such as R_k, R_s, C_k, C_s, for a lower half-power frequency of 20 cps. Use a transformer phase splitter.

8.6 Using linear circuit theory, calculate the midfrequency gain of the circuit shown in Fig. 8.23 if $R = 2$ K. Use one-half of a 12AX7 as the triode.

8.7 In Fig. 8.23, calculate the output impedance at A and B in the midfrequency range. Put the answer in terms of tube and circuit parameters.

8.8   In Fig. 8.24, a 12AX7 is used for V_1 and V_2. $R_L = 100$ K for each tube. Determine R_1 and R_2 for proper operation, and the voltage gain.

8.9   In Fig. 8.25, using a 12AU7 and an R value of 33 K, determine the proper value of R_k so that the gain at A and B is the same. Determine the voltage gain.

8.10 The following conditions must be met: An input voltage with a maximum peak value of 0.1 volt will be available from a phono-crystal. It is desired to operate an 8-ohm loudspeaker at a maximum ac power level of ≥ 2 watts at a second-harmonic distortion level of 5 percent. Using a 12AX7 voltage amplifier and a 6AQ5 power tube, design a circuit that will perform this task. The lower half-power is to be 30 cps and a paraphase amplifier is to be used as a phase splitter. Determine all resistance and capacitance values. Use $E_{bb} = 300$ volts.

9.1 Oscillators

Basically an oscillator is an electronic circuit whose output is a voltage varying cyclically with time. The oscillator output is sustained by the use of feedback, wherein a portion of the output is fed back to the input through a phase-changing network. The output voltage wave form can be sinusoidal or nonsinusoidal, depending upon the type of feedback. Oscillators have been designed to produce cyclic voltages with frequencies from cycles per second ($f < 0.01$ cps) to several thousand megacycles per second ($f > 10^{10}$ cps).

There is a wide range of uses for oscillators. They are used in laboratories for testing other circuits. They are used to generate electromagnetic radiation and accompanying transmission of intelligence, to generate and synchronize complex voltages in television and radar sets, and to measure very small time spans, of the order of 10^{-10} sec.

An oscillator and its feedback circuit is shown in block diagram form in Fig. 9.1. As can be seen, part of the output voltage e_3 is fed back to the input through a feedback network whose amplification is given by A_f. For passive networks, A_f is less than 1. The input to the amplifier, e_2, is made up of two components: the signal input e_1 and the feedback voltage e_4. Thus

$$e_2 = e_1 + e_4 \tag{9.1}$$

The output voltage e_3 is given by,

$$e_3 = A_v e_2 \tag{9.2}$$

and a fraction of the output voltage is fed back:

$$e_4 = A_f e_3 \tag{9.3}$$

Eliminating e_2 from Eqs. (9.1) and (9.2),

$$e_3 = A_v (e_1 + e_4) \tag{9.4}$$

Eliminating e_4 from Eqs. (9.3) and (9.4) and solving the output voltage for e_3,

$$e_3 = \left(\frac{A_v}{1 - A_v A_f} \right) e_1 \tag{9.5}$$

The denominator is frequency-dependent because of the reactive character of the feedback network. Therefore a frequency can exist where $1 - A_v A_f$ can equal zero,

$$1 - A_v A_f = 0 \tag{9.6}$$

In this case the ratio $A_v/(1 - A_v A_f) \rightarrow \infty$ so that e_3 can have a finite value even though $e_1 \rightarrow 0$. Mathematically, it can be seen that it is possible to have an output e_3 even though the input e_1 is zero. In effect this is the definition of an oscillator circuit. For

Oscillators

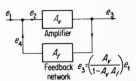

Fig. 9.1 Block diagram of a voltage amplifier utilizing feedback.

zero signal input, there is a finite output e_3. The condition on A_v and A_f in Eq. (9.6) is a *Barkhausen condition* for oscillation,

$$A_v A_f = 1 \qquad (9.7)$$

Furthermore, for sinusoidal oscillations, the regenerative feedback must be frequency selective.

9.2 Phase-Shift Oscillator

A good example of an oscillator circuit that produces a sinusoidal voltage output by producing a 180-deg phase shift for one frequency only is the phase-shift oscillator. The phase-shift feedback network consists of three capacitor-resistor voltage dividers in series. A circuit diagram for a phase-shift oscillator is shown in Fig. 9.2(a). The feedback network is composed of C_1, C_2, C_3, R_1, R_2, and R_3. The voltage input into the network is the voltage across the tube. The output of the network is the voltage drop across R_3. This output voltage appears on the control grid of the oscillator tube and is amplified. The resistor R_L is a load resistor for the circuit, and the sinusoidal voltage developed across the tube is coupled to another circuit through capacitor C_c. The resistor R_k is usually a 6-volt filament lamp. Its function is to keep constant the amplitude of oscillations. The filament bulb has a resistance that increases as the current through it increases. Thus, if the current through the oscillator tube tends to increase, R_k increases in value, biasing the tube more negatively and re-reducing the output to the former level.

To show why the feedback occurs at a definite frequency, consider the voltage output of one RC voltage divider C_1 and R_1. This is diagrammed in Fig. 9.2(b). A phasor diagram relating E_{R_1} and E_1 is drawn in Fig. 9.2(c). Since the circuit is capacitive, the total current I_1 will lead the total voltage E_1. This is shown in the phasor diagram. The voltage E_{R_1} is in phase with the cur-

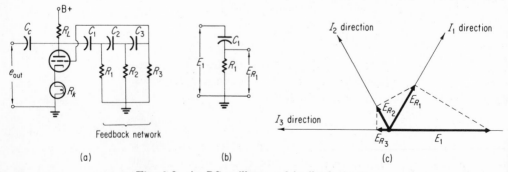

Fig. 9.2 An RC oscillator and feedback circuit.

rent I_1 and the phasor for E_{R_1} will be along I_1. With E_1 known, E_{R_1} is found by dropping a perpendicular from the tip of E_1 to the I_1 axis. For the other sections, the same construction is used to find E_{R_2} and E_{R_3}. For only one given frequency, E_{R_3} will be found to be 180 deg out of phase with E_1. This condition is shown in the phasor diagram. Even qualitatively, it can be seen that E_{R_3} is much smaller than E_1. In order to sustain oscillations, the gain of the tube circuit must be high to offset this reduction of the signal by the feedback circuit.

Example 9.1 Calculate the frequency of oscillation and find A_f for the circuit below (left).

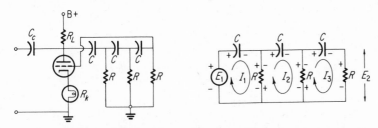

1. If the voltage across the tube is called E_1 and the output of the feedback circuit is called E_2, as shown, a three-loop circuit is formed. This neglects any loading effects due to the tube, R_L and any circuits on the other side of C_c.

2. The circuit equations become

$$I_1(R - jX_C) \quad - I_2R \qquad\qquad\qquad = E_1$$
$$I_1R \qquad - I_2(2R - jX_C) + I_3R = 0$$
$$I_2R - I_3(2R - jX_C) = 0$$

3. Solving for I_3, so that the product $I_3R = E_2$ can be formed,

$$I_3 = \cfrac{\begin{vmatrix} R - jX_C & -R & E_1 \\ R & -(2R - jX_C) & 0 \\ 0 & R & 0 \end{vmatrix}}{\begin{vmatrix} R - jX_C & -R & 0 \\ R & -(2R - jX_C) & R \\ 0 & R & -(2R - jX_C) \end{vmatrix}}$$

$$= \frac{E_1R^2}{(R^3 - 5RX_C^2) + j(X_C^3 - 6R^2X_C)}$$

4. Solving for E_2 and rationalizing,
$$E_2 = I_3 R$$

$$= \frac{E_1 R^3}{(R^3 - 5RX_C{}^2) + j(X_C{}^3 - 6R^2 X_C)}$$

$$= \frac{E_1 R^3}{(R^3 - 5RX_C{}^2)^2 + (X_C{}^3 - 6R^2 X_C)^2}$$

$$\times \left[(R^3 - 5RX_C{}^2) - j(X_C{}^3 - 6R^2 X_C) \right]$$

5. In order for E_2 to be 180 deg out of phase with E_1, the j term must equal zero:

$$X_C{}^3 - 6R^2 X_C = 0$$

$$X_C{}^2 = 6R^2$$

$$\frac{1}{4\pi^2 f_R{}^2 C^2} = 6R^2$$

$$f_R = \frac{1}{2\pi \sqrt{6} RC}$$

6. Putting this value of f_R back into the general expression relating E_2 to E_1, and solving for the ratio E_2/E_1,

$$\frac{E_2}{E_1} = A_f = -\frac{1}{29}$$

Thus a voltage gain $A_v \geq 29$ is needed to sustain oscillations:

$$A_v A_f = 1$$

$$A_v = \frac{1}{A_f} = -29$$

An oscillator using the circuit shown in Fig. 9.2, can be made to have a variable-frequency output by making the resistors R_1, R_2, and R_3 variable. The three variable resistors can be mounted on a single shaft so that as one resistor is varied, the others are varied with it, Fig. 9.3. This changes the value of R in the equation

$$f_R = \frac{1}{2\pi \sqrt{6} RC} \tag{9.8}$$

which in turn changes the frequency of oscillation. For simplicity of design and operation, $C_1 = C_2 = C_3$ and $R_1 = R_2 = R_3$. For coarse changes in frequency, different values of C can be switched

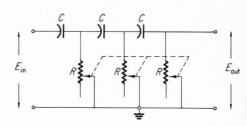

Fig. 9.3 Varying R changes the frequency of feedback.

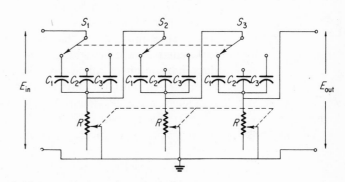

Fig. 9.4 Coarse and fine frequency control are produced by varying C and R.

in. The feedback network will then appear as shown in Fig. 9.4 above.

The capacitor switches are "ganged" also; that is, S_1, S_2, and S_3 are mounted on a common shaft. Switching capacitors changes the range of oscillation, while varying the resistors labeled R varies the frequency within a range.

9.3 Tickler-Coil Oscillator

One of the simplest types of sinusoidal oscillators utilizing a tuned circuit for frequency selection is the tickler-coil oscillator. The frequency-determining network is a parallel tuned circuit, and the phase shifting from plate to grid is accomplished by coupling between the primary and secondary windings of a transformer. The oscillator circuit is shown in Fig. 9.5(a). A random change in plate current due to noise will produce a changing magnetic field in L_1. This changing field will intercept L and induce a voltage across the tuned circuit.

The voltage across the tuned circuit depends upon the impedance of the tuned circuit. Because the impedance characteristics for a high-Q tuned circuit change so rapidly about the resonant frequency, the condition expressed by Eq. (9.7) is satisfied within a very narrow band of frequencies centered about the resonant frequency. Thus the tuned circuit oscillator has good frequency stability, and the frequency of oscillation is given by

$$f = \frac{1}{2\pi\sqrt{LC}} \tag{9.9}$$

Thus the grid voltage depends upon the parallel tuned circuit in the grid circuit. The grid voltage in turn modulates the plate current, which produces the changing magnetic field in L_1. This coupling between plate and grid occurs because of the interaction of L_1 and L through a transformer action.

A simplified calculation involving the principles of Fig. 9.5(b) would be of interest. In this calculation, the minimum value of μ for oscillation will be determined. It is assumed that the coupling does not seriously affect the resonant frequency of the tuned circuit L–C. This is justified by the fact that, in practice, the frequency of oscillation is predicted by Eq. (9.9) to within a fraction of a percent. The circuit of Fig. 9.5(a) must be reduced to an ac voltage-equivalent circuit, and calculations will be made on the equivalent circuit. Figure 9.5(a) is shown redrawn in Fig. 9.5(c), with all components having negligible ac voltage drops shorted out. A further rearrangement of components produces Fig. 9.5(d). This circuit shows the transformer-like coupling between plate and grid circuit. The quantities e_3 and e_2 of Fig. 9.5(b) are indicated in the ac equivalent circuit.

Since it is assumed that the circuit is operating at the resonant frequency, the impedance of the L–C–R_1 parallel circuit is purely resistive and given by $Z = R_1$. This resistance appears as an effective resistance R in the primary, or plate, circuit. This is shown in Fig. 9.5(e). R is given by

$$R = K^2 \left(\frac{N_1}{N_2}\right)^2 R_1 \tag{9.10}$$

where N_1/N_2 is the turns ratio of the coils L_1 and L and K is the coupling factor for the system. Because all the flux produced by L_1 does not intercept L, since L_1 is removed from L by a small distance, the ratio of flux intercepted to flux produced, K, is less than 1. Since there is no external signal input, $e_1 = 0$ and $e_2 = e_4$.

Comparing Figs. 9.5(b) and (d),

$$e_4 = A_f e_3$$

$$= -K \left(\frac{N_2}{N_1}\right) e_3$$

Therefore

$$A_f = -K \left(\frac{N_2}{N_1}\right) \tag{9.11}$$

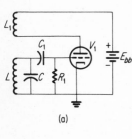

(a)

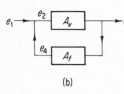

(b)

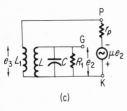

(c)

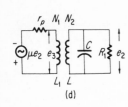

(d)

(e)

Fig. 9.5 The reduction of a tickler coil oscillator into an ac equivalent circuit.

Comparing Figs. 9.5(b) and (e),

$$e_3 = A_v e_2$$

$$= -\mu \left(\frac{R}{R + r_p} \right) e_2$$

Therefore

$$A_v = -\mu \left(\frac{R}{R + r_p} \right) \qquad (9.12)$$

Using the Barkhausen condition for oscillation, $A_v A_f - 1 = 0$,

$$\mu = \left(\frac{R + r_p}{KR} \right) \left(\frac{N_1}{N_2} \right) \qquad (9.13)$$

Eliminating R between Eqs. (9.10) and (9.13),

$$\mu = \left(\frac{N_1}{N_2} \right) \frac{1}{K} + \left(\frac{r_p}{R_1} \right) \left(\frac{N_2}{N_1} \right) \frac{1}{K^3} \qquad (9.14)$$

To obtain a numerical value for μ, the following typical values for the parameters in Eq. (9.14), will be used:

$$\frac{N_1}{N_2} = \frac{1}{3} \qquad r_p = 30 \text{ K}$$

$$K = 0.4 \qquad R_1 = 30 \text{ K}$$

Therefore $\mu \cong 50$.

As can be seen from Eq. (9.14), as K is made greater, μ can be smaller and still sustain oscillations. Actually, the value for μ is a minimum value. Greater values of μ than that given by Eq. (9.14) still produce oscillations. There is a natural limiting action to the oscillations due to the finite cutoff and saturation values for a given tube. Therefore a tube with μ less than the required value will not sustain oscillations, whereas tubes with μ greater than calculated will work very well.

An interesting demonstration of the onset of oscillations occurs when L_1 is moved toward L until a distance is reached where oscillations are produced. Moving L_1 away stops the oscillations; moving L_1 closer increases the amplitude of oscillations up to a maximum value, and then further coupling has no effect on the amplitude.

This buildup and natural limitation of oscillations is shown in Fig. 9.6 with the aid of a transfer curve. In this example the coupling is fairly large, $K \cong 0.5$. As the oscillations begin, there will be grid current due to the positive-going peaks of the grid voltage. This causes the R_1 and C_1 combination to produce a

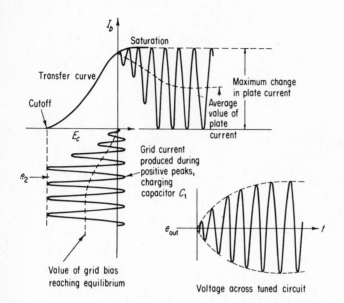

Fig. 9.6 Limitation of oscillation amplitude due to saturation and cut off.

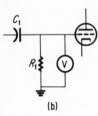

(a)

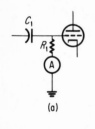

(b)

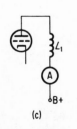

(c)

Fig. 9.7 Methods for detecting oscillation.

negative voltage between grid and cathode, owing to grid-leak bias. Each succeeding cycle is increased in amplitude as a result of the charging of the capacitor in the tuned circuit. As the amplitude of oscillation increases, the average grid-to-cathode voltage E_c becomes more negative until an equilibrium value is produced. This is due to the fact that as e_2 increases, the amplitude of the plate current oscillations increases to a natural limit imposed by the tube characteristics. The limiting is due to tube cutoff and saturation, as shown by the transfer curve. When this point is reached, the grid voltage cannot change in amplitude because the amplitude of the magnetic field change is now constant. The voltage across the tuned circuit R_1–C–L increases in amplitude until the equilibrium value is reached and maintained. This voltage is sinusoidal and is used as the output voltage for the circuit.

The onset of oscillations can be indicated in several ways. As can be seen in Fig. 9.6, as the oscillations increase, the average plate current I_b decreases, and the average voltage E_c between grid and cathode becomes negative because of current flow through R_1. Thus a dc ammeter can be used to indicate oscillation by placing it in the grid circuit or plate circuit, Figs. 9.7(a) and (c). A quick check to see if an oscillator circuit is oscillating is to place a voltmeter across R_1 and note that a negative voltage is present on the grid, Fig. 9.7(b). Also, there is the direct method of simply connecting an oscilloscope across the tuned circuit and observing the sinusoidal voltage that will be present when the circuit is oscillating.

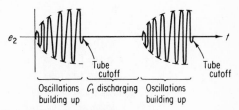

Fig. 9.8 Intermittant oscillation due to incorrect grid leak bias.

The values of R_1 and C_1 are critical in that if the time constant is made too large, intermittent oscillations can occur. This could happen if the tube were to bias itself beyond cutoff as a result of a slow discharge of C_1 through R_1. In that case, the voltage across the tuned circuit would look much like the representation of Fig. 9.8. In normal use of an oscillator circuit, this condition is to be avoided.

9.4 Hartley Oscillator

The Hartley oscillator is one of the most common oscillator circuits. It is a logical extension of the tickler-coil oscillator. The schematic diagram is shown in Fig. 9.9(a). The circuit is redrawn in Fig. 9.9(b) to show the coupling circuit more clearly. The tuned circuit is formed by L, C, and R_1. The grid-leak bias is produced by the R_1 and C_1 combination. In the plate circuit, RFC stands for a radio frequency choke (that is, an inductor having appreciable inductance at the frequency of the oscillator) and is the load for the amplifier. In many cases, a resistor of about 10 to 30 K is found in place of the RFC. The capacitor C_2 is made to have a low reactance at the oscillator frequency and is a coupling capacitor. Although it appears not to couple anything, a glance at Fig. 9.9(b) shows that it couples the plate voltage to the control grid through the LC circuit. The coupling actually is of a transformer type, with the primary consisting of a portion of L be-

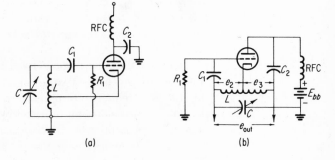

Fig. 9.9 The Hartley oscillator.

(a)   (b)

tween cathode and ground, and the secondary consisting of the portion of L between grid and cathode.

The voltage e_3 is stepped up by transformer action into the voltage e_2, since the number of turns between grid and cathode is usually about three times the number between cathode and ground. A sinusoidal voltage output is taken between the two points indicated in Fig. 9.9(b). As indicated in the diagram, capacitor C can be a variable capacitor. This will enable one to change the frequency of the output voltage by a simple change in capacitance.

Whenever a fixed frequency with good stability is desired, the components determining the frequency (that is, L and C) would be kept at a constant temperature. This would be necessary because a change in physical size of L or C would change the frequency of oscillation.

The frequency of the Hartley oscillator is determined by the tuned circuit and is given by

$$f = \frac{1}{2\pi\sqrt{LC}}$$

9.5 Crystal Oscillator—The Pierce Oscillator

In some applications, extreme stability is required. Then, instead of using tuned circuits composed of inductors and capacitors as the frequency-determining components, the piezoelectric effect of crystal structures such as quartz is utilized in conjunction with the resonant mode of mechanical vibration of a solid. Quartz crystals, cut in a predetermined manner and mounted so that two opposing faces have electrical contacts, will compress or expand when an electric field is impressed across the faces. The rate at which the crystal can change its shape depends upon its physical parameters, such as thickness and density. If the applied field is cyclic in nature, it will cause the crystal to vibrate. The amplitude of vibration of the crystal falls off very rapidly as the applied electric-field frequency deviates from the resonant frequency of the crystal. Thus the crystal acts like a tuned circuit having a very high Q.

Values of the order $Q = 10^5$ are not uncommon for crystals. Because of the high-Q value and the fact that the crystal has inertia, the frequency of oscillation cannot be easily changed and is a very pure sinusoid. This makes for a very stable oscillator. To increase stability still more, the crystal is usually mounted in a small temperature-controlled oven so as to reduce the frequency drift to values as small as 1 part per 10 million over long time spans.

One of the simplest circuits utilizing the crystal is the Pierce oscillator, shown in schematic form in Fig. 9.10(a). The capacitor C_2 is a dc blocking capacitor whose function is to keep the large dc voltage at the plate from the crystal. Its value is such that its reactance is low at the resonant frequency of the crystal. Grid-leak bias is produced by the combination R_1 and the effective capacitance of the crystal plates. The choke in the plate circuit is the plate load, and C_c is the coupling capacitor between the oscillator and other circuits. Figure 9.10(b) shows the Pierce oscillator circuit redrawn in order to illustrate how feedback occurs. The plate voltage appears across C_p, composed of inter-electrode and wiring capacitances. The voltage-divider action is such that a fraction of this voltage appears across C_G and is amplified. Only at the resonant frequency of the crystal is the voltage developed across C_G of the correct phase to sustain oscillations. A detailed examination would show that the amplification of the tube circuit would have to be at least C_G/C_p in order to sustain oscillations:

$$A_v \geq \frac{C_G}{C_p}$$

The frequency-determining component is the crystal itself. In overall effect it acts like a parallel tuned circuit of extremely high Q.

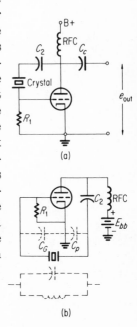

Fig. 9.10 The Pierce crystal oscillator.

9.6 Multivibrator—Square-Wave Oscillator

The multivibrator is an example of a nonsinusoidal oscillator. The oscillations depend upon a feedback circuit that is independent of frequency. Also, the feedback is such that the tubes involved switch suddenly from nonconductive to fully conductive states. Intermediate states of operation do not occur.

The simplest multivibrator circuit uses two tubes, and the output of one tube is connected to the input of the next tube, as shown in Fig. 9.11(a). This diagram is used simply to illustrate that the output of V_1 is connected through the coupling capacitor C_1 to the grid of V_2, and that the output of V_2 is connected through C_2 to the grid of V_1. The commonly used diagram to illustrate a multivibrator is shown in Fig. 9.11(b). The output voltage is taken across either V_1 or V_2 or, in some cases, from both tubes. The two outputs are shown in Fig. 9.11(c). During time t_1, tube V_1 is nonconducting and its plate voltage is equal to E_{bb}, and tube V_2 is conducting and its plate voltage is less than E_{bb}. At time τ, the tubes switch so that tube V_1 is conducting and tube V_2 is non-

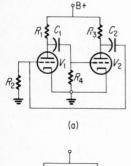

(a)

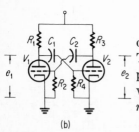

(b)

Fig. 9.11 The multi-vibrator circuit and output voltages.

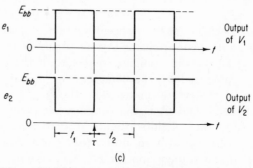

(c)

conducting. The switching occurs in a fraction of a microsecond. This cyclic switching from one state to another and back again produces the "square wave" voltage waveforms shown. A multivibrator that oscillates without a signal input is called a *free-running multivibrator*.

A detailed explanation of multivibrator action will be given by resorting to a numerical example. Consider the circuit and tube characteristics given in Fig. 9.12. When the plate voltage is applied, one of the tubes will conduct more than the other; this produces a negative-going voltage at the plate of the tube that is conducting more. This negative-going signal is applied to the control grid of the second tube, making it conduct less. This makes the plate voltage of the second tube rise. This positive-going signal is coupled to the control grid of the first tube, making it conduct still more. This action drives the second tube to cutoff and beyond, and causes the first tube to conduct heavily. Assume that this has happened and that V_1 is conducting and V_2 is at cutoff. The grid voltage on V_1 will be $E_c = 0$, and the voltage on the grid of V_2 will be $E_c = -6.5$ volts, the cutoff voltage for the tube.

The other voltages are obtained by drawing in the load lines for V_1 and V_2 and using point A, Fig. 9.12(a). Thus V_1 will have a dc resistance of about 45.2 K and a plate voltage of 190 volts, Fig. 9.12(b), while the plate voltage for V_2 will be $E_b = E_{bb} = 400$ volts, since the tube is at cutoff. The capacitors C_1 and C_2 will be charged to 196.5 volts and 400 volts, respectively. Capacitor C_1 is discharging to 190 volts, but does not reach this value because when the value of the voltage drop across R_4 reaches -6.5 volts, V_2 starts to conduct, dropping its plate voltage. This negative-going signal is coupled to the control grid of V_1, lowering its current. Within a fraction of a microsecond, V_1 switches from a conducting to a nonconducting state and V_2 goes from a non-conducting to a conducting state.

Immediately after switchover, the voltages in the circuit are as shown in Fig. 9.12(c). Capacitor C_2 must now discharge to

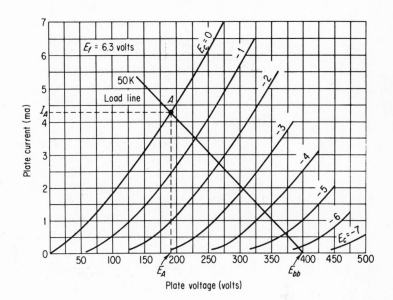

Fig. 9.12 Analytical determination of the output voltage waveform.

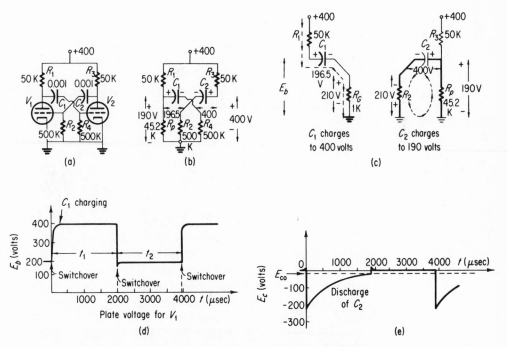

190 volts, producing a negative voltage drop across R_2. Capacitor C_1 must charge up to 400 volts, producing a positive voltage drop across R_4. This causes the control grid of V_2 to act like the

plate of a diode, and grid current flows to charge C_1. The effective resistance of the cathode-to-grid circuit is about 1 K. Thus C_2 discharges through the path $R_p = 45.2$ K and $R_2 = 500$ K, while C_1 charges through the path $R_G = 1$ K and $R_1 = 50$ K. The time constant for the charging of C_1 is given by

$$\tau = (R_1 + R_G)C_1 = 51 \ \mu\text{sec} \tag{9.15}$$

where $C_1 = 0.001 \ \mu\text{f}$.

Thus, in $5\tau \cong 255 \ \mu\text{sec}$, C_1 is charged to 400 volts and the current will stop in the R_1, C_1, R_G circuit. Tube V_1 will then be operating with $E_c = 0$ volt. The charging of C_1 determines the plate voltage of V_1. This is shown in Fig. 9.12(d). At the same time, C_2 is discharging. As C_2 discharges, the grid voltage for V_1 rises exponentially as the current through R_2 decreases with time. This is shown in Fig. 9.12(e). Tube V_1 will be held in the nonconducting state until the grid-to-cathode voltage reaches the cutoff value of -6.5 volts; then switchover occurs again. The time t_1 depends upon the discharge time of C_1. The time constant for the C_2 discharge circuit is

$$\tau_1 = (R_2 + R_p)C_2$$

$$= 545.2 \ \mu\text{sec} \tag{9.16}$$

where $C_2 = 0.001 \ \mu\text{f}$. The time t_1 will be given by solving the following equation:

$$E_{co} = E'_c e^{-(t/\tau_1)} \tag{9.17}$$

where $E'_c = -210$ volts, $E_{co} = -6.5$ volts, and $\tau_1 = 545.2 \ \mu\text{sec}$;

$$t_1 = \tau_1 \ln\left(\frac{E'_c}{E_{co}}\right)$$

$$\cong 1900 \ \mu\text{sec} \tag{9.18}$$

Thus, approximately 1900 μsec after the first switchover, V_1 starts to conduct again because switchover occurs again. For the time t_2, V_2 is driven below cutoff, tube V_1 conducts, and the voltages on the capacitors C_1 and C_2 will be reversed from that shown in Fig. 9.12(c). Since the two tube circuits are identical, V_2 reaches cutoff 1900 later, starts to conduct, and switchover occurs again. This process continues as long the plate-supply voltage $E_{bb} = 400$ volts is applied.

Equation (9.18) gives the duration of the first half-cycle of oscillation. Since the two tube circuits are symmetrical, $t_1 = t_2$ and the period of oscillation would be 3800 μsec. The frequency would then be about 263 cps. If the two grid circuits were to have different time constants, then the plate-voltage wave forms would be asymmetrical. Such an oscillator output is shown in Fig. 9.13. V_1 is cut off for a longer period of time than V_2 because of the difference in grid resistor values.

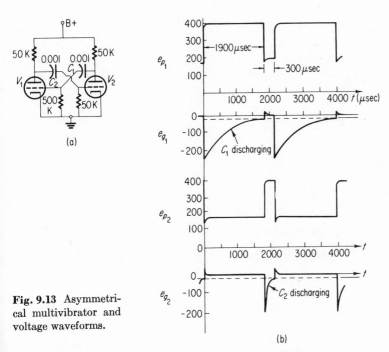

Fig. 9.13 Asymmetrical multivibrator and voltage waveforms.

The period of oscillation is given by the sum of the half-periods, where each is of the form given by Eq. (9.18). Thus

$$T = t_1 + t_2 \tag{9.19}$$

where T is the period of oscillation and t_1 and t_2 are given by

$$t_1 = \tau_1 \ln\left(\frac{E'_c}{E_{co}}\right) \tag{9.20}$$

$$t_2 = \tau_2 \ln\left(\frac{E'_c}{E_{co}}\right) \tag{9.21}$$

$$T = (\tau_1 + \tau_2) \ln\left(\frac{E'_c}{E_{co}}\right) \tag{9.22}$$

In these equations, τ_1 and τ_2 are the time constants for the *discharge* of C_1 and C_2, respectively. E_c is the peak negative voltage that appears at the grid of each tube immediately after switchover, and E_{co} is the value of the cutoff voltage. If the plate voltage during conduction is called E_A, given by point A in Fig. 9.12(a), then

$$E'_c = E_{bb} - E_A \tag{9.23}$$

The value of R_p for $E_c = 0$ is given by

$$R_p = \frac{E_A}{I_A} \tag{9.24}$$

Using these values, Eq. (9.22) can be put into an operational form. The period of oscillation for a free-running multivibrator is then given by

$$T = \left[R_{g_1}C_1 + R_{g_2}C_2 + \frac{E_A}{I_A}\,(C_1 + C_2) \right] \ln\left(\frac{E_{bb} - E_A}{E_{co}} \right) \tag{9.25}$$

where:

T = period of oscillation

R_{g_1} = grid resistor for V_1

C_1 = grid capacitor for V_1

R_{g_2} = grid resistor for V_2

C_2 = grid resistor for V_2

E_A = plate voltage for V_1 and V_2 at $E_c = 0$

I_A = plate current for V_1 and V_2 at $E_c = 0$

E_{bb} = plate-supply voltage

E_{co} = cutoff voltage for V_1 and V_2

9.7 Blocking Oscillator

A blocking oscillator produces a very narrow positive or negative pulse at a selected repetition frequency. The feedback is provided by transformer coupling between plate and control grid, but the frequency is determined by the discharge of a capacitor in the control grid circuit, much like that for a multivibrator. The blocking oscillator circuit is shown in Fig. 9.14.

The feedback circuit consists of L_1 and L_2. The extra winding L_3 is simply a convenient way of producing the output voltage. L_1 is wound in such a sense that when the plate current through L_2 is increasing, the voltage on the control grid will be positive. Thus, when the plate voltage is applied, the plate current I_b will increase; this produces a positive voltage on the grid, and thus a

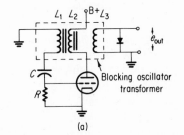

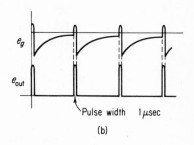

Fig. 9.14 Blocking oscillator and wave forms.

(a)

(b)

greater increase in current. This action very quickly (in 10^{-8} sec) brings the tube near saturation. At this point the rate of change of current decreases and the voltage coupled to the control grid becomes negative-going. This causes the plate current to decrease.

Figure 9.15(a) shows what would happen if capacitor C were not present. From A to B the current increases and the voltage at the grid of the tube would be as shown. Where di/dt is constant, the voltage would be constant. At B, saturation of the tube is approached and therefore e_g drops. This causes a cumulative effect, which causes I_b to decrease from B to C. The voltage e_g on the grid will now change to a negative value, since di/dt is negative.

Charging of C during t_1

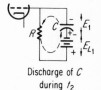

Discharge of C during t_2

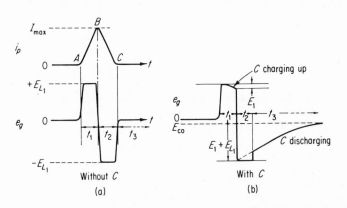

Without C

(a)

With C

(b)

Discharge of C during t_3

(c)

Fig. 9.15 Blocking oscillator grid bias and output wave forms.

This would cut the tube cutoff until the plate current (now being furnished by the interaction of a collapsing magnetic field on L_2) becomes zero. Then the cycle would repeat itself. However, with C in the circuit as shown, when the grid voltage becomes positive, C will charge up to some value E_1 as a result of grid-current flow during time t_1, Fig. 9.15(c). This alters the grid-voltage wave form somewhat, Fig. 9.15(b). When the switchover occurs (point B on the I_b curve), the grid voltage is negative and equal to E_{L_1} in series with E_1, as shown in Fig. 9.15(c). The capacitor C now discharges through R, a large time constant in relation to the pulse width.

After time t_2, the grid voltage cannot return to zero and repeat the cycle because the voltage e_g is below the cutoff voltage of the tube and remains so until capacitor C has discharged to the value E_{co}. Thereupon, plate current starts and the rapid pulse cycle t_1–t_2 is repeated again. Because capacitor C acquires a charge during the positive half-cycle of the pulse, which keeps the tube cut off for a long period of time (t_3), the output of the blocking oscillator is a series of very narrow pulses. In order to produce only a positive half-cycle, a semiconductor diode is placed across the output winding, Fig. 9.14(a). The diode is an open circuit for the positive half-cycle and a short circuit during the second half-cycle. This ensures that the output will be a sharp positive pulse of very small time duration or "width." The output pulse can be either positive or negative, depending upon which of the two output leads is grounded. As shown, the output is positive with respect to ground.

It is difficult to ascertain the pulse width for a given transformer unless all properties for the transformer are determined. For this reason the repetition frequency is not calculated here, since E_1 would be a function of the particular blocking transformer used.

9.8 Sawtooth Oscillator

The sawtooth oscillator is a nonsinusoidal oscillator producing an output voltage that repeats itself cyclically. It is different from the oscillator circuits described so far because it has no feedback mechanism. Instead, it depends on the action of an electronic switch, which closes and opens periodically with time. The switch used in this example is the thyratron whose characteristics were discussed in Sec. 4.13 The circuit for a sawtooth oscillator is shown in Fig. 9.16(a). The circuit consists of a dc source, E_{bb}, a resistor R_1, and a capacitor C. Shunting the capacitor is the thyratron and associated biasing voltage E_c. If the thyratron were not present, the capacitor C would charge up to E_{bb} and remain at that potential. Because there is a thyratron across capacitor C, however, the voltage across the capacitor is also the plate voltage for the thyratron. As the voltage rises because of the charging of capacitor C, the firing potential for the thyratron is reached, E_F in Fig. 9.16(b). Up until this time, the thyratron was an open circuit. Now the tube ionizes in about 0.2 μsec, becomes a small resistance, and discharges capacitor C very quickly. The voltage rapidly decreases to the point where the deionizing potential for the thyratron is reached, E_D in Fig. 9.16(b). The tube suddenly becomes nonconducting again, and capacitor C recharges through R. The plate voltage again rises along an exponential curve until the firing potential is reached and the thyratron ionizes, repeating the cycle.

Thus the thyratron acts like a voltage-sensitive switch, closing when E_F is reached, and opening when E_D is reached. During the period when the tube is nonconducting, capacitor C charges through R_1. When the tube fires, the capacitor discharges through the small resistance the tube presents when it is conducting. Comparing the current through the tube with the voltage across the tube, in Fig. 9.16(b), it can be seen that the tube conducts in pulses.

Fig. 9.16 Thyratron sawtooth oscillator.

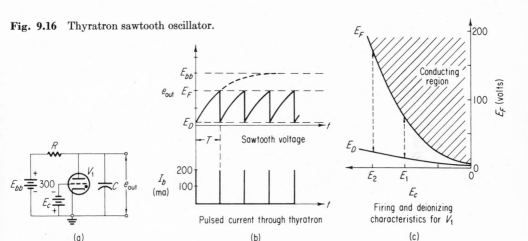

(a)   (b)   (c)

Sawtooth voltage

Pulsed current through thyratron

Firing and deionizing characteristics for V_1

In normal circumstances, the charge time is much larger than the discharge time for the capacitor, and the discharge time is neglected in calculations. Figure 9.16(c) shows a typical set of firing characteristics for a thyratron, that is, E_F as a function of the grid bias E_c. The deionizing potential is included on the same graph. The deionizing potential changes very little with change in E_c. If the grid bias is set at $E_c = E_1$, then the firing potential will be about 75 volts and the deionizing potential will be about 15 volts. The amplitude of oscillation will be given by the difference of these numbers, or 60 volts. If E_c is changed to $E_c = E_2$, the firing potential will be increased to $E_F = 168$ volts, and the deionizing potential will be $E_D = 20$ volts. The amplitude of oscillation will now be 148 volts. The frequency of oscillation will have changed also, since it will take a longer time for C to charge to 168 volts than to 75 volts. If the amplitude is kept small, so that only a small part of the charging curve is used, as shown in Fig. 9.16(b), the sawtooth voltage will be fairly linear and can be used as a linear time base in an oscilloscope. With E_c fixed, the frequency can be changed by varying R and C, thus changing the charging-time·constant.

The period T for one cycle can be obtained analytically, with

E_{bb}, E_F, E_D, and the time constant known. Since the charging curve starts from E_D, the voltage across the capacitor is given by,

$$e_{out} = E_{bb} + (E_D - E_{bb})e^{-(t/RC)} \qquad (9.26)$$

For the period T, $e_{out} = E_F$ and Eq. (9.26) becomes

$$E_F = E_{bb} + (E_D - E_{bb})e^{-(T/RC)} \qquad (9.27)$$

Solving Eq. (9.27) for T,

$$T = RC \ln\left[\frac{E_{bb} - E_D}{E_{bb} - E_F}\right] \qquad (9.28)$$

For the two numerical cases shown in Fig. 9.16(c), T becomes:
(a)   For $E_c = E_1$,

$$T = RC \ln\left(\frac{300 - 15}{300 - 75}\right)$$

$$= 0.231\ RC$$

(b)   For $E_c = E_2$,

$$T = RC \ln\left(\frac{300 - 20}{300 - 168}\right)$$

$$= 0.751\ RC$$

Equation (9.28) suggests that the frequency of oscillation can be easily changed by changing the values of R or C, or both. A circuit that allows one to change the frequency of a sawtooth oscillator is shown in Fig. 9.17. Switching in different capacitors

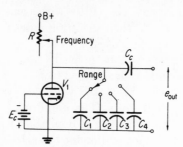

Fig. 9.17   Variable-frequency sawtooth oscillator.

changes the frequency in large steps. Varying the value of R provides a continuous change in frequency. In most of the simpler oscilloscopes, this is the type of sawtooth oscillator used to provide a voltage that changes linearly with time. It is applied to the horizontal deflection plates and produces a linear time base. This is shown in more detail in Chapter 14.

Problems

9.1 In the circuit of Fig. 9.2 (a), if $R_1 = R_2 = R_3 = 100$ K and $C_1 = C_2 = C_3$ are variable capacitors that are connected mechanically so that each of their capacities vary simultaneously from 50 pf to 300 pf, find the following:
(a) The frequency range of oscillation.
(b) The value of A_v to sustain oscillation over this range.

9.2 In a tickler-coil oscillator, $C = 200$ pf and the frequency desired is 3 Mc. Find L.

9.3 The oscillator of Fig. 9.9(a) is to operate at 1,000,000 cps (1 Mc). If L has a Q of 100 and a dc resistance of 5 ohms, find L and C. What is the impedance of L at resonance?

9.4 Using a 12AU7 vacuum tube in a symmetrical multivibrator, as shown, determine the oscillating frequency and the amplitude of the square wave. Use $E_{bb} = 400$ volts.

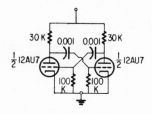

9.5 Do Problem 9.4, using a 12AX7 tube. Use $E_{bb} = 150$ volts.

9.6 For the sawtooth oscillator shown in Fig. 9.16(a), and the switching characteristics shown, let $E_C = -3$ volts, $C = 0.001$ μf. Find (a) R for a frequency of 1000 cps and (b) the amplitude of the oscillations, and (c) plot the wave form for two cycles. Assume the deionization potential is zero volts and that the resistance of the thyratron is negligible when ionized.

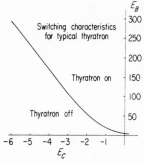

9.7 Change E_C to -6 volts in Problem 9.6 and find the new frequency of oscillation. What is the new amplitude of oscillation?

10.1 Clipping Circuits

In time-measuring, counting, and image-forming devices such as oscilloscopes, electronic timers, radar sets, and television sets, nonsinusoidal voltage wave forms are the basic ingredients. In many applications the output of a sinusoidal oscillator is purposely shaped and distorted to produce timing pulses that in turn initiate other circuits. The purpose of this chapter is to introduce the concepts used in various pulse- and wave-shaping circuits.

The clipping circuit is used to purposely distort a given voltage wave form. There are two basic clipping circuits utilizing diodes: *shunt* and *series*. In the series diode clipper, the diode is placed as shown in Fig. 10.1. The input voltage e_{in} is assumed to be

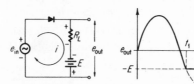

Fig. 10.1 Series clipping circuit with negative bias.

sinusoidal. During the first half-cycle, the diode is forward-biased and conducts. The output voltage e_{out} will be given by

$$e_{out} = iR_L - E \tag{10.1}$$

Adding up the voltage drops in the circuit, i is found to be

$$e_{in} - iR_L + E = 0$$

$$i = \frac{e_{in} + E}{R_L} \tag{10.2}$$

Equation (10.1) becomes

$$e_{out} = \left(\frac{e_{in} + E}{R_L}\right) R_L - E$$

$$= e_{in} \tag{10.3}$$

This will continue to be true even during the negative half-cycle until time t_1, when the input voltage becomes more negative than E. During the time interval t_1 to t_2, the diode will be reverse-biased and in a nonconducting state. The output voltage will remain at $e_{out} = -E$, the biasing battery voltage. At t_2, the diode again begins to conduct and Eq. (10.3) will apply again. Thus a portion of the negative half-cycle of the sine wave is clipped. If the biasing voltage E is made more positive (for example, $E = 0$), the resulting circuit is a half-wave rectifier, Fig. 10.2(b). Making the bias voltage E positive, as in Fig. 10.2(c), keeps the diode reverse-biased for a longer portion of the period. Only during the time t_1 to t_2, is the diode forward-biased. The net result of biasing

308

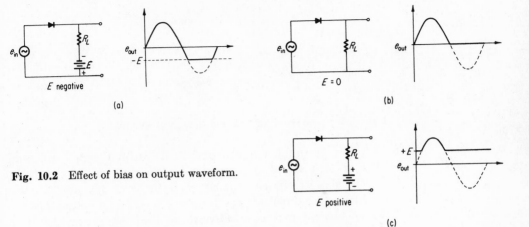

Fig. 10.2 Effect of bias on output waveform.

a diode clipper circuit is that any portion of a cycle can be passed undistorted while the remainder is clipped off.

Example 10.1 Using the diode clipping circuit shown below and a sinusoidal voltage input with a peak voltage of 10 volts, deduce the output voltage.

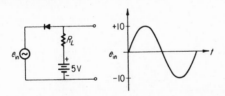

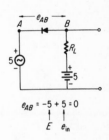

$e_{AB} = -5 + 5 = 0$

$E \quad e_{in}$

1. In the circuit above, right, the diode will be forward-biased during the first half-cycle, until the input voltage reaches $e_{in} = +5$ volts. At this point, the voltage across the diode will be $e_{AB} = 0$, as shown at right (middle).

2. During the time $e_{in} < +5$ volts, the diode conducts and the resultant current flow through R_L produces a voltage drop in series with E.

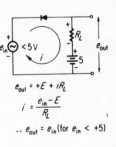

$e_{out} = +E + iR_L$

$i = \dfrac{e_{in} - E}{R_L}$

$\therefore e_{out} = e_{in} \text{ (for } e_{in} < +5)$

3. During the portion of the first half-cycle when $e_{in} > +5$ volts, the diode will be reverse-biased and will not conduct, and the output voltage will remain constant at $e_{out} = +5$ volts for the remainder of the first half-cycle and during the second half-cycle.

4. Thus the output voltage will look like the illustration at right (bottom).

Comparing this result with the circuit shown in Fig. 10.2(c), it can be seen that the portion passed or clipped depends upon the

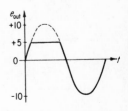

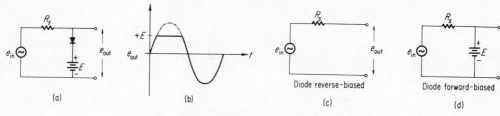

Fig. 10.3 Basic shunt clipping and equivalent circuits.

direction of the diode. The circuits shown in Example 10.1 and Fig. 10.2(c) are identical except for the direction of the diode.

The shunt diode clipping circuit is shown in Fig. 10.3. In this particular example, for the time when $e_{in} < +E$, the diode is reverse-biased and is effectively an open circuit, as shown in Fig. 10.3(c). During the time $e_{in} > +E$, the diode conducts and is effectively a short circuit. The equivalent circuit is shown in Fig. 10.3(d). Thus the output voltage is as shown in Fig. 10.3(b). The input voltage above E is clipped. Figure 10.4 shows how the

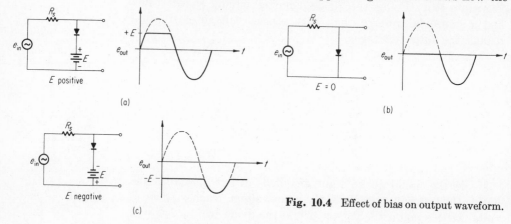

Fig. 10.4 Effect of bias on output waveform.

output voltage depends on the biasing voltage E. With the diode directed as shown, the input above E is clipped in all three cases. If the diode is reversed, it will be found that the input voltage below E will be clipped in all three cases.

An interesting combination of parallel biased diodes allows one to clip both the top and bottom of a wave form, as shown in Fig. 10.5. When the input voltage is between $+E_1$ and $-E_2$, the diodes

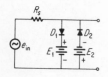

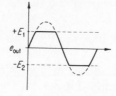

Fig. 10.5 Effect of two shunt clipping circuits in parallel.

D_1 and D_2 are each reverse-biased. Thus the output voltage will be equal to the input voltage. When e_{in} exceeds E_1, diode D_1 conducts and the output will be equal to $+E_1$. When the input voltage goes below $-E_2$, diode D_2 conducts, and the output voltage remains constant at $e_{out} = -E_2$.

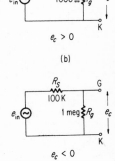

(a)

10.2 Grid Limiting

In grid limiting, the grid performs the same function as the plate in diode clipping. An essential part of the grid-limiting circuit is the series resistor R_S. The circuit for a typical grid limiter is shown in Fig. 10.6(a). The combination of R_S and R_g forms a voltage divider. The voltage to be amplified by the tube V_1 is the grid-to-cathode voltage e_c. Notice that the cathode is grounded. Therefore the grid bias is initially zero.

During the first half-cycle of the input wave form, the grid is positive with respect to the cathode and will conduct, much like a diode. The grid-to-cathode resistance r_g, when the grid is conducting, is about 1000 ohms. The voltage divider during this half-cycle is as shown in Fig. 10.6(b). The grid-to-cathode voltage is given by

$$e_c = e_{in}\left(\frac{r_g}{R_S + r_g}\right) \qquad (10.4)$$

For the example, $r_g \cong 1$ K, $R_S = 100$ K, and

$$e_c \cong 0.01e_{in}$$

Thus the positive half-cycle is effectively clipped as a result of cathode-to-grid conduction. During the second half-cycle the grid is negative with respect to the cathode, and there is no grid current. The input circuit is now shown by Fig. 10.6(c).

$$e_c = e_{in}\left(\frac{R_g}{R_S + R_g}\right) \qquad (10.5)$$

For the example, $R_g = 1$ meg, $R_S = 100$ K, and

$$e_c = 0.91e_{in}$$

The resulting plate current and plate voltage are shown in Fig. 10.6(d). The voltage E_b is an amplified version of the grid-to-cathode voltage e_c. Also, the amplifier produces a 180-deg phase shift between e_c and e_b.

An overdriven grid-limiting amplifier can be used to produce a square wave output from a sine wave input. Consider the circuit of Fig. 10.6(a) and a sinusoidal input voltage with a peak-to-

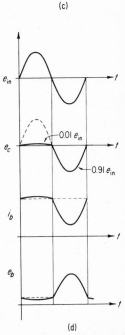

(b)

(c)

(d)

Fig. 10.6 Basic grid limiter and equivalent circuits.

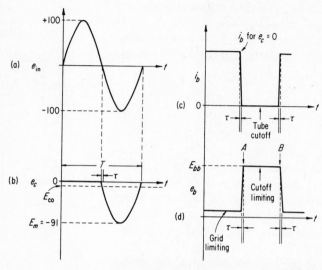

Fig. 10.7 Production of a square wave from a sine wave input by grid clipping and overdriving.

peak value of 200 volts. The grid-limiting action cuts off the positive half-cycle of the input voltage, Fig. 10.7(b). The resulting plate current during this half-cycle is the value for $E_c = 0$, Fig. 10.7(c).

For the negative half-cycle of the input voltage, the tube is cut off during most of the half-cycle because the amplitude of the input voltage, Fig. 10.7(b) and (c) exceeds the cutoff voltage. The resulting plate voltage is a square wave with a switching time equal to the time taken for the input voltage e_c to go from $E_c = 0$ to $E_c = E_{co}$, the cutoff voltage. This time is labeled τ in Fig. 10.7(b). When the amplitude of e_{in} is made large, τ is a very small fraction of the total period. For this case, the slope of the e_c curve can be considered linear and is given by

$$\left(\frac{de_c}{dt}\right)_{t=T/2} = \frac{d}{dt}\left(E_m \sin \omega t\right)_{t=\tau/2} \tag{10.6}$$

$$\frac{de_c}{dt} = -\omega E_m \tag{10.7}$$

Integrating Eq. (10.7) with respect to dt,

$$\int_0^{E_{co}} de_c = -\omega E_m \int_0^\tau dt \tag{10.8}$$

$$E_{co} = -\omega E_m \tau \tag{10.9}$$

and

$$\tau = -\frac{1}{\omega}\frac{E_{co}}{E_m} \tag{10.10}$$

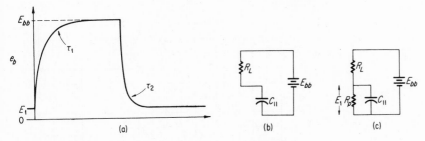

Fig. 10.8 Effect of circuit and interelectrode capacitance on square waves.

As a numerical example, let $E_{co} = -6$ volts, $E_m = 91$ volts, and $f = 1000$ cps.

$$\tau \cong 10 \; \mu\text{sec}$$

As a further refinement, the output voltage e_b can be put into a similar amplifier and the process repeated. This will make τ even smaller. It should be noted that e_b varies from E_{bb} to a value near zero. Thus the peak-to-peak value of e_b is several hundred volts. There is a limiting value for τ, for a given tube. This is due to interelectrode and wiring capacitance.

The voltage across the plate can change only as fast as the shunt capacitance charges up. An estimate of the time constants involved can be obtained by using $R = R_L$ and $C = C_{\|}$ for the positive slope, point A in Fig. 10.7(d), and by using $R = R_{\|}$ and $C = C_{\|}$ for the negative slope, point B in Fig. 10.7(d), where $R_{\|}$ is the parallel combination of R_L and R_p. During the positive slope, the tube is cut off and $C_{\|}$ charges through R_L, Fig. 10.8(b). During the negative slope, the tube is conducting, and $C_{\|}$ discharges through the voltage divider R_L and R_p, where R_p is the dc resistance of the conducting tube, Fig. 10.8(c). Assuming that a perfect square wave is impressed on the grid of the tube, the resultant plate voltage waveform will resemble that of Fig. 10.8(a). The time constant τ_1 is greater than τ_2, since R_L is greater than $R_{\|}$.

Example 10.2 For the tube circuit and characteristics shown, at the top of page 314, estimate τ_1, τ_2, and the output wave form for a 20-μsec square-wave pulse input.

1. The shunt capacitance $C_{\|} = 10$ pf does not include the input capacitance of the next stage. Thus the output voltage deduced here is that of a minimum case. The increased value for $C_{\|}$ that is due to input capacitance will increase the charging time.

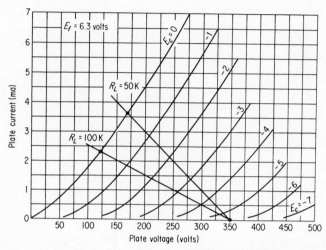

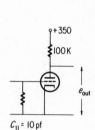

2. From the plate characteristics and the 100-K load line, estimate for:

$$E_c = 0 \text{ volts}$$

$$E_b = 120 \text{ volts}$$

$$I_b = 2.3 \text{ ma}$$

$$R_p = E_b/I_b = 52 \text{ K}$$

3. Thus

$$\tau_1 = R_L C_{||} = 1.0 \ \mu\text{sec}$$

and

$$\tau_2 = R_{||} C_{||} = 0.34 \ \mu\text{sec}$$

4. The plate voltage will change from 120 volts at $E_c = 0$ to 350 volts at cutoff. The resulting output pulse is as shown.

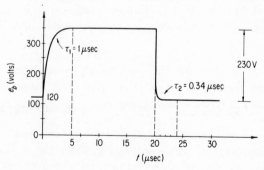

5. As can be seen, the shunt capacitance produces a serious limit on the "squareness" of the output voltage wave form. For

this reason, pulse amplifiers must be wired carefully to keep wiring and stray capacitances down to a minimum.

Example 10.3 Using the circuit and characteristics shown in Example 10.2, change R_L to $R_L = 50$ K and determine the new output voltage as follows:

1. For the load line $R_L = 50K$ at:

$$E_c = 0$$

$$E_b = 168 \text{ volts}$$

$$I_b = 3.6 \text{ ma}$$

$$R_p = \frac{168}{3.6 \times 10^{-3}} = 46.7 \text{ K}$$

2. Thus $\tau_1 = R_L C_{||} = 0.5$ μsec and $\tau_2 = R_{||} C_{||} = 0.24$ μsec.
3. The plate voltage will vary from 168 volts at $E_c = 0$ to 350 volts at cutoff, a peak-to-peak value of 182 volts.
4. Thus, by decreasing the plate load resistor by a factor of one-half, the time constants are reduced by approximately this factor, while the peak-to-peak voltage is reduced only by about 40 percent. The overall effect is a sharpening of the pulse at the expense of amplification.

10.3 Clamping

Until now, the effect of a coupling capacitor has not been considered. For cases where there is zero grid bias, or where diodes are purposely placed across the grid resistor, a phenomenon called *clamping* can take place. It is useful in establishing a dc level for complicated wave forms. A simplified clamping circuit is shown in Fig. 10.9(a). The diode D is called the *clamping diode,* and the time constant for the RC circuit is made large compared to the period of the input-voltage wave form to be clamped.

In the example, consider an input square wave such as might be expected from the output of a pulse amplifier, Fig. 10.9(b). The square wave has an amplitude of 150 volts and varies from +350 volts to +200 volts. The equivalent circuit used to explain the operation of the clamping circuit is shown in Fig. 10.9(c). The square wave is represented by switching between a 350-volt battery and a 200-volt battery. Assume that C is uncharged initially and that the 350-volt battery is switched into the circuit. Since the initial voltage on C is zero, the full 350 volts appears across the diode and biases it in the forward direction. Thus capacitor C charges to the applied voltage through the small resistance offered by the diode, and the output voltage quickly falls to zero. This is the origin of the positive spike in Fig. 10.9(d).

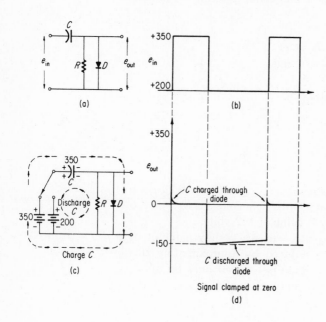

Fig. 10.9 Basic clamping circuit, signal clamped at zero volts.

The capacitor remains charged to 350 volts with the polarity shown in Fig. 10.9(c). As the input voltage is switched to +200 volts, the voltage across the diode will be the sum of the capacitor and battery voltages, or −150 volts. Thus, at switchover, the output voltage goes negative by 150 volts, as shown. The voltage biases the diode in the reverse direction so that the capacitor must discharge through R. The time constant RC is purposely made large, and C does not discharge appreciably. When the input voltage jumps to +350 volts again, the voltage across the output terminals will be the sum of the capacitor voltage and the battery voltage. This will be slightly positive due to the slight discharge of the capacitor.

Diode D conducts and quickly charges the capacitor to 350 volts, bringing the output voltage to zero. This action is now repeated as long as the input voltage is present. Except for the initial transient produced in charging C, the output wave form is of the same shape as the input wave form, but is displaced so that the top of the output voltage wave form is clamped at zero. With this circuit in general, any input-voltage wave form would be displaced so that its top would be clamped at zero.

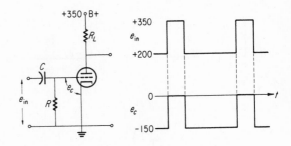

Fig. 10.10 Clamping due to grid leak current.

Grid-leak bias is an example of such clamping action. In the circuit of Fig. 10.10, the plate of the clamping diode is the control grid. Since there is zero bias, any positive-going signal will cause grid current to charge capacitor C during the positive peaks. Capacitor C will discharge during the remainder of the cycle. The net result is that the top of the signal impressed between grid and cathode will be clamped at zero, as shown. In this particular example, then, the tube will alternate between points $E_c = 0$ and $E_c = E_{co}$ on the load line, since the -150 volts is more than enough to cut off the tube. Circuits that make use of this type of clamping, called *grid clamping*, are audio amplifiers and oscillators utilizing grid-leak bias (see Fig. 9.6) as well as the pulse amplifiers presently being discussed.

10.4 RC Coupling of Square Waves

In some applications, a square wave must be passed from one amplifier to another without distortion. In other applications, the square wave is intentionally distorted in its passage from one circuit to another. The coupling between circuits is conventional RC coupling. The purpose of this section is to see what effect the RC-coupling circuit has on the square wave.

Consider the square wave in Fig. 10.11. For simplicity, it is a symmetrical square wave (that is, each alternation is one half-cycle, $T/2$ seconds in duration, where T is the period). For this example, assume that the time constant for the coupling circuit, Fig. 10.11(b), is equal to $T/10$.

$$RC = \frac{T}{10}$$

The square wave can be represented by the switch and battery shown in Fig. 10.11(c). The equations derived in Chapter 5 for

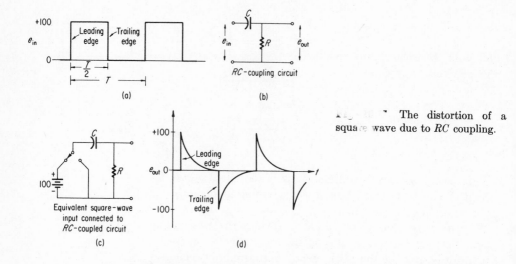

The distortion of a square wave due to RC coupling.

the charge and discharge of a capacitor in an RC circuit show that a capacitor can be considered fully charged or discharged in 5 RC time constants. For the circuit and parameters given in Fig. 10.11, $RC = T/10$, and $5RC = T/2$.

During the first half-cycle of the square wave input, there is a constant potential of 100 volts across the RC combination. Initially, the potential across the capacitor is zero and the full 100 volts appears across R. As C charges up, the voltage across R decreases, as shown in Fig. 10.11(d). The capacitor charges to the applied voltage in $T/2$ sec, and the resultant voltage across R decreases to zero in $T/2$ sec as shown. At the beginning of the second half-cycle, the square wave input drops from $+100$ volts to zero, the trailing edge. During the trailing edge, the capacitor is effectively connected across R, with a polarity such that the voltage across R is negative by 100 volts. The capacitor now discharges in $T/2$ sec and the potential across R decays from -100 volts to zero in $T/2$ sec.

The output does not resemble the square wave input. It consists of a series of positive and negative spikes with vertical leading edges and exponentially shaped trailing edges. The concept of C charging and discharging through R during the first and second half-cycles of the square wave is used to deduce the output wave forms shown in Fig. 10.12. In this figure, the parameter RC is varied from $RC = T/100$ to $RC = 5T$, to show how this affects the output-voltage wave form. For the case where $RC = T/100$, the capacitor is fully charged in $5RC = T/20$. Thus the initial voltage across R is decreased to zero in one-twentieth of the period. The result is a series of very sharp positive and negative pulses, corresponding to the leading and trailing edges of the square wave input, Fig. 10.12(b). As can be seen, as RC is made larger, it takes longer for C to charge and discharge.

In Fig. 10.12(c), $5RC = T/10$, while in Fig. 10.12(d), $5RC = T/2$. As the time constant is made larger, the capacitor does not fully charge or discharge in one half-period. In Fig. 10.12(e), for example, at switchover there is still about 8.2 volts across R. The voltage across the capacitor adds to this voltage, producing a negative leading edge of 100-volt amplitude. In Fig. 10.12(f), $RC = T$, and C charges to about 39.4 volts in one half-period, leaving 60.6 volts across R. In Fig. 10.12(g), the time constant is so large compared with $T/2$ that the capacitor charges only to 9.5 volts before switchover. The results in Fig. 10.12 show that as the time constant RC is made larger and larger, the output wave form approaches the input wave form.

Thus, if one wishes to convert a square wave into positive and negative pulses, a time constant $RC \ll T$ is used. This type of circuit is called *a peaking circuit*. For the case where the square wave must be passed through the coupling network undistorted, $RC \gg T$ must be used.

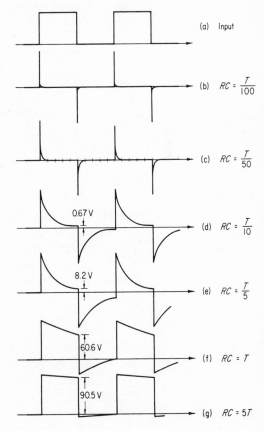

Fig. 10.12   Effect of the coupling time constant on a square wave input.

Example 10.4 For a 10-kc symmetrical square-wave input, determine an RC-coupling circuit that will produce sharp positive and negative pulses.

1. The period T for a 10-kc square wave is given by

$$T = \frac{1}{f} = \frac{1}{10^4} = 100 \ \mu\text{sec}$$

Thus $T/2 = 50 \ \mu\text{sec}$. To ensure good peaking, RC must be a small fraction of $T/2 = 50 \ \mu\text{sec}$.

2. Let

$$RC = \frac{1}{100} \times \frac{T}{2}$$

$$= \frac{1}{100} \times 100 \ \mu\text{sec}$$

$$= 1 \ \mu\text{sec}$$

3. To select values of R and C such that the product is equal to 1 μsec, one can construct a table of values:

R, K		C, pf		RC, μsec
500	$\times$	2	=	1
100	$\times$	10	=	1
50	$\times$	20	=	1
10	$\times$	100	=	1
5	$\times$	200	=	1
1	$\times$	1000	=	1

Since the resistance R is the grid resistor of the next stage, it will help determine the maximum rise time of the preceding circuit. Thus R should be relatively small. However, the preceding amplifier circuit has an internal impedance, so that the voltage across R depends upon the ratio

$$e_R = \left(\frac{R}{R + R_0} \right) e_0$$

where R_0 is the output impedance of the preceding circuit and e_0 is the emf of the preceding circuit.

For maximum voltage transfer, R must be as large as possible. The two conditions are contradictory. An optimum value for R

cannot be settled upon explicitly because of the diversity of circuits encountered. However, some values of R can be omitted from the table given above because of the small value of C required. For example, a 2-pf capacitor is not a practical unit because of the extremely small physical size required. Practical values of C generally start at 5 pf. Thus, without further knowledge of the requirements, values of $R \leq 100$ K can be used to form the coupling circuit.

4. Each of the circuits shown will produce the same effect on the square wave, since the time constant RC is the same for each circuit.

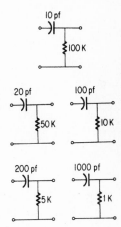

Example 10.5 For a 10-kc square wave and a value for R of 50K, design a peaking circuit that will retain *only* positive pulses.

1. For a 10-kc square wave,

$$T = \frac{1}{f} = 100 \ \mu\text{sec}$$

To obtain good peaking,

$$RC = \frac{1}{100} \frac{T}{2} = 0.5 \ \mu\text{sec}$$

For $R = 50$ K, $C = 10$ pf.

2. The peaking circuit and its output becomes as shown below.

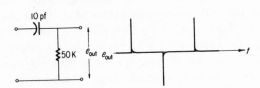

3. To eliminate the negative pulses, a clipping circuit must be used. A simple type would incorporate shunt clipping, as shown at right.

The diode will conduct during the negative pulses and present a very small resistance in parallel with the 50-K resistor, effectively shorting the output during the negative peaks.

The output becomes a series of positive pulses.

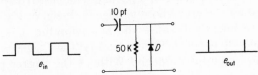

10.5   Triggered Free-Running Multivibrator

As discussed in Sec. 9.6, the frequency of a multivibrator depends upon time constants involving the coupling capacitors and grid resistors, and a natural-log functional dependence on the circuit voltages, Eq. (9.25). The switchover occurs when the grid voltage of the nonconducting tube reaches cutoff. This point is shown in Fig. 10.13. Because the slope of the e_c curve is rather shallow as e_c approaches E_{co}, the switchover point is susceptible to small

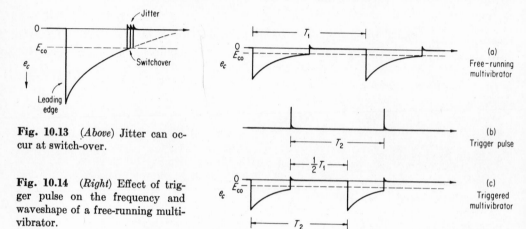

Fig. 10.13   (*Above*) Jitter can occur at switch-over.

Fig. 10.14   (*Right*) Effect of trigger pulse on the frequency and waveshape of a free-running multivibrator.

voltage variations in the grid circuit. This indeterminateness in switchover produces a *jitter* in the resulting square wave. Figure 10.13 shows three superimposed cycles of e_c for a free-running multivibrator. Because of the action just described, switchover is not the same for each cycle, with respect to the negative leading edge. The frequency of the square wave is thus not constant, but deviates at random over a small range. For most pulse circuits this cannot be allowed.

Multivibrator frequency can be accurately stabilized by triggering the multivibrator with a train of pulses having the required frequency stability. This train of trigger pulses can be derived from a crystal-controlled oscillator by proper pulse shaping. For correct triggering, the natural frequency of the multivibrator must be lower than the driving frequency of the trigger pulses. The free-running multivibrator period is T_1 and the period of the trigger pulse is T_2, where $T_1 > T_2$. The positive trigger pulse is impressed on the grid whose voltage is shown in Fig. 10.14(a). The positive pulse will cause the multivibrator to switch over sooner. The switchover is definite because the trigger pulse drives the tube above cutoff suddenly. Without the trigger pulses, the

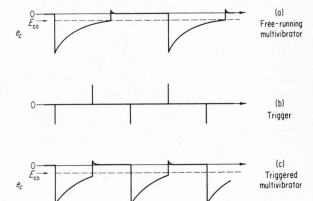

Fig. 10.15 Use of positive and negative pulses in triggering a multivibrator.

multivibrator oscillates at its natural frequency. As soon as the trigger pulses are applied, the multivibrator "locks in" with the frequency of the trigger pulses.

Although the frequency of the triggered multivibrator will now be equal to the frequency of the trigger pulses, there will still be a jitter in the duration of the two half-cycles because the second tube still approaches cutoff along an exponential curve. This can be corrected by impressing a chain of alternate positive and negative trigger pulses on the control grid of one of the tubes. The positive pulse will drive the tube above cutoff, producing switchover, and the negative pulse will drive the tube below cutoff, initiating switchover again. This is shown in Fig. 10.15. This type of triggering of a free-running multivibrator produces a stable, jitter-free output.

10.6 Single-Shot Multivibrator

In some applications, a square wave is needed only when a trigger pulse is applied, and until this pulse is applied, there is to be no output. For this requirement, a single-shot multivibrator is used. The important characteristic of a single-shot multivibrator is that the circuit remains quiescent until a trigger pulse is applied. The most common circuit is the cathode-coupled multivibrator. The circuit is shown in Fig. 10.16. The grid of V_1 is grounded and the grid of V_2 is connected to the cathode. Thus V_1 is biased by the voltage drop across R_k and V_2 is operating at zero bias. R_k is selected so that the bias voltage E_k is enough to keep V_1 below cutoff. In the quiescent state, V_1 is normally cut off and V_2 is normally conducting.

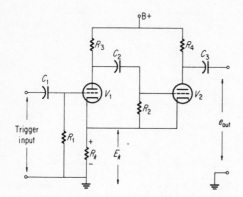

Fig. 10.16 Basic cathode-coupled multivibrator.

To start the multivibrator action, a positive pulse of sufficient amplitude is applied to the control grid of V_1. This causes V_1 to start conducting, and a switchover is effected. V_1 will conduct and V_2 will remain at cutoff for a period of time, determined essentially by the discharge of C_2. The detailed explanation of the cathode-coupled multivibrator is shown in Fig. 10.17. In Fig. 10.17(a), the voltage wave forms at the various grids and plates are shown. Initially, V_1 is nonconducting and V_2 is conducting. Capacitor C_2 is charged to a potential given by $E_{bb} - E_k$, where the magnitude of E_k is enough to keep V_1 cut off, Fig. 10.17(b). With the application of a positive trigger pulse on the grid of V_1, tube V_1 is driven above cutoff and begins to conduct. This pro-

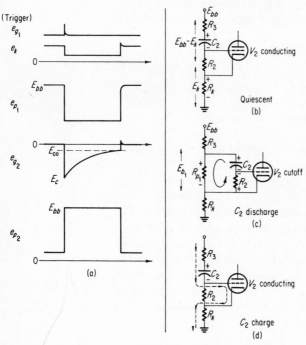

Fig. 10.17 Waveshapes and charge and discharge circuits in the cathode-coupled multivibrator.

duces a voltage drop at the plate of V_1, which is coupled by C_2 to the grid of V_2 as a negative-going signal. This causes V_2 to conduct less, reducing E_k. This in turn causes V_1 to conduct more. This type of regenerative action causes V_1 to switch to the conducting state while V_2 is driven to cutoff by the voltage drop across R_2 due to the initial charge on C_2, Fig. 10.17(c). With V_1 conducting, E_{b_1} is less than the voltage across C_2, and C_2 discharges, as shown. Immediately after switchover, the voltage across R_2 is given by

$$E_{R_2} = -E_{C_2} + E_{b_1} \qquad (10.11)$$

where E_{b_1} is the voltage drop across V_1 when in the conducting state. E_{R_2} is negative enough to drive the grid of V_2 well below cutoff. As C_2 discharges, the grid voltage e_{g_2} approaches cutoff, Fig. 10.17(a). When cutoff is reached, the tubes switch back to the quiescent state, and C_2 charges up to its quiescent voltage through R_3, R_2, and R_k, Fig. 10.17(d).

Example 10.6   For the circuit shown, obtain the magnitudes of the voltages illustrated in Fig. 10.17(a). Also determine the width of the pulse at the plate of V_2.

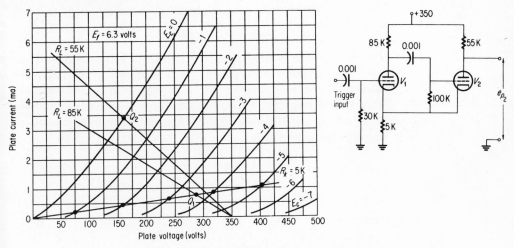

1.   To obtain the quiescent voltages for the circuit, the load line for V_2 is drawn in on the tube characteristics as shown.

$$I_{b_2} = \frac{E_{bb} - E_{b_2}}{R_{L_2} + R_k}$$

Neglecting R_k, the working equation becomes

$$I_{b_2} = \frac{350 - E_{b_2}}{55} \quad ma$$

This curve is labeled $R_L = 55$ K.

2. The quiescent voltages, as obtained from the $R_L = 55$ K load line, are as shown.

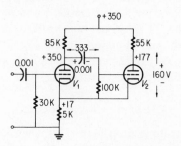

3. With the application of a positive trigger pulse on the grid of V_1, tube V_1 conducts and V_2 is cut off. To determine the voltage drops for this condition, the operating point for V_1 must be determined. The load line for V_1 must first be drawn. The equation for the load line is

$$I_{b_1} = \frac{E_{bb} - E_{b_1}}{R_{L_1}}$$

$$= \frac{350 - E_{b_1}}{85} \quad \text{ma}$$

This load line is labeled $R_L = 85$ K on the tube characteristics.

4. The Q point for V_1 must be determined graphically. Because of the voltage drop across R_k, V_1 has a bias other than zero while conducting. The Q point is found by the intersection of the load line and the following curve, obtained by summing up voltage drops in the grid circuit:

$$E_{c_1} = I_{b_1} R_k$$

$$= 5I_{b_1} \quad \text{volts} \qquad (I_{b_1} \text{ in ma})$$

E_{c_1} (volts)	I_{b_1} (ma)
0	0
−1	0.2
−2	0.4
−3	0.6
−4	0.8
−5	1.0

This curve is drawn in on the characteristics and is labeled $R_k = 5$ K. Q_2 is established at $E_{c_1} = -3.6$ volts, $E_{b_1} = 288$ volts, and $I_{b_1} = 0.7$ ma.

5. The voltages in the circuit immediately after switchover are shown below.

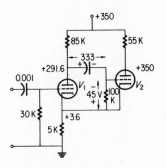

As can be seen, the initial grid-to-cathode voltage for V_2 is -45 volts, as obtained from Eq. (10.11).

6. Cutoff for the tubes is estimated as $E_{co} = -6$ volts. With the results of steps 2 and 5, the voltage wave forms can be drawn in.

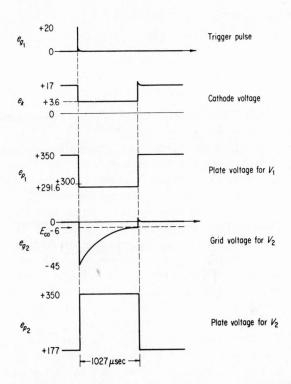

Since the initial bias on V_1 is $-E_k = E_{C_1} = -17$ volts, the trigger pulse must be greater than 11 volts in amplitude in order to bring V_1 above cutoff. A pulse of 20 volts amplitude is shown.

7. The width t of the positive pulse at the plate of V_2 is de-

termined by the time it takes e_{g_2} to change from -45 volts to -6 volts. This is given by

$$e_{g_2} = -45e^{-(t/RC)}$$

where R is given by Fig. 10.17(c):

$$R = R_2 + R_{p_1}$$

From the Q point for V_1, R_{p_1} is given by

$$R_{p_1} = \frac{E_{b_1}}{I_{b_1}} = \frac{288}{0.7 \times 10^{-3}} \cong 411 \text{ K}$$

Thus the time constant is given by

$$RC = (R_2 + R_{p_1})C_2$$

$$= 511 \ \mu\text{sec}$$

The equation for t becomes,

$$-6 = -45e^{-(t/511)}$$

where t is in microseconds. Solving for t,

$$t = 511 \ln \left(\tfrac{4.5}{6}\right) \ \mu\text{sec}$$

$$= 1027 \ \mu\text{sec}$$

10.7 Single-Shot Multivibrator (Positive Grid Return)

In the cathode-coupled, single-shot multivibrator shown in Fig. 10.16, R_2 is connected to the cathode, and tube V_2 has zero bias when conducting. Thus, when C_2 begins to discharge immediately after switchover, e_{g_2} approaches zero asymptotically, Fig. 10.17(a). Therefore the same problem occurs here as in the free-running multivibrator. The switchover when e_{g_2} reaches E_{co} will have some jitter in it and, as a consequence, succeeding pulses will not have exactly the same widths. This drawback can be obviated by connecting R_2 to the plate-supply voltage E_{bb} rather than to the cathode. This circuit modification is shown in Fig. 10.18.

The grid resistor R_2 is connected from the grid of V_2 to the plate-supply voltage E_{bb}. Essentially, this does not change the voltages in the circuit. For all practical purposes the voltage at the grid of V_2 will still be zero when V_2 is conducting. This is due to the fact that the grid will draw current that produces a bias on the tube, as shown in Fig. 10.19(a). Due to conduction between cathode and grid, the conduction resistance R_G of the cathode-to-grid "diode" is of the order of 500 ohms. The voltage drop across R_G is the bias for tube V_2. It is obtained from the series network of R_2, R_G, and R_K.

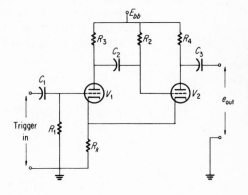

Fig. 10.18 Cathode-coupled multivibrator with positive grid return.

$$E_G = \left(\frac{R_G}{R_2 + R_K + R_G}\right) E_{bb} \qquad (10.12)$$

Putting in representative values for R_G, R_2, and R_K,

$$R_G = 500 \text{ ohms}$$

$$R_2 = 1 \text{ meg}$$

$$\left. \right\} \qquad E_G \cong +0.17 \text{ volt}$$

$$R_K = 5 \text{ K}$$

$$E_{bb} = 350 \text{ volts}$$

Therefore, to a good approximation, $E_G \cong 0$ when V_2 is conducting. Although the bias remains essentially the same, the discharge action of C_2 is different. C_2 now discharges to E_{bb} rather than zero. The two cases are shown in Fig. 10.19(b) and (c). For positive grid return, the discharge curve crosses E_{co} at a large angle, making the switchover time more precise. For a given value of R_2, $t_1 > t_2$.

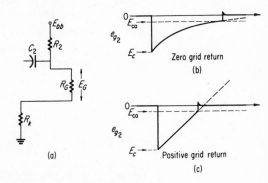

Fig. 10.19 Charge path for C_2 in positive grid return.

To determine the width of the rectangular pulse produced at the plate of V_2, the time taken for the voltage on the grid of V_2 to change from E_c to E_{co} must be calculated. Where E_c is the initial negative voltage on the grid of V_2 immediately after switch-over and E_{co} is the cutoff voltage for V_2, the curve tries to rise exponentially from E_c to E_{bb}, Fig. 10.20. The equation for the discharge curve is of the form

$$e_{g_1} = A + Be^{-(t/RC)} \tag{10.13}$$

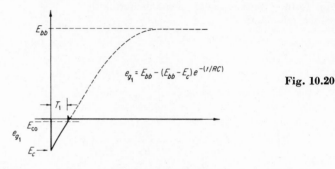

Fig. 10.20

To evaluate A and B, the following conditions can be used:

$$\text{At } t = 0, \quad e_{g_1} = E_c; \quad \text{at } t = \infty, \quad e_{g_1} = E_{bb} \tag{10.14}$$

Putting these into Eq. (10.13), one obtains

$$E_c = A + B$$
$$E_{bb} = A \tag{10.15}$$

Equation (10.13) becomes

$$e_{g_1} = E_{bb} - (E_{bb} - E_c)e^{-t/(RC)} \tag{10.16}$$

To obtain t_1, the width of the output pulse, set $e_{g_1} = E_{co}$ and solve Eq. (10.16) for t_1:

$$t_1 = RC \ln\left(\frac{E_{bb} - E_c}{E_{bb} - E_{co}}\right) \tag{10.17}$$

If one realizes that E_{bb} plays the role of the voltage that C_2 charges to asymptotically, then Eq. (10.17) can be reduced to the one necessary to calculate t_1 for the case where R_2 is returned to the cathode by replacing E_{bb} by zero in Eq. (10.17). For R_2 returned to zero, then

$$t_1 = RC \ln\left(\frac{E_c}{E_{co}}\right) \tag{10.18}$$

10.8   Production of Delayed Pulses

The single-shot multivibrator provides a simple way of producing a delayed pulse. It also allows the choice of a positive or negative delayed pulse. The delayed pulse is produced by peaking and clipping the output voltage. For example, in the circuit of Fig. 10.18, the output can be taken, as shown, from the plate of V_2, or it can be taken from the plate of V_1. Both outputs will be square waves. The V_2 output will be a positive square wave, Fig. 10.21(b), and the V_1 output will be a negative square wave, Fig. 10.21(c).

If either of these square waves is passed through an RC peaking circuit, the output will consist of very narrow positive and negative pulses, Fig. 10.21(d) and (f). These can be clipped in turn with a shunt or series diode circuit so that only the pulse due to the trailing edge is left, Fig. 10.21(e) and (g). In this way, the initial pulse, Fig. 10.21(a), is effectively delayed by the width of the rectangular pulse produced by the single-shot multivibrator. As noted, depending upon the plate from which the output is taken, the delayed pulse can be of either polarity. This type of circuit can produce delays from microseconds to several thousand microseconds, depending upon the time constants involved.

A continuously variable delay can be produced by replacing R_2 with a variable resistor, as shown in Fig. 10.22. The waveforms for several points in the circuit are also shown. The value of R_2 determines the time constant, and therefore the pulse width of waveform (B). The combination $R_5 - C_3$ is a peaking circuit converting waveform (B) into positive and negative spikes, (C). The diode D_1 produces clipping, allowing only the negative pulse through, (D).

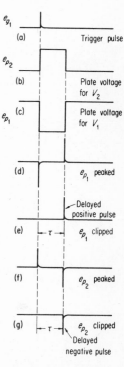

(a) e_{g_1} Trigger pulse

(b) e_{p_2} Plate voltage for V_2

(c) e_{p_1} Plate voltage for V_1

(d) e_{p_1} peaked

(e) τ Delayed positive pulse e_{p_1} clipped

(f) e_{p_2} peaked

(g) τ e_{p_2} clipped   Delayed negative pulse

Fig. 10.21 Production of positive and negative delayed pulses by peaking and clipping the multivibrator output.

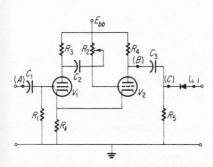

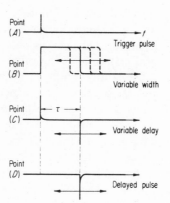

Point (A)   Trigger pulse

Point (B)   Variable width

Point (C) τ Variable delay

Point (D)   Delayed pulse

Fig. 10.22 Production of a variable delayed pulse.

10.9 Gate Voltages

The rectangular output of a multivibrator can be used, as is, to turn another circuit on or off, depending upon the polarity of the rectangular pulse. The circuit will stay on or off for the duration of the pulse.

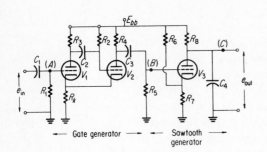

Gate generator — · — Sawtooth —
generator

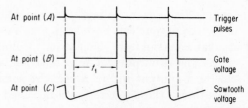

Fig. 10.23 Production of a cyclic sawtooth by gating a vacuum-tube sawtooth generator circuit.

Figure 10.23 shows an application where a gate voltage turns a tube on, causing it to conduct heavily. The tube is across a capacitor in an RC circuit. The action is similar to that of the sawtooth oscillator, except that in this case an auxiliary circuit (the single-shot multivibrator) turns on tube V_3 every time a trigger pulse appears on the grid of V_1. The output of V_3 will be a sawtooth voltage whose duration is equal to the time t_1 in Fig. 10.23. The time constant R_2C_2 is such that the output of V_2 is an asymmetrical square wave. The positive half-cycle is of short duration compared to that of the period. This positive rectangular pulse in (B) is the gate voltage. It will cause tube V_3 to conduct for the duration of the positive pulse. The coupling network R_5C_3 has a time constant large compared with the period of voltage (B). Thus the gate voltage is not distorted by the coupling circuit. The tube V_3 is normally held at cutoff by means of the voltage divider R_6 and R_7. The voltage drop across R_7 is positive. This makes the cathode positive with respect to the grid and cuts off the tube.

With V_3 cut off, C_4 begins to charge to E_{bb} through R_8. With the application of a trigger pulse (point A), the multivibrator (now called a *gate generator*) produces a positive gate that drives V_3 above cutoff. It conducts heavily and discharges C_4 rapidly. The voltage across C_4 remains low until the gate is removed, V_3 is again cutoff, and C_4 charges through R_8 again. The result is a sawtooth output across C_4. For the successive sawtooths to be identical, the trigger pulses must occur periodically.

The circuit will now be modified to produce a given sawtooth with every application of a trigger pulse. In this case the trigger pulses need not occur periodically. See Fig. 10.24. In this application the output is taken from the plate of V_1. This output is a negative gate. Tube V_3 has zero bias, so it is normally conducting. This keeps the plate voltage of V_3 low. With the application of the negative gate on the grid of V_3, V_3 is cutoff for the duration of the gate. During this time, C_4 begins to charge to the applied voltage E_{bb} through R_8. With the removal of the gate voltage, V_3 conducts and discharges C_4 rapidly. The resultant output consists

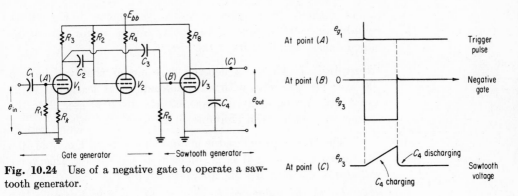

Fig. 10.24 Use of a negative gate to operate a sawtooth generator.

of a predetermined sawtooth with every application of a trigger pulse to the circuit.

The two applications described here, illustrate how a positive gate can be used to turn on a tube for a period of time and how a negative gate can be used to turn off a tube for a period of time. The gated circuit, in both cases, is a sawtooth generator. Note that the sawtooth generator is biased differently in the two applications.

10.10   Delayed Gate

If two single-shot multivibrators are put in series, the first one can be used to produce a delayed trigger and the second can be used to produce a gate voltage. The result is a delayed gate voltage. The setup is shown in block diagram form in Fig. 10.25. The actual circuit diagram and voltage wave forms are also shown in Fig. 10.25. In this way a gate voltage with a fixed delay can be produced. If a variable-delay gate voltage is required, then R_2

Fig. 10.25 Production of a fixed delay gate voltage by using two multivibrators in series.

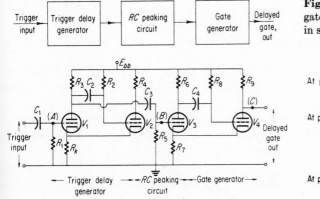

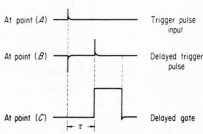

can be replaced by a variable resistor, as shown in Fig. 10.22. By adjusting the value of R_2, the delay can be set for a predetermined value, within limits of the circuit.

Example 10.7 Design a delayed gate circuit that will have a 500 μsec positive gate with an amplitude of at least 150 volts, and which can be delayed from 500 μsec to 2000 μsec. Use the characteristics shown in Example 10.6.

1. The circuit to be used is given in Fig. 10.25. Example 10.6 has shown that $R_k = 5$ K is sufficient to cut off V_1. This value will also be used for R_7, Fig. 10.25, to cut off V_3.

2. Using the values of R_3 and R_4 as given in Example 10.6 (that is, $R_3 = 85$ K, $R_4 = 55$ K), the value of the RC time constant, R_2C_2, must be determined for the minimum delay of 500 μsec and the maximum delay of 2000 μsec. Using Eq. (10.17),

$$t = RC \ln \left(\frac{E_{bb} - E_c}{E_{bb} - E_{co}} \right)$$

and selecting $E_{bb} = +350$ volts, E_{co} will be -6 volts; from Example 10.6, $E_c = -45$ volts. Thus $t = 0.13R_2C_2$.

Now R_2 must be at least 1 meg to ensure that the grid voltage on V_2 is of the order of $E_c = 0$ volt, Eq. (10.12). If R_2 is less than 1 meg, then a new Q point must be found on the load line and the voltages for V_2 must be redetermined, since E_c would now be appreciable positive. Thus R_2 is being restricted to values of 1 meg and higher so that the voltages determined in Example 10.6 can be used "as is". For $R_2 = 1$ meg, the minimum condition on t will be satisfied, and this allows for the calculation of C_2:

$$500 \times 10^{-6} = 0.13 \times 1 \times 10^6 C_2$$

$$C_2 \cong 3800 \text{ pf} = 0.0038 \text{ μf}$$

For this value of C_2, the maximum value of R_2 can be determined for $t = 2000 \times 10^{-6}$ sec:

$$2000 \times 10^{-6} = 0.13 \times 3.8 \times 10^{-9} \times R_2 \text{ (max)}$$

$$R_2 \text{ (max)} \cong 4.0 \times 10^6 \text{ ohms}$$

Thus, for a C_2 value of 0.0038 μf, R_2 must vary from 1 meg to 4.0 meg. If R_2 is constructed as shown, this will meet the requirements.

3. The coupling circuit R_5 and C_3 must be a peaking circuit. Therefore the time constant $R_5C_3 \ll t$. Using $t = 500$ μsec, the minimum pulse width,

$$R_5C_3 = \frac{1}{100} t \text{ (min)} = 5 \text{ μsec}$$

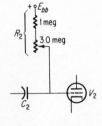

Choosing a value of $C_3 = 100$ pf, R_5 can be determined as $R_5 = 10$ K.

4. In the gate generator circuit, using the values found in Example 10.6 for R_6, R_9, and R_7, C_4 can be determined by using Eq. (10.17):

$$t = R_8 C_4 \ln \left(\frac{E_{bb} - E_c}{E_{bb} - E_{co}} \right)$$

where

$$t = 500 \ \mu\text{sec}$$

$$R_8 = 2 \ \text{meg}$$

$$E_{bb} = +350 \ \text{volts}$$

$$E_c = -45 \ \text{volts}$$

$$E_{co} = -6 \ \text{volts}$$

Then

$$500 \times 10^{-6} = 2 \times 10^6 \times C_4 \times 0.13$$

$$C_4 \cong 1900 \ \text{pf} = 0.0019 \ \mu\text{f}$$

5. The circuit values and voltage wave forms are as shown.

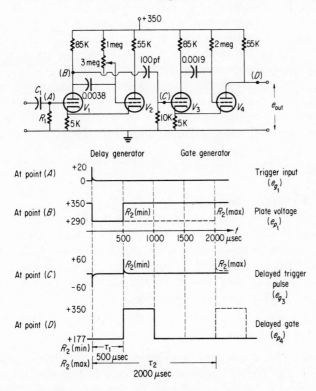

10.11   Coincidence Circuits

An important process in pulse circuitry is the selection of a desired voltage wave form, or portion thereof, from a train of incoming signals. The essential ingredients are an incoming signal, a gate voltage, and an appropriate circuit, called a *coincidence circuit*.

Diode

The simplest type of coincidence circuit employs a diode, and is shown in Fig. 10.26. The diode D_1 is biased in the reverse direction by the positive voltage E. The voltage E is selected as greater than the signal voltage at all times. Thus, with the application of the signal voltage only, there is no output because $E > e$ signal.

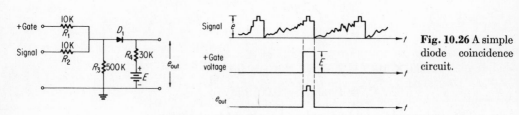

Fig. 10.26 A simple diode coincidence circuit.

With the application of a gate voltage, where the amplitude of the gate voltage is equal to the bias voltage E, the diode will be at zero bias for the duration of the gate and will conduct during this time, as shown in Fig. 10.26. Resistors R_1 and R_2 are used to isolate the gate circuit from the signal circuit. R_3 is the load for the signal and gate circuits when D_1 is not conducting, and furnishes a dc path between the negative terminal of E and the diode. R_4 is the load for the circuit when D_1 is conducting. The output is reduced by the voltage divider R_2 and R_4. Thus

$$e_{\text{out}} = e_{\text{signal}} \left(\frac{R_4}{R_2 + R_4} \right)$$

when diode D_1 is conducting.

Representative values for R_1, R_2, R_3, and R_4 are shown. The gate voltage can be seen selecting a pulse from the input signal. If the gate amplitude is not made equal to the bias voltage E, a well or pedestal will be produced, depending upon whether the gate amplitude is less than or greater than E. This is usually an undesirable effect and is to be avoided.

Pentode

Pentodes with a separate connection to the suppressor grid can be used effectively in coincidence circuits. In this application the control grid is held at cutoff and the suppressor grid is held at a negative potential of about -20 to -40 volts. The pentode will not conduct until there is simultaneously a positive-going signal on the control grid and a gate voltage on the suppressor grid that drives it to zero potential, or a slightly positive potential.

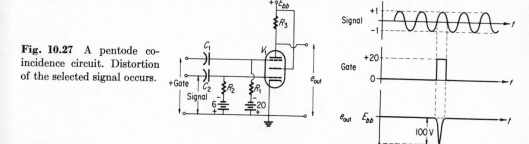

Fig. 10.27 A pentode coincidence circuit. Distortion of the selected signal occurs.

Such a coincidence circuit is shown in Fig. 10.27. The signal is applied to the control grid through the coupling circuit R_2 and C_2, and the gate voltage is applied to the suppressor grid through the coupling circuit R_1 and C_1. The plate load for the pentode circuit is R_3. Notice that the screen grid is held at E_{bb}. This is done to increase the transconductance of the tube so that when the tube is momentarily turned on, there will be a sufficient supply of electrons to produce normal action. In the circuit shown, Fig. 10.27, there is amplification as well as coincidence. However, the output is distorted because the tube is operating on the nonlinear portion of its transfer curve. For signal voltages consisting of pulses and square waves, this distortion does not change the signal appreciably. For other types of signal such as sawtooths and portions of sine waves, the distortion produced can appreciably alter the signal voltage.

TRIODE

A triode can be used in much the same way to produce a linear response during the coincidence time. It is known from Chapter 6 that a triode with a large value of R_L has a fairly linear transfer curve.

The circuit is shown in Fig. 10.28. The combination of R_4 and R_2 forms a voltage divider that keeps the cathode at 14 volts, 4 volts below cutoff. It is assumed that the input signal amplitude is kept to less than 4 volts. The plate-load resistor is large, 100 K, ensuring that the transfer curve will be quite linear. The gate voltage is applied to the cathode in this case, and must be a negative-going voltage, to reduce the bias on V_1. To ensure that no pedestal is produced, the gate voltage should be sufficient to bring the tube to cutoff. R_3 and R_2, a voltage divider, are used to reduce the negative gate amplitude to the desired value.

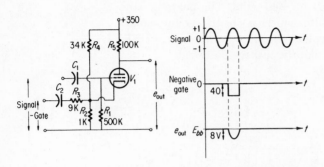

Fig. 10.28 A triode coincidence circuit. Distortion can be minimized.

The output will be a linear reproduction of the input selected by the negative gate. Due to the unbypassed cathode resistor R_2 and the fact that normal current cannot start to flow instantaneously with the application of the gate and signal voltages, the gain for the circuit will be low, of the order of 5 to 10. This is quite acceptable, since the transfer curve is linear. One must remember that there is also a phase reversal due to the amplifier.

An interesting and simple application of coincidence circuitry is the *electronic switch*. It is used in conjunction with an oscilloscope to show two voltages simultaneously. In this way the two voltages can be compared as to phase, wave shape, or distortion. In essence, each wave form is sampled many times per second and the sampling is shown on the oscilloscope screen. The cyclic sampling is produced by switching on and off a circuit, such as shown in the triode coincidence circuit, by means of a multivibrator.

The schematic for a simple electronic switch is shown in Fig. 10.29. V_3 and V_4 form a plate-coupled multivibrator circuit. When V_3 is conducting, there is a voltage drop across R_3, and when V_4 is conducting, there is a voltage drop across R_4. R_3 is the common cathode resistor for V_1 and V_3, and R_4 is the common cathode resistor for V_2 and V_4. The values are so chosen that when V_3 is conducting, V_1 will be cut off by the voltage drop across R_3. And when V_3 is nonconducting, R_3 biases V_1 to the middle of the linear portion of its transfer curve. The same explanation suffices for V_2, V_4, and R_4. V_1 and V_2 have a common plate-load resistor R_7. First V_1 and then V_2 alternately conduct through R_7. The resultant voltage across R_7 is shown in Fig. 10.29(b).

The net effect is the superimposition of both wave forms so that the relative phase shift between voltages can be seen on the oscilloscope. To show the voltages clearly, as can be seen, the sampling rate must be substantially higher than the frequency of the voltages being sampled. A higher sampling rate is shown in Fig. 10.29(c). The wave forms are more distinct. In practice, the multivibrator frequency is made variable by switching in different values of C_3 and C_4 and by making R_5 and R_6 variable.

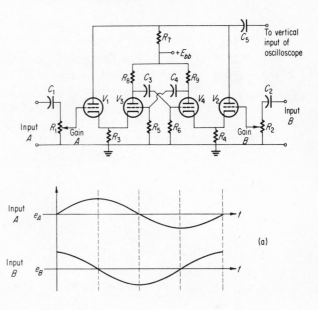

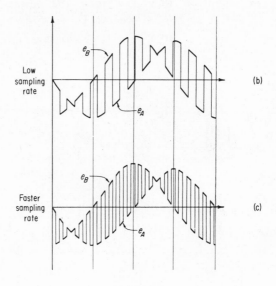

Fig. 10.29 A simplified electronic switch and two sample outputs, one at a low sampling rate, and the other at a higher sampling rate.

Problems

10.1 For $e_{in} = \sin \omega t$, find e_{out} for the circuit shown.

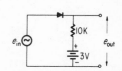

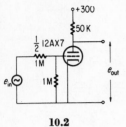

10.2

10.2 Determine e_{out} for the following grid limiter. Let $e_{in} = 5$ sin ωt volts. Assume $r_g = 1$ K.

10.3 For the grid-limiting circuit of Problem 10.2, let $e_{in} = 50$ sin $2\pi ft$, where $f = 10$ kc.
   (a) Find the amplitude of e_{out}.
   (b) Draw the wave form for 2 cps.
   (c) Determine the switching time.

10.4 Determine e_{out} if $e_{in} = 1$ sin $2\pi ft$, where $f = 1000$ cps. Draw the wave forms expected at points A, B, C, and e_{out}.

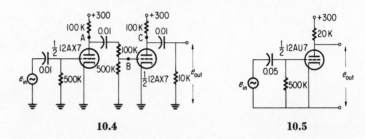

10.4   **10.5**

10.5 Determine the output of this circuit. (*Hint*: Clamping occurs). Let $e_{in} = 5$ sin ωt volts, where $\omega = 2000 \pi$.

10.6 For the following input voltage, determine the output of the clamping circuits shown.

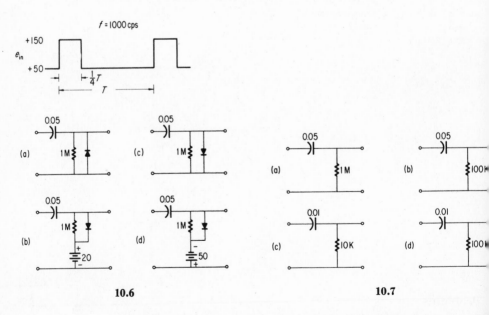

10.6   **10.7**

10.7 For the input voltage shown in Problem 10.6, determine the output voltages for the coupling circuits shown.

10.8 For the single-shot multivibrator shown in Fig. 10.16, use the following circuit values:

$R_3 = 50$ K $E_{bb} = 300$ volts

$R_4 = 50$ K V_1 and $V_2 = $ 12AU7

$C_2 = 0.001$ μf

$R_2 = 100$ K

Determine the minimum value of R_k for oscillation.

10.9 If the feedback capacitor and grid resistor in Example 10.6 are changed to $C_2 = 0.01$ μf and $R_2 = 50$ K, find the pulse width at the plate of V_2. Use the characteristics given in the example.

10.10 Draw the schematic suggested by the following block diagram.

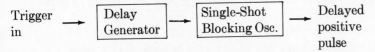

Trigger in → Delay Generator → Single-Shot Blocking Osc. → Delayed positive pulse

11.1 Transistor Action

The transistor is a solid-state device. It differs from a vacuum tube in several respects. The charge carriers move in the solid, so there is no need for a vacuum. Also, the charge carriers are always available, so there is no need for a filament. Thus no filament power or warm-up time is required for a transistor. The transistor is much smaller than vacuum tubes. This allows for the manufacture of compact electronic equipment. Transistors are used in all applications previously utilizing tubes, such as, radio-frequency and audio amplifiers, power amplifiers, oscillators, and pulse generators.

In order to explain transistor action, the transistor is depicted as two diodes in series, Fig. 11.1(a), with the emitter and base forming one diode, and the base and collector forming the other diode. The transistor shown is called a PNP transistor because of the nature of the semiconductor layers forming the transistor. Three leads are taken off the transistor, one lead from each of the semiconductor materials used. The lead connections are resistive that is, there is no rectification at the lead connections. Rectificacation does occur at the emitter-base junction and at the base-collector junction. The base region is made very thin, of the order of 10^{-3} cm. It should be noted that the applied electric fields are strongest at the emitter-base junction and base-collector junction. The remainder of the semiconductor is relatively field-free. Therefore the flow of injected carriers is governed chiefly by diffusion until the carriers approach a junction. Here they can acquire drift velocities due to the electric field.

In Fig. 11.1(a), the emitter-base circuit is biased in the forward direction by battery E_1, while the base-collector circuit is biased in the reverse direction by battery E_2. One would expect I_E to equal I_B, since this is in the forward-biased emitter-base circuit, and one could expect the collector current I_C to be negligible, since this circuit is reverse-biased. What is actually observed is that I_C is almost equal to I_E and I_B is almost zero. This seemingly anomolous effect comprises the transistor action. Instead of the holes in the emitter flowing through the base to B, they pass into the collector and flow back to the emitter through the external collector circuit and batteries E_2 and E_1.

This action is explained qualitatively in the following way: Holes are swept across the emitter-base junction as a result of the field impressed by battery E_1. Normally, they would combine with electrons in the N-type material of the base. However, the base has been purposely made so thin that an appreciable number of holes drift by diffusion to the base-collector junction instead of to the base lead. The electric field at the base-collector junction repels electrons in the N-type material, but is of the correct polarity to sweep the holes into the collector. They continue their

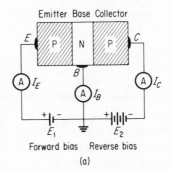

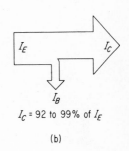

Fig. 11.1 (a) Bias and test circuit for a PNP transistor. (b) Most of the emitter current flows into the collector. (c) Simplified physical construction of a transistor (not to scale).

I_C = 92 to 99% of I_E

(b)

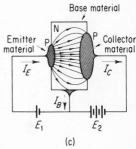

(c)

journey through the P-type material of the collector and produce a net current in the external collector circuit. Thus the holes swept across from emitter to base drift through the base with negligible recombination and are swept into the collector, contributing to collector current and subtracting from base current. From 92 to 99 percent of the emitter current passes through the base into the collector circuit. This process is illustrated in Fig. 11.1(c).

The diagram approximates the construction geometry for a PNP transistor. The biggest piece of material is N type and forms the base. Diffused into opposite sides of the base material is the P type of material. The smaller one is the emitter and the larger one is the collector. The lines drawn in between the emitter and the collector are the flow lines for the holes injected into the base material by the emitter. Essentially, the holes move away from the emitter by diffusion. Thus the lines radiate away from the emitter. To make the efficiency of the transfer of holes high, the collector is made larger than the emitter so that it intercepts most of the flow lines. Thus the collector collects most of the holes. Very little of this hole current drifts into the base connection.

Since the emitter-base junction is biased in the forward direction, a small value of E_1 will produce a large value of I_E. Consequently a large value of I_C is produced, even though the base-collector junction is reverse-biased. This control action is similar to that of a vacuum tube where a small voltage on the control grid controls the current in the plate circuit. The similarity ends here, since in a transistor the input current is not negligible. A measure of the efficiency of transfer of emitter current into the collector circuit is given by the ratio of change in collector current to change in emitter current, keeping the collector-base voltage constant. This is essentially a current amplification factor and is given by

$$\alpha = \left(\frac{\Delta I_C}{\Delta I_E}\right)_{E_C} \qquad (11.1)$$

For "good" transistors, α will be of the order of 0.99; that is, 99 percent of the emitter current passes through the collector circuit.

11.2   Practical Considerations

Two types of transistor are possible. Figure 11.2 shows that the base region can be either P type or N type of material. The two transistors thus formed are called NPN and PNP transistors, respectively. The symbols for these transistors are also shown.

The symbols used in practice are still in a slight state of flux. The symbols shown seem to be the ones that have survived the evolutionary process. One type differs from the other only in the inclusion of a circle around the elements. In each symbol, the emitter is identified as the lead with the arrow head. In the PNP transistor the arrow points toward the base, and in the NPN transistor, the arrow points away from the base. The arrow indicates the direction of positive current flow when the emitter-base junction is biased in the forward direction. Thus, glancing at the symbols, it is easy to remember the biasing convention. The NPN transistor would have to have the emitter negative with respect to the base, for the arrow to indicate positive current, and the PNP transistor would have to have the emitter positive with respect to the base, for the arrow to be in the correct direction. Care must be taken that the transistor not be biased so that the collector-to-base voltage is in the forward direction. Since the collector voltage is in volts, of the order of 5 to 10 volts, such a large voltage would produce an excessive current and destroy the transistor by overheating.

Both types of transistors operate in the same general manner. In the NPN type it is the electrons that are swept from emitter to collector, while in the PNP type, the carriers producing the transistor action are the holes. It is also evident that to bias the NPN transistor properly so that the emitter-base junction is forward-biased and the base-collector junction is reverse-biased, the battery polarities will be reversed from that shown in Fig. 11.1 for the PNP transistor. Either transistor type can be used in a particular application, and the two types can be used in conjunction with each other to produce circuits not possible with vacuum tubes. Figure 11.3 shows some typical transistors along with the physical dimensions and base connections. The placement of the leads emerging from the case have been standardized to a degree.

The most common type of transistor is shown in Fig. 11.3(a) and (b). It is essentially a rectangular parallelepiped of less than $\frac{1}{2}$ sq in. on the large face. These transistors can pass as much

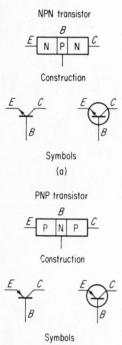

NPN transistor

Construction

Symbols
(a)

PNP transistor

Construction

Symbols
(b)

Fig. 11.2 Construction and schematic symbols for the NPN and PNP transistors.

current as the conventional vacuum tubes used in voltage amplifiers, but are considerably smaller. The transistors come from the
manufacturer with leads of about $1\frac{1}{2}$ in., as shown. Thus the
transistors can either be wired into a circuit or the leads clipped
to a length of $\frac{3}{8}$ in. and the transistors inserted into sockets especially designed for them.

Fig. 11.3 Some typ-
ical transistor external
construction and base
wiring.

(c)

(d)

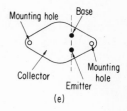

(e)

Care must be taken when soldering transistors into circuits so
that they will not be overheated. This would ruin the transistor,
converting it into a rather expensive but useless resistor. A heat
sink must be applied to the lead being soldered. A convenient way
to apply a heat sink is to clamp the transistor lead with a pair of
long-nosed pliers so that the pliers are between the transistor and
the soldering point. In this way the pliers must heat up before
the heat can pass to the transistor. Even so, it is advisable that
the soldering be done quickly.

Table 11.1 shows the magnitudes of voltage and current that
are to be expected for a typical transistor. The current is not much
different from that expected for a typical vacuum tube. However,
the voltage on the collector is much lower than the corresponding
plate voltage for a tube. The transistor is a low-voltage device.
The specifications show that transistors are highly temperature
dependent. The collector power dissipation is usually given at
25 °C, ambient air temperature, and this value must be decreased
if the transistor is to be operated at higher temperatures. This
is specified in the footnote of Table 11.1: "Derate 1.1 mw/°C
increase in ambient temperature." With this factor, the transistor
cannot be operated above 84 °C. At this temperature it can dissipate zero watts. The storage temperature indicates that a
transistor may be damaged by simply overheating it when not
in use. This is due mainly to the fact that diffusion is accelerated
by temperature increases. Impurities on the surface may diffuse into the body of the transistor and the P and N types of
materials will diffuse into each other, changing the character of
the transistor.

TABLE 11.1   Transistor Voltage and Current

Collector to emitter voltage (base open)	15 volts
Collector to base voltage (emitter open)	15 volts
Collector current	20 ma
Emitter current	−20 ma
Collector dissipation (25 C)*	65 mw
Storage temperature	85°C

* Derate 1.1 mw/°C increase in ambient temperature.
SOURCE: General Electric Co.

11.3   DC Characteristics for the Grounded Emitter Configuration

Transistors differ from tubes in that the input circuit draws current. There are four variables that have to be considered in a transistor. In the grounded emitter circuit shown in Fig. 11.4(a), the four variables are I_B, the base current; I_C, the collector current; E_B, the base-to-emitter voltage; and E_C, the collector-to-emitter voltage. The base circuit is a low-current circuit, of the order of microamperes, and the collector circuit is a relatively high-current circuit, Fig. 11.4(b). Due to the four variables, one must not only use output characteristics, but also input characteristics in order to determine the transfer curves.

Figure 11.4(c) shows a typical set of input characteristics. They are obtained by keeping E_C constant, varying E_B, and measuring I_B. Then E_B is plotted as a function of I_B. The input current increases with the increase in E_B ,in a nonlinear fashion. Also, the input characteristics are slightly dependent upon the collector voltage E_C. Three curves are shown for $E_{CE} = 0$, $E_C = 10$ volts, and $E_C = 20$ volts. The spacing of the three input curves shows that the higher-voltage curves are crowded together. It would be of interest to determine the approximate value of the resistance of the emitter-base junction. For example, in Fig. 11.4(c) at an I_B value of 200 μa, and using the $E_C = 10$-volt curve, the dc input resistance becomes

$$R_i = \frac{0.195}{200 \times 10^{-6}} \cong 975 \text{ ohms}$$

This is quite different from the input resistance of a triode, which is considered infinite because no grid current flows. For $I_B = 200$ μa, R_i varies from 900 to 1000 ohms, depending upon the value of E_C. To complicate matters further, the base current also depends upon temperature. For a given value of E_B, the base current increases

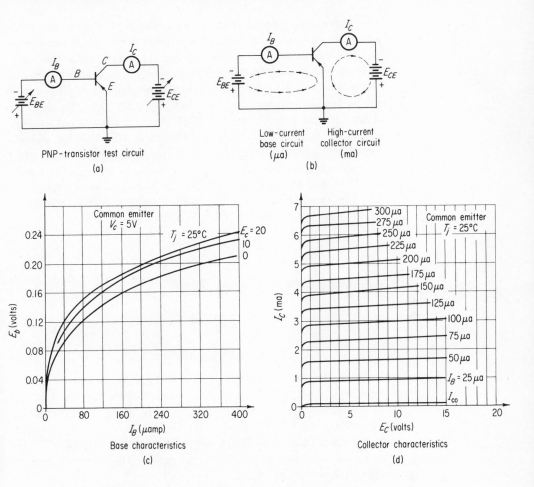

Fig. 11.4 Typical base and collector characteristics for a transistor.

with increase in temperature. Base-current dependence on temperature must be compensated for or distortion may result from a shift in the operating point. The curves for a transistor are usually shown for an ambient air temperature of 25°C.

In Fig. 11.4(d), the output characteristics are seen to resemble the curves found for the plate characteristics of pentodes. The linearity is much better for a transistor, that is, the spacing between successive curves is more nearly the same. Notice that I_B is used as the input parameter rather than E_B. Thus the curves are linear with respect to an input current change rather than to an input voltage change. A change in input voltage will not produce a proportional change in output current because of the nonlinear characteristic of the input resistance. At $I_B = 0$, the current through the collector circuit does not go to zero, even though the collector circuit may have a relatively large reverse bias.

This current is proportional to I_{CEO}. The CEO stands for "collector to emitter with the base current zero." It is sometimes abbreviated as I_{CO}, as is done in Fig. 11.4(d) and in this text. This current is due to the intrinsic conduction inherent in semiconductors and is highly temperature-dependent. It is often called the *saturation current*. To show how I_{CEO} fits into the collector characteristics given in Fig. 11.4(d), consider the current amplification for constant E_C. It is evident from the collector characteristics that there is a current amplification in the grounded emitter configuration. The input current, I_B is measured in microamperes, and the output current I_C is measured in milliamperes. For the grounded emitter, the symbol for this current amplification is the Greek letter β, which is defined as

$$\beta = \left(\frac{\Delta I_C}{\Delta I_B}\right)_{E_C} \tag{11.2}$$

Integrating this equation, one obtains

$$\int_{I_{CO}}^{I} dI_C = \beta \int_{0}^{I_B} dI_B \tag{11.3}$$

$$I_C = \beta I_B + I_{CO} \tag{11.4}$$

assuming that β is constant. Since the $I_B = \text{const}$ curves are not equally spaced, β does vary from point to point in the collector characteristics. However, for a large area of the characteristics, an average value for β works very well. Also, I_{CO} is the collector current when $I_B = 0$. Equation (11.4) can be used over a large range of collector characteristics to predict the collector current, once I_B has been assumed.

Example 11.1 For the collector characteristics shown in Fig. 11.4(d), find β for the point $E_C = 5$ volts, $I_B = 50$ μa. Check Eq. (11.4) for several values of I_B. For example, $I_B = 25, 50, 75, 100, 125$ μa.

1. For the determination of β, Eq. (11.2) will be used.

$$\beta = \left(\frac{\Delta I_C}{\Delta I_B}\right)_{E_C}$$

Using $E_C = 5$ volts, and allowing I_B to vary from 25 to 50 μa, the corresponding values of I_C will be

$$I_B = 25 \text{ μa}, 50 \text{ μa} \qquad I_C = 0.9 \text{ ma}, 1.6 \text{ ma}$$

Thus β becomes

$$\beta = \frac{(1.6 - 0.9) \times 10^{-3}}{(50 - 25) \times 10^{-6}} = 28$$

2. I_{CO} is determined from the collector characteristics for $I_B = 0$ as $I_{CO} = 0.1$ ma.

3. Equation (11.4) becomes

$$I_C = (28I_B + 0.1) \text{ ma}$$

where I_B is expressed in milliamperes.

4. Checking I_C versus I_B for several values of I_B at $E_C = 5$ volts, we find

I_B, ma	Calculated I_C, ma	Graph I_C, ma
25	0.8	0.9
50	1.5	1.6
75	2.2	2.3
100	2.9	2.9
125	3.6	3.45

Comparing the calculated values to the graphical ones, it can be seen that Eq. (11.4) works out fairly well. It is a useful relationship for the grounded emitter configuration.

Example 11.2   For the collector characteristics shown, find β as a function of E_C. Use $\Delta I_B = 50\ \mu a$, from $I_B = 100$ to $I_B = 50\ \mu a$. Let E_C vary from 0 to 30 volts.

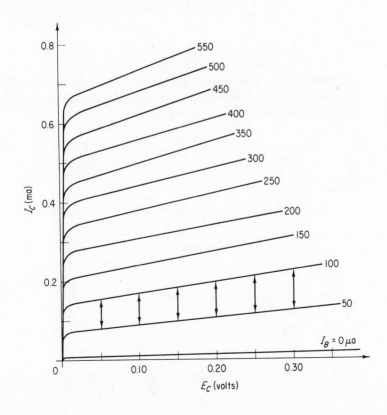

1. The quantity β is defined as

$$\beta = \left(\frac{\Delta I_C}{\Delta I_B}\right)_{E_C}$$

Qualitatively, the graphical approach consists of taking a change in I_B as determined by the family of curves, as shown above, and then tracing the two ends points of I_B to the ordinate and obtaining the corresponding change in I_C, ΔI_C. β is now obtained from the ratio of these two changes. As can be seen, the slopes of the two I_B curves used are different. Therefore, ΔI_C will change even though ΔI_B remains fixed.

2. Performing the indicated operation on the collector characteristics supplied, the following β values are obtained:

E_C, volts	ΔI_B, μa	ΔI_C, ma	β
5	50	0.74	14.8
10	50	0.80	16.0
15	50	0.85	17.0
20	50	0.91	18.2
25	50	0.97	19.4
30	50	1.02	20.4

3. These values of β can be graphed as a function of E_C. As is quite evident from the graph, the value of β is a linear function of E_C. For small changes in E_C, β can be considered constant. However, for large anticipated changes in E_C, β can be expressed as a simple linear equation:

$$\beta = a + bE_C$$

where a and b are constants that can be determined from the graph.

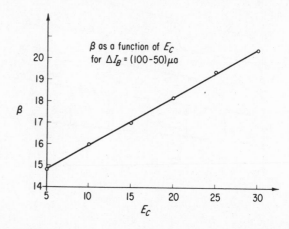

β as a function of E_C for $\Delta I_B = (100-50)\mu a$

11.4 Graphical Construction for the Grounded Emitter Configuration

The grounded emitter circuit has some general characteristics that are usually compared to the grounded cathode-vacuum-tube circuit. First, there is an appreciable voltage gain, and second, there is a 180-deg phase shift between input and output voltage. The graphical construction will bring these points out in detail. The basic, simplified circuit for a grounded emitter is shown in Fig. 11.5(a). The input voltage is between base and ground and the output voltage is taken between collector and ground. A PNP transistor is being used in this example.

In order to show voltage amplification, it would be helpful to develop the input-voltage/output-current transfer curve for this circuit. The procedure is not so simple as it was for the vacuum tube; first, because the output characteristics are shown with respect to I_B, and second, because the input characteristics are nonlinear. The construction will involve the simultaneous use of input and output characteristics. The load-line equation for the collector circuit of Fig. 11.5(a) is

$$-E_{CE} + I_C R_C + E_C = 0 \qquad (11.5)$$

where E_C is the voltage from collector to emitter. This can be solved for I_C:

$$I_C = \frac{E_{CE} - E_C}{R_C} \qquad (11.6)$$

The two variables are I_C and E_C. Thus the load-line circuit for the transistor is exactly similar to that for a vacuum tube. Equation (11.6) is plotted on the output characteristics shown in Fig. 11.5(b) for an R_C value of 2000 ohms. A transfer curve relating output current I_C to input current I_B can be constructed from this load line, as shown in Fig. 11.5(c). This curve is labeled "Current Transfer Curve." It shows that the output current change will be relatively linear with respect to an input current change. However, because the input resistance of a transistor is not linear, Fig. 11.4(c), the input voltage from the preceding circuit will not in general produce a base current proportional to the signal voltage. The saving grace is that when the input and output characteristics are combined to produce an E_B and I_C transfer curve, a small portion of the characteristics will be linear. This will then enable one to use the transistor as a small signal amplifier.

The procedure for the formation of the E_B and I_C transfer curve is shown in Fig. 11.5(c). The current transfer curve, obtained from the load-line points, and the resultant input characteristics, also obtained from the load-line points plotted on the family of

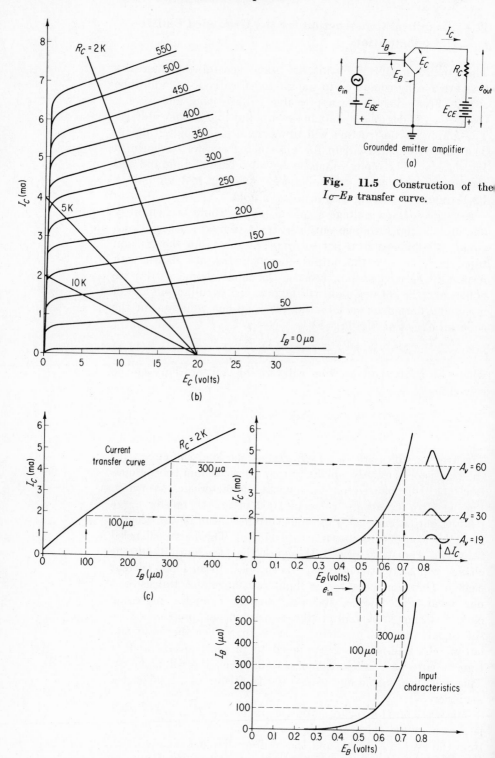

Fig. 11.5 Construction of the I_C–E_B transfer curve.

curves of I_B versus E_B as a function of E_C, such as shown in Fig. 11.4(c), are used. The corresponding points of each curve are transferred to an E_B and I_C coordinate system. The resultant curve is nowhere linear, but for small input voltages the curve can be considered linear over segments. Three such cases are shown. A signal voltage variation of 0.04 volt peak to peak is shown at $E_B = 0.5$ volt, $E_B = 0.6$ volts, and $E_B = 0.7$ volt. The corresponding changes in I_C are also shown. Since $E_C = E_{CE} - I_C R_C$, the output voltage can be constructed and the voltage gain can be obtained.

Example 11.3 For the characteristics shown in Fig. 11.5, determine the voltage gain at various points on the transfer curve, that is, $E_B = 0.5$, 0.6, and 0.7 volt. Use an input peak-to-peak voltage of 0.04 volt.

1. For $E_B = 0.5$ volt,

$e_{in} = 0.04$ volts pp

$e_{out} = \Delta I_C R_C = 0.38 \times 10^{-3} \times 2 \times 10^3 = 0.76$ volts pp

$A_v = \dfrac{e_{out}}{e_{in}} = \dfrac{0.76}{0.04} \cong 19$

2. For $E_B = 0.6$ volts,

$e_{in} = 0.04$ volts pp

$e_{out} = \Delta I_C R_C = 0.6 \times 10^{-3} \times 2 \times 10^3 = 1.2$ volts pp

$A_v = \dfrac{e_{out}}{e_{in}} = \dfrac{1.2}{0.04} = 30$

3. For $E_B = 0.7$ volt,

$e_{in} = 0.04$ volt pp

$e_{out} = \Delta I_C R_C = 1.2 \times 10^{-3} \times 2 \times 10^3 = 2.4$ volts pp

$A_v = \dfrac{e_{out}}{e_{in}} = \dfrac{2.4}{0.04} = 60$

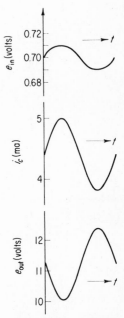

Fig. 11.6 Voltages and currents in a grounded emitter. Note phase reversal between e_{in} and e_{out}.

As would be expected from an inspection of the E_B–I_C transfer curve, the voltage gain is greatest where the slope of the curve is greatest. For the case where $E_B = 0.7$ volt, the quiescent values of I_B, I_C, and E_C can be determined as $I_B = 300$ μa, $I_C = 4.4$ ma, and $E_C = 11.2$ volts. Using these values as the Q point, then the input voltage, collector current, and output voltage can be graphed. The resulting graphs are shown in Fig. 11.6. The output voltage is 180 deg out of phase with the input voltage, and there is a substantial voltage gain.

Example 11.4 Show the effect of R_C, the load resistance, on the voltage gain at a given value of bias. Use $R_C = 2, 5$, and 10 K; $E_B = 0.5$ volt; and $E_{CE} \doteq 20$ volts.

1. The E_B–I_C transfer curves for $R_C = 5$ K and $R_C = 10$ K must be constructed. The case for $R_C = 2$ K has been done in Example 11.2.

2. For $R_C = 5$ K, the load-line equation becomes

$$I_C = \left(\frac{20 - E_C}{5} \right) \text{ma}$$

This is plotted in Fig. 11.5(b). From this curve and its intersection with the collector characteristics, one can construct the I_B–I_C current transfer curve. It will have a saturation region because the load line falls to the left of the knee of the collector curves. The resultant I_B and I_C curve is shown below.

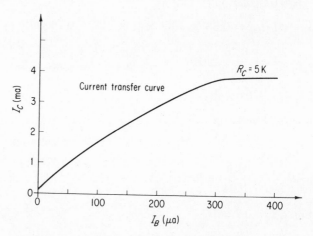

3. Next, the E_B and I_C transfer curve is obtained by utilizing the I_B and I_C curve, as above, and the input characteristics of Fig. 11.5(c). The result is an E_B and I_C transfer curve with a saturation region, as shown in Fig. 11.7. Below the saturation region, that is, for values of $E_B < 0.7$ volt, the transfer curve does not differ much from that for the $R_C = 2$ K case because the slopes of the I_B constant curves in Fig. 11.5(b) are almost zero.

4. For an input voltage of 0.04 volt pp, as shown in Fig. 11.7, the resultant current change $\Delta I_C = 0.3$ ma:

$R_C = 5$ K

$e_{\text{in}} = 0.04$ volt pp

$e_{\text{out}} = I_C R_C = 0.3 \times 10^{-3} \times 5 \times 10^3 = 1.5$ volts pp

$A_v = 37.5$

5. For $R_C = 10$ K, the load-line equation becomes

$$I_C = \left(\frac{20 - E_C}{10}\right) \text{ma}$$

This is plotted in Fig. 11.5(b).

6. From this graph, the I_B and I_C transfer curve can be determined and is as shown.

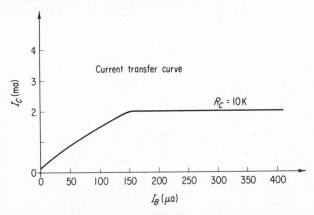

The saturation is more pronounced because the 10-K load line intersects the $E_C = 0$ axis at $I_C = 2$ ma.

7. Following the procedure illustrated in Fig. 11.5(c), the E_B and I_C transfer curve can be constructed. The result is shown in Fig. 11.7, and is labeled "$R_C = 10$ K."

8. For an input voltage of 0.04 volt pp, the resultant current change is $\Delta I_C = 0.22$ ma:

$R_C = 10$ K

$e_{\text{in}} = 0.04$ volt pp

$e_{\text{out}} = I_C R_C = 0.22 \times 10^{-3} \times 10^4 = 2.2$ volts pp

$A_v = 55$

9. The results are as graphed and show that as R_C increases, A_v increases for a given bias point.

The transfer curves do not change very much about the point $E_B = 0.5$ volt for the values of R_C selected. This is why A_v increases with increase of R_C.

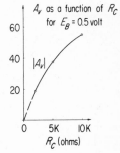

For the higher values of R_C, care must be taken not to overdrive the transistor; otherwise, severe limiting due to saturation and cutoff can occur.

To operate the transistor at the point shown in Fig. 11.7, the quiescent base voltage must be 0.5 volt. The input characteristics show that the base current at this voltage will be 50 μa.

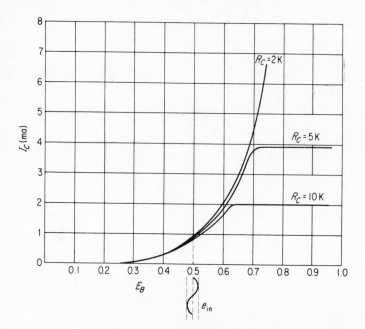

Fig. 11.7 Transfer curve as a function of collector load for the grounded emitter.

Thus, as in vacuum tubes, the transistor must have the proper dc-biasing voltages applied to it in order to operate on the linear portion of the transfer curve. The proper Q point can be obtained by setting either $E_B = 0.5$ volt, or $I_B = 50$ μa. The two conditions are synonymous. The values of the batteries in the simplified circuit of Fig. 11.5(a) have thus been determined as $E_{CE} = 20$ volts and $E_{BE} = 0.5$ volt, with the polarities as shown for the case illustrated in Example 11.3.

APPROXIMATE METHOD FOR OBTAINING THE VOLTAGE GAIN

In this method, one only needs the collector characteristics, Fig. 11.5(b), and a knowledge of the ac input resistance. If one has the input characteristics, Fig. 11.5(c), then the ac input resistance is known explicitly. It is given by the slope of the E_B and I_B curve at any point. Obviously, the ac input resistance will vary with input current. Illustrating the method is the best way to explain it. In Example 11.3, a load line of 2 K was used. The bias point was changed to show how voltage gain varied with the choice of bias. This calculation will be repeated, using the approximate method.

1. For $E_B = 0.5$ volt (the bias point), the input resistance as given by Fig. 11.5(c) is about 2000 ohms. This is obtained by finding the slope of the input curve for the point $E_B = 0.5$ volt, $I_B = 50$ μa. Now, moving to the output characteristics, at $I_B = 50$

μa, and along the load line, assume a 50-μa peak signal swing. The resultant change in collector current will be $\Delta I_C = 1.7$ ma. The voltage gain can now be approximated by comparing the change in input and output voltages for this case.

(a) For the output circuit,

$$\Delta E_C = R_C \, \Delta I_C = 2 \times 10^3 \times 1.7 \times 10^{-3} = 3.4 \text{ volts}$$

(b) For the input circuit,

$$\Delta E_B = r_{\text{in}} \, \Delta I_B = 2 \times 10^3 \times 100 \times 10^{-6} = 0.2 \text{ volt}$$

(c) The voltage gain A_v is then given approximately by

$$A_v = \frac{\Delta E_C}{\Delta E_B} = \frac{3.4}{0.2} = 17$$

(d) The actual value, as obtained in Example 11.3, is $A_v = 19$.

2. For $E_B = 0.6$ volts (the new bias point), the new ac resistance can be obtained from Fig. 11.5(c). The point at which r_{in} is calculated is $E_B = 0.6$ volt, $I_B = 110$ μa. To obtain this value graphically, one draws a line tangent to the input curve at this point and then deduces the slope, which is given by

$$r_{\text{in}} = \frac{\Delta E_B}{\Delta I_B}$$

The value found here is $r_{\text{in}} = 857$ ohms. The changes in voltage for input and output circuit can now be calculated.

(a) For the output circuit, using a variation I_B of 100 μa about the point $I_B = 100$ μa,

$$\Delta E_C = R_C \, \Delta I_C = 2 \times 10^3 \times 1.6 \times 10^{-3} = 3.2 \text{ volts}$$

(b) For the input circuit,

$$\Delta E_B = r_{\text{in}} \, \Delta I_B = 857 \times 100 \times 10^{-6} = 0.0857 \text{ volts}$$

(c) The voltage gain A_v becomes

$$A_v = \frac{\Delta E_C}{\Delta E_B} = \frac{3.2}{0.0857} = 37$$

(d) The value obtained in Example 11.3 was $A_v = 30$.

3. For $E_B = 0.7$ volt, the input resistance is found for the point $E_B = 0.7$ volt, $I_B = 300$ μa:

$$r_{\text{in}} = 333 \text{ ohms}$$

The input and output voltage changes can now be calculated.

(a) For the output circuit, using a variation in I_B of 100 μa about the point $I_B = 300$ μa,

$$\Delta E_C = R_C \, \Delta I_C = 2 \times 10^3 \times 1.11 \times 10^{-3} = 2.22 \text{ volts}$$

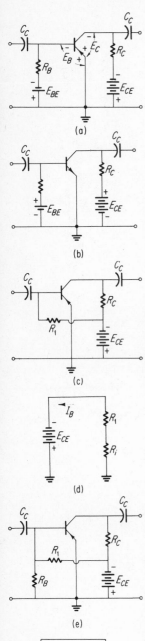

(a)

(b)

(c)

(d)

(e)

(f)

Fig. 11.8 Various biasing methods for the grounded emitter.

(b) For the input circuit,

$$\Delta E_B = r_{\text{in}} \, \Delta I_B = 333 \times 100 \times 10^{-6} = 0.0333 \text{ volt}$$

(c) The voltage gain A_v becomes

$$A_v = \frac{\Delta E_C}{\Delta E_B} = \frac{2.22}{0.0333} = 66.7$$

(d) The actual value, as found in Example 11.3, is $A_v = 60$.

The method just described assumes a small arbitrarily chosen change in collector current along the load line and converts that to a change in input voltage by multiplying the input current change by a resistance; in this case, the ac input resistance. Knowing the output current change is enough to calculate the output voltage change. The ratio of these two voltage changes is the desired voltage gain. If the changes used were smaller (for example, choosing ΔI_B to be 50 μa in each case), then the answers would approach the values found in Example 11.3. There is a point of diminishing returns, since making ΔI_B too small will introduce errors in estimating on the graph, which will decrease the precision. However, if carefully done, values for A_v can be obtained that are within 5 percent of the actual values. One deficiency of this method is that the transfer curve is never obtained, so that one is essentially working in the dark and cannot predict the maximum signal input before distortion will become appreciable. However, for small signal work, it is a handy, fast way of getting the answer.

11.5 Biasing the Grounded Emitter

A transistor must have two dc voltages applied to its elements in order to make it operate in the proper region of the transfer curve. The emitter-base junction must be forward-biased, and the collector-base junction must be reverse-biased. The simplest method is to use two batteries, as shown in Fig. 11.8(a). The transistor shown is a PNP type in the grounded emitter configuration. The collector must be negative with respect to the base, and the emitter must be positive with respect to the base. From the dc characteristics of Sec. 11.4 it is evident that E_C will be of the order of volts, and E_B will be of the order of tenths of volts. In summing up voltages from the collector to base, E_B is in opposition to E_C. Since $E_B \ll E_C$, the collector is negative with respect to the base, as required.

In this circuit, the capacitors shown are coupling capacitors that serve to isolate the dc voltages of one stage from those of the next stage. The base resistor R_B serves the same function as the grid resistor in vacuum tube circuits. The ac signal to be amplified is developed across R_B. The collector resistor, R_C, is the load resistor across which the amplified ac signal is developed.

For the NPN transistor shown in Fig. 11.8(b), the circuit is exactly the same as the one used for the PNP type, except for the polarities of the batteries used (they are reversed with respect to the PNP type), and the transistor symbol is that used for the NPN type. The bias voltages E_B and E_C are both of the same sign with respect to ground. This means that a single battery in conjunction with the proper voltage divider can be used to produce the two bias voltages required. Such circuits are shown in Fig. 11.8(c) and (e). For the two-battery bias circuit, it is necessary for R_B to be larger than the input resistance so that temperature variations do not displace the Q point significantly. R_B cannot be too small, for otherwise it will affect the input resistance of the transistor and reduce the overall voltage gain. A rule of thumb is to make R_B equal to approximately ten times the input resistance. Since the input resistance is of the order of 1000 ohms, an R_B value of 10 K would be satisfactory. The battery E_{BE} essentially has R_B and R_i in series across it, as shown in Fig. 11.9. Because $R_B > R_i$ for this circuit,

Fig. 11.9 DC equivalent circuit for the base bias circuit in Fig. 11.8(a).

$$I_B = \frac{E_{BE}}{R_B + R_i} \tag{11.7}$$

For the condition $R_B > R_i$, Eq. (11.7) becomes

$$I_B = \frac{E_{BE}}{R_B} \tag{11.8}$$

E_{BE} is determined as

$$E_{BE} = I_B R_B \tag{11.9}$$

Thus, knowing the base bias current and the value of R_B, E_{BE} can be determined.

Single-Battery Bias—Series Resistance

In the simplest type of single-battery bias, Fig. 11.8(c), a resistor R_1 is placed in series with the base and supply voltage E_{CE}. Since E_{CE} is of the order of 10 volts and I_B must be of the order of microamperes, R_1 will be large. The base bias circuit can be represented by the simple series circuit shown in Fig. 11.8(d). From this circuit,

$$I_B = \frac{E_{CE}}{R_1 + R_i} \tag{11.10}$$

where R_i represents the dc input resistance of the transistor. It turns out that $R_1 > R_i$. Thus, to a good approximation,

$$E_{CE} = I_B R_1 \tag{11.11}$$

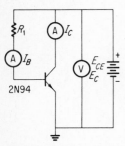

Fig. 11.10 The effect of R_1 on base bias current.

To show that this is indeed an accurate relation, a set of typical values of I_B versus E_{CE} has been obtained experimentally for a 2N94 transistor. The test circuit used is shown in Fig. 11.10. the value of E_{CE} was changed in 5-volt steps from 5 to 15 volts, and the resultant values of I_B and I_C (as measured by the ammeters) have been tabulated for two values of R_1 in Table 11.2. The first thing to be noticed is that the value of I_B does determine the operating point of the transistor. In Table 11.2 for $R_1 = 470$ K, when I_B is equal to 10 μa, $I_C = 0.5$ ma. For the case where $R_1 = 1$ meg and E_{CE} is adjusted so that I_B is again 10 μa, it should be noted that $I_C = 0.5$ ma again.

TABLE 11.2   Values of I_B and I_C

	E_C, volts	I_C, ma	I_B, μa
$R_1 = 470$ K	15	2.3	32
	10	1.27	21.2
	5	0.5	10.04←
$R_1 = 1$ meg	15	0.92	15.8
	10	0.50	10.06←
	5	0.21	5.2

To check Eq. (11.11), the value of I_B determined experimentally is compared with the value calculated by $I_B = E_{CE}/R_1$, Table 11.3. As can be seen, the agreement is excellent.

TABLE 11.3   Experimentally Determined Values of I_B Compared with Values Calculated from Eq. (11.11).

E_{CE}, volts	$R_1 = 470$ K		$R_1 = 1$ meg	
	Experimental, μa	Calculated, μa	Experimental, μa	Calculated, μa
15	32	31.9	15.8	15.0
10	21.2	21.3	10.06	10.0
5	10.04	10.65	5.2	5.0

SINGLE-BATTERY BIAS—VOLTAGE DIVIDER

Another commonly used bias circuit is shown in Fig. 11.8(e). R_B is chosen to be about ten times the ac input resistance of the transistor, so that it does not reduce the ac input resistance appreciably. It forms a relatively low-resistance dc return to ground and reduces the time constant for the coupling circuit. Due to the small ac input resistance of a transistor, the capacitor must be large in order to have normal low-frequency response. But the time constant for dc voltages and currents can be large, as would be the case for the series resistance bias, Fig. 11.8(c). This could cause blocking of the amplifier in much the same way that excess grid-leak bias in a vacuum tube can block a tube. With the voltage divider consisting of R_1 and R_B, the ac input resistance is not affected, but the dc resistance from base to ground is significantly reduced.

The effective dc circuit for this type of bias is shown in Fig. 11.8(f). Since $R_i < R_B$, and $R_i \ll R_1$, Eq. (11.11) can be used here also for determining the base bias current.

Example 11.5   Determine the values of R_1 and R_B for the circuit shown. Use the operating point established in Example 11.4 for $R_C = 10$ K and the characteristics shown in Fig. 11.5.

1.   From the input curve, the ac input resistance can be calculated from the following definition:

$$r_i = \left(\frac{\Delta E_B}{\Delta I_B}\right)_{EC}$$

This is given by the slope of the input curve at the Q point. An inspection of Fig. 11.7 shows that the Q point should be $E_B = 0.5$ volt. Moving to Fig. 11.5(c), the input characteristics, a line is drawn tangent to the curve at $E_B = 0.5$ volt. The construction is indicated in the illustration.

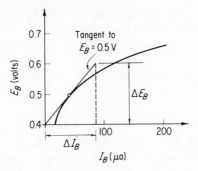

Putting in values, r_i becomes

$$r_i = \frac{(0.6 - 0.4)\ \text{volt}}{(90 - 0)\ \times\ 10^{-6}\ \text{amp}} \cong 2.2\ \text{K}$$

2. R_B is determined by the condition $R_B > r_i$; letting $R_B = 10r_i$,

$$R_B = 22 \text{ K}$$

3. R_1 is determined independently of R_B, and is given by Eq. (11.11):

$$R_1 = \frac{E_{CE}}{I_B}$$

From the input characteristics, Fig. 11.5(c), $I_B = 40$ μa at $E_B = 0.5$ volt:

$$R_1 = \frac{20 \text{ volts}}{40 \times 10^{-6} \text{ amp}} = 500 \text{ K}$$

4. The circuit, with correct values, is indicated in the accompanying illustration.

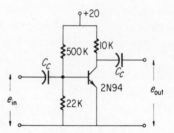

SINGLE-BATTERY BIAS WITH CURRENT FEEDBACK

Transistors are highly temperature-dependent. In fact most of the sophisticated circuits are usually complicated because of the efforts made to nullify the effects of temperature changes on the operation of the circuit. From elementary semiconductor theory, one realizes that the intrinsic current increases exponentially with temperature. A large enough change in temperature can produce an increase in intrinsic current that begins to affect the collector current. In fact, any small change in current in the base circuit is increased by a factor β in the collector circuit. Because of the small region of linear characteristics in the transfer curve, any change in dc parameters due to temperature variations can produce severe distortion.

The circuit shown in Fig. 11.11 tends to nullify temperature effects by means of negative feedback. The circuit differs from the previous ones because of the inclusion of R_E in the emitter circuit. It is bypassed by C_E, so that the combination produces no appreciable ac voltage drop. R_E affects only the dc parameters.

The voltage drop across R_E produces a voltage E_E with polarity as shown. Thus the emitter resistor produces a voltage drop that tends to make the emitter negative with respect to the base. The biasing network, consisting of R_1 and R_B, produces a voltage drop E_1 that tends to make the emitter positive with respect to the base. The total bias on the emitter-base junction E_B is given by the sum of the voltage drops across R_1 and R_E:

$$E_B = E_1 - E_E$$

where E_B is the voltage at the emitter with respect to the base. If the collector current I_C tends to increase because of an increase in temperature, this means that there is a corresponding increase in I_E. Thus E_E increases, and as the equation for E_B shows, E_B will be reduced. Reducing the bias voltage on a transistor makes it conduct less. The net result is that, owing to current feedback through R_E, the tendency for the collector current to increase with temperature is reduced.

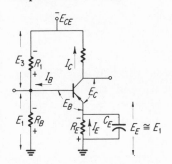

Fig. 11.11 Grounded emitter configuration utilizing an emitter resistor R_E for temperature compensation.

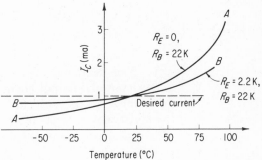

Fig. 11.12 Effect of R_E on overall temperature stabilization of I_C.

The qualitative results are shown in Fig. 11.12. They show that as R_E increases, the stabilization increases. For the single-battery case there is a limit to the value of R_E, since the greater R_E, the smaller R_1 will have to be for a given bias. Since R_1 and R_B will be in parallel with r_i in the ac equivalent circuit, too low a value of R_1 will affect the input circuit. As a rule of thumb, R_E is usually chosen equal to r_i, the equivalent ac input resistance.

To obtain the proper dc bias for the base, note the voltages shown in Fig. 11.11. R_B is again large compared with R_i and does not enter into the dc bias calculations. The important dc circuit is shown in Fig. 11.13. Note that the current through R_E is due to I_E, while the current through R_1 is the necessary bias current I_B. The current through R_1 is due to the difference in potential across R_1. Although the voltage drop across R_i is insignificant, as before, the voltage drop across R_E is important. It is given by

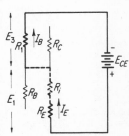

Fig. 11.13 Biasing current circuit path.

$$E_1 = I_E R_E \qquad (11.12)$$

The current should be put in terms of I_C so that, once the Q point has been determined, the collector characteristics can be used to determine the parameters. Using the definition for α,

$$\alpha I_E = I_C \tag{11.13}$$

Thus E_1 can be put in the form

$$E_1 = \frac{I_C}{\alpha} R_E \tag{11.14}$$

The voltage drop across R_1 is given by the difference between E_{CE} and E_1 (neglecting the voltage drop across R_i):

$$E_3 = E_{CE} - E_1 \tag{11.15}$$

$$I_B R_1 = E_{CE} - \frac{I_C}{\alpha} R_E \tag{11.16}$$

$$R_1 = \frac{E_{CE}}{I_B} - \left(\frac{I_C}{I_B}\right)\frac{1}{\alpha} R_E \tag{11.17}$$

This equation can be used as it stands, which means that both input and output characteristics must be used; that is, I_B and I_C must be determined. Sometimes the equation for R_1 is written entirely in terms of the collector characteristics by using the relationship between I_C and I_B, Eq. (11.4):

$$I_C = \beta I_B + I_0 \tag{11.18}$$

This allows one to put Eq. (11.17) into the form

$$R_1 = \left(\frac{\beta}{\alpha}\right)\left(\frac{I_C}{I_C - I_0}\right)\left[\alpha \frac{E_{CE}}{I_C} - R_E\right] \tag{11.19}$$

The values of α, β, and r_i are usually given by the manufacturer in his specification of the transistor. Thus, once a load line is drawn and a Q point selected, I_C, E_{CE}, and I_0 can be obtained from the collector characteristics and used in Eq. (11.19).

Example 11.6 Determine the resistor values needed to bias the transistor shown in the diagram. Assume $E_{CE} = 20$ volts, $R_C = 10$ K, and $E_B = 0.5$ volt (as in Example 11.5).

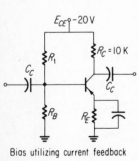

$E_{CE} - 20\,V$
$R_C = 10\,K$
C_C
R_1
C_C
R_B
R_E

Bias utilizing current feedback

1. R_E is selected from the relation

$$R_E = r_i$$
$$= 2.2 \text{ K}$$

2. R_B is selected from the relation

$$R_B = 10r_i = 22 \text{ K}$$

3.   Using the collector characteristics shown in Fig. 11.5(b), but drawing in a load line given by the equation

$$I_C = \frac{E_{CE} - E_C}{R_C + R_E}$$

Since R_E is not insignificant compared with R_C, it is included in the denominator:

$$I_C = \left(\frac{20 - E_C}{12.2}\right) \text{ ma}$$

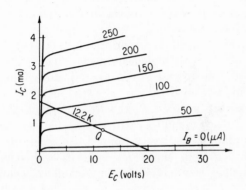

The Q point is given as $E_B = 0.5$ volt, which corresponds to $I_B = 40 \ \mu a$, Fig. 11.5(c). This point is indicated on the load line. From this, I_C can be determined as

$$I_B = 0.7 \text{ ma}$$

4.   Using Eq. (11.17) to determine R_1,

$$R_1 = \frac{E_{CE}}{I_B} - \left(\frac{I_C}{I_B}\right)\frac{1}{\alpha} R_E$$

α must be known. This is always given by the manufacturer. For the transistor used here, assume $\alpha = 0.97$:

$$R_1 = \left(\frac{20}{40 \times 10^{-6}} - \frac{0.7 \times 10^{-3}}{40 \times 10^{-6}} \times \frac{1}{0.97} \times 2.2 \times 10^3\right)$$

$$\cong 460 \text{ K}$$

5.   The circuit becomes as shown in the accompanying diagram.

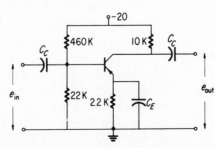

SINGLE-BATTERY BIAS WITH VOLTAGE FEEDBACK

Voltage feedback can be used in conjunction with current feedback to reduce further the tendency for I_C to increase with temperature. The circuit that will accomplish this is shown in Fig. 11.14(a) and (b). In (a), the circuit operation is as follows: As I_C increases because of increase in ambient temperature, the voltage at the collector tends to decrease. This in turn lowers the bias voltage across R_B, which in turn reduces I_C. The net effect is that I_C is stabilized so that the tendency for I_C to increase with temperature is reduced considerably. Also, R_E is in the circuit, producing its own retarding effect on the tendency of I_C to increase with temperature.

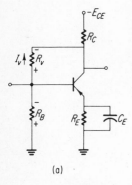

If the circuit is used as shown in Fig. 11.14(a), there will be feedback at the signal frequencies also, reducing the overall gain of the circuit. Where this is not desired, a decoupling circuit is inserted. This is shown in Fig. 11.14(b). R_V has been split up into two resistors in series, R_1 and R_2. The junction of R_1 and R_2 is connected to ac ground by capacitor C_V. Thus, ac feedback from collector to plate is effectively stopped. However, in the ac equivalent circuit, R_2 will shunt R_B, and R_1 will shunt R_C. This will also produce a reduction in the overall voltage gain for the circuit.

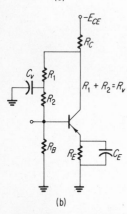

Fig. 11.14 Two voltage feedback temperature compensation circuits.

To calculate R_1 and R_2, one wants the bias current to flow through R_1 and R_2 as a result of the difference in potential across R_1 and R_2. This condition can be written down as

$$R_1 + R_2 = \frac{E_C - E_E}{I_B} \tag{11.20}$$

where E_C is the collector-to-emitter voltage, E_E is the voltage drop across R_E, and I_B is the desired base bias current. In terms of α and β, I_E can be written as

$$\alpha I_E = I_C = \beta I_B$$

$$I_E = \frac{\beta}{\alpha} I_B \tag{11.21}$$

Putting this in Eq. (11.20),

$$R_1 + R_2 = \frac{E_C}{I_B} - \frac{\beta}{\alpha} R_E \tag{11.22}$$

The values for R_1 and R_2 should be such that $R_1 > R_C$ and $R_2 > R_B$. However, $R_1 + R_2$ is usually of the order of 250 K. This does not allow one to meet the conditions stated above. The optimum course is to divide R: into two equal parts and take the consequences in the form of reduced gain.

11.6 DC Characteristics for the Grounded Base Configuration

This configuration has been likened to the vacuum-tube grounded-grid circuit. The reasons for this are: First, the output voltage is in phase with the input voltage; second, there is a voltage gain greater than 1; and third, the input ac impedance is low. For the transistor circuit, the input signal is impressed between emitter and ground, and the output signal is taken between collector and ground, Fig. 11.15. The battery bias is used to keep the circuit simple so that the fundamental characteristics can be seen. Replacing the two batteries by a single battery plus voltage dividers and inserting temperature-compensation circuitry will make the final circuit more complicated. Note that for the PNP type of transistor used, the emitter must be positive with respect to the base and that the collector must be negative with respect to the base. In order to make this circuit amenable to single-battery operation, the forward bias is usually applied between base and ground, as shown in Fig. 11.15(b). Now both E_{BB} and E_{CB} are negative with respect to ground, but the emitter is still positive with respect to the base.

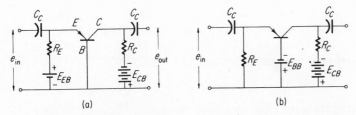

(a)   (b)

Fig. 11.15 Grounded base amplifier showing two biasing configurations.

In order to discuss the amplifying properties of the grounded base configuration, the input and output properties must first be obtained graphically. The test circuit is shown in Fig. 11.16(a). To obtain the output (or collector) characteristics, I_C is measured as E_C is varied. It will be found that to keep I_E constant, E_E will have to be varied as E_C is changed. In this manner, the family of curves shown in Fig. 11.16(b) can be obtained. A most distinctive feature is the extreme linearity of the collector characteristics in the common base configuration. As can be seen, I_E is slightly less than I_C. In fact the difference is a measure of α, given by Eq. (11.1), that is,

$$\alpha = \left(\frac{\Delta I_E}{\Delta I_C}\right)_{E_C} \tag{11.1}$$

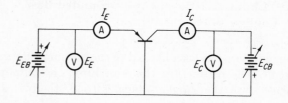

Fig. 11.16(a)

Example 11.7 Find α graphically for the transistor shown in Fig. 11.16(b) at the point $E_C = -10$, $I_C = 1$ ma.

1. Using Fig. 11.16(b) and allowing I_E to change from 1 to 2 ma along $E_C = -10$ volts,

$$\Delta I_E = (2 - 1) \quad \text{ma}$$

The corresponding change in I_C becomes

$$\Delta I_C = (1.92 - 0.95) \quad \text{ma}$$

2. Applying Eq. (11.1),

$$\alpha = \left(\frac{1.92 - 0.95}{2 - 1}\right) = 0.97$$

The usual value for α is anywhere from 0.90 to 0.99, depending upon the transistor type. The value for α can also be used to determine dc currents. A simple integration of Eq. (11.1) yields

$$I_E = \alpha I_C \tag{11.23}$$

where the zero value for each current is taken as equal to zero. This is not quite correct because, in actuality, there is a small collector current even though $I_E = 0$. This is due to the intrinsic current in the reverse-biased collector junction circuit. This current is low (of the order of 0.1 μa) and can usually be neglected in dc calculations at normal operating temperatures.

A typical set of input characteristics relating I_E to E_E is shown in Fig. 11.16(c). There is very little difference between the curves for $E_C = 0$ and $E_C = -20$ volts. Also, something not shown is the fact that the $E_C = $ const curves are fairly evenly spaced from $E_C = 0$ to $E_C = -20$ volts. The ac input resistance, given by the slope of an I_E and E_E curve is not constant and depends upon the values of I_E and E_E selected. For a range of I_E, however, r_{eb} can be considered essentially constant.

Example 11.8 For the $E_C = -20$ curve, find r_{eb} for $I_E = 1, 2, 3, 4,$ and 5 ma. Use the curve shown in Fig. 11.16(c).

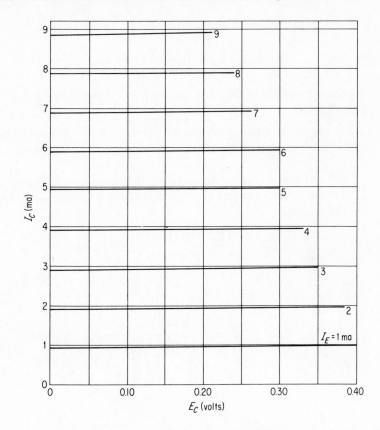

Fig. 11.16(b)

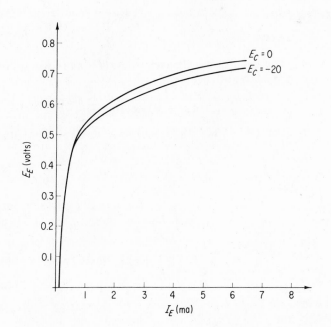

Fig. 11.16(c)

1. The ac resistance is given by

$$r_{eb} = \left(\frac{\Delta E_C}{\Delta I_C}\right)_{E_C}$$

which corresponds to the slope of the E_C = const curve shown in Fig. 11.16(c).

2. For I_E = 1 ma, a portion of the input curve is reproduced below (left) along with the construction necessary to determine r_{eb}.

First, a line is drawn tangent to the $E_C = -20$ curve at I_E = 1 ma. Then a triangle is constructed from which ΔI_E and ΔE_E are determined:

$$r_{eb} = \frac{0.72 - 0.42}{(2.4 - 0) \times 10^{-3}} = 125 \text{ ohms}$$

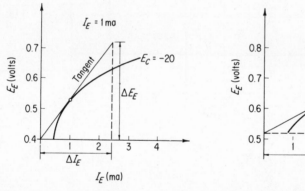
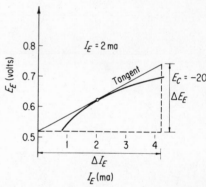

3. For I_E = 2 ma, the construction is as shown above (right).

$$r_{eb} = \frac{0.74 - 0.52}{(4.2 - 0) \times 10^{-3}} \cong 52 \text{ ohms}$$

4. For I_E = 3 ma, the construction is as shown below.

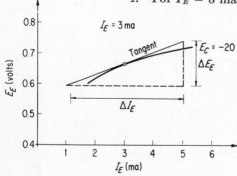

$$r_{eb} = \frac{0.74 - 0.59}{(5.0 - 0) \times 10^{-3}} = 30 \text{ ohms}$$

5. Similar constructions for I_E = 4 and 5 ma produce the following values for r_{eb}:

$$I_E = 4 \text{ ma:} \quad r_{eb} = 29 \text{ ohms}$$

$$I_E = 5 \text{ ma:} \quad r_{eb} = 22 \text{ ohms}$$

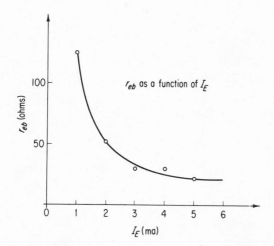

6. A curve showing r_{eb} as a function of I_E is shown above.

For a value of $I_E \geq 3$ ma, r_{eb} can be considered a constant for linear circuit approximations. As mentioned before, the ac input resistance for the grounded base configuration is rather small here, of the order of 30 ohms.

11.7 Graphical Construction for the Grounded Base Configuration

The grounded base configuration has an appreciable voltage gain. To explore this property, it would be useful to construct the transfer curve relating input voltage and output current, that is, an I_C and E_E curve. The procedure is similar to that used in the grounded emitter configuration. Using the collector characteristics and a load line, an I_C and I_E current transfer curve can be constructed.

The collector characteristics are shown in Fig. 11.17. Three load lines are drawn in for a supply voltage $E_{CB} = 20$ volts. They are $R_C = 2$ K, $R_C = 5$ K, and $R_C = 10$ K. For the load line $R_C = 2$ K, the I_C and I_E transfer curve shown in Fig. 11.18 can be constructed. As predicted from the appearance of the collector characteristics, the current transfer curve is extremely linear. This, coupled with the input characteristics, produces an I_C and E_E transfer curve that looks very much like the input curve. The procedure for obtaining the I_C and E_E transfer curve is illustrated in Fig. 11.19. Redrawing the input curve so that I_E is along the ordinate, the points A, B, C, D, $\cdots$ are transferred from the I_C and I_E transfer curve and the I_E and E_E input curve

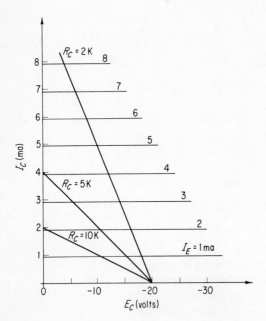

Fig. 11.17 Several load lines drawn in on the collector characteristics.

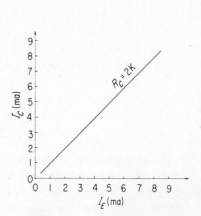

Fig. 11.18 Current transfer characteristics are linear in the grounded base configuration.

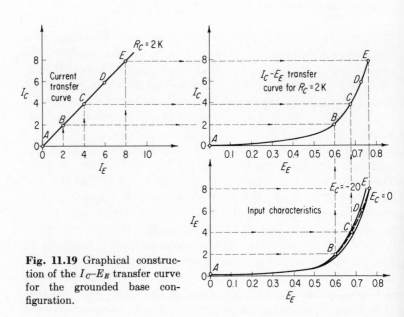

Fig. 11.19 Graphical construction of the I_C–E_E transfer curve for the grounded base configuration.

to produce the I_C and E_E curve. Because the I_C and I_E curve is so linear, the desired transfer curve, I_C and E_E, is a reproduction of the input characteristics as far as general shape is concerned. This will be true only for the case where the load line is to the right of the knee of the I_C and E_C curves. For the case of the higher-value load lines, the I_C and I_E curve will have a saturation plateau. The I_C and E_E transfer curves for $R_C = 2$ K, $R_C = 5$ K, and $R_C = 10$ K are shown in Fig. 11.20. As can be seen, the transfer curves for $R_C = 5$ K and 10 K have saturation plateaus. For a 0.12-volt pp input signal at A, the peak-to-peak output-current variation is about 1.48 ma. The voltage gains for the various load resistors become

1. $R_C = 2$ K; $A_v = 24.6$.

2. $R_C = 5$ K; $A_v = 61.5$.

3. $R_C = 10$ K; $A_v = 123$.

If one stays on the $R_C = 2$ K load line and biases the emitter to 0.72 volt, as shown for input voltage B, then operation takes place on the steep part of the I_C and E_E curve. In this case, the voltage gain becomes

$$R_C = 2 \text{ K}; \qquad A_v = 90$$

To show the phase relationships between input and output signals, consider the input voltage A, Fig. 11.20, and $R_C = 10$ K. As the voltage goes more positive, I_C increases. This will cause the voltage from collector to base to go more positive; see Fig. 11.21(a). For example, when E_E changes from 0.54 volt to 0.60 volt, the collector voltage given by

$$E_C = -E_{CB} + I_C R_C \qquad (11.24)$$

changes from -8.0 volts to 0.0 volt, Fig. 11.21(b). When the emitter voltage changes from 0.06 volt to 0.48 volt, the collector changes from 0.0 volt to -14.8 volts. The results are shown in Fig. 11.21(b). As the input voltage goes positive, the output voltage goes positive. Thus there is no phase change between input and output for a grounded base amplifier.

11.8 Biasing the Grounded Base

For a PNP type of transistor, the emitter must be positive with respect to the base and the collector must be negative with respect to the base. This can be effected by a two-battery bias, as shown in Fig. 11.22(a). The battery E_{EB} is of the order of 0.5 volt and E_{CB} is of the order of 10 to 20 volts. For the NPN type of tran-

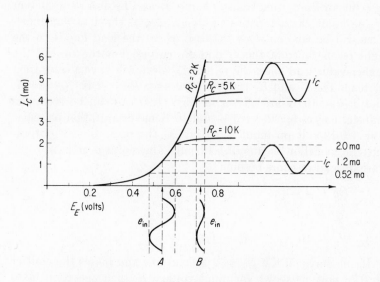

Fig. 11.20 Effect of R_C on the I_C–E_E transfer curve.

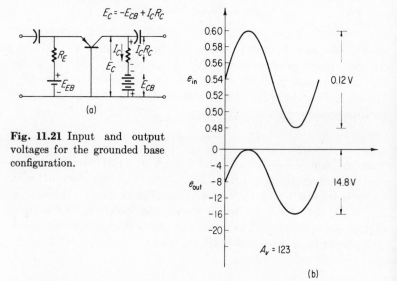

Fig. 11.21 Input and output voltages for the grounded base configuration.

sistor, the circuit is the same; only the battery polarities are reversed, as shown in Fig. 11.22(b).

A variation of the two-battery bias is shown in Fig. 11.22(c). The base is negative with respect to ground. This effectively makes the emitter positive with respect to the base. More important, both voltages, E_{EB} and E_{CB}, are negative with respect

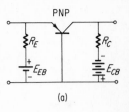

(a)

to ground. This makes the use of a single battery possible. Voltage E_{EB} is obtained from E_{CB} by means of a voltage divider, as shown in Fig. 11.22(d). The bias voltage is developed across the R_2 and C_2 combination. R_2 is usually made of the order of 100 ohms. The series combination, emitter-to-base resistance plus R_E, is in parallel with R_2.

Since R_E is of the order of 1000 ohms, by purposely making R_2 of the order of 100 ohms, $R_2 < R_E + R_i$, and the bias voltage E_B can be obtained by a simple voltage-divider calculation on R_1 and R_2.

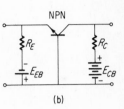

(b)

Example 11.9 Find R_1 for the conditions $E_{CB} = -20$ volts, $E_B = 0.5$ volt

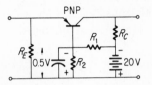

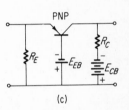

(c)

1. Let $R_2 = 100$ ohms. Then, from the voltage divider R_1 and R_2, the equation relating E_{CB} and E_B can be written down:

$$E_B = E_{CB}\left(\frac{R_2}{R_1 + R_2}\right)$$

Solving for R_1,

$$R_1 = R_2\left(\frac{E_{CB} - E_B}{E_B}\right)$$

2. Substituting the values given,

$$R_1 = 100\left(\frac{20 - 0.5}{0.5}\right)$$

$$= 3900 \text{ ohms}$$

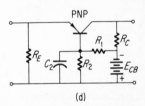

(d)

Fig. **11.22** Various biasing configurations for the grounded base.

3. R_1 is rather low; however, the circuit is not loaded down in any way because R_1 will appear across R_2 in the ac equivalent circuit. Since R_2 is shunted by C_2, R_1 and R_2 will not appear in the ac equivalent circuit.

11.9 Graphical Construction for the Grounded Collector Configuration

In the grounded collector, the input signal is applied between base and ground, and the output signal voltage is developed between emitter and ground. Although it is called a *grounded* collector, for biasing reasons, the collector is not at dc ground, although in the ac voltage-equivalent circuit, the collector would be grounded. Hence the collector is at ac ground but not usually at dc ground. The reason for doing this will be made clear in the section on biasing of the grounded collector.

The test circuit for determining the dc characteristics is shown in Fig. 11.23. The emitter must be slightly positive with respect to the base in order to secure proper forward biasing for the PNP transistor shown. Therefore $E_E > E_B$, but by only a few tenths of a volt. I_E will be in the milliampere range, while I_B will be of the order of 100 μa. As is usual, the output characteristics I_E and E_E are plotted as a function of the input current I_B. The results for a typical transistor are shown in Fig. 11.24(a). The I_B = const curves are plotted in increments of 50 μa and are uniformly spaced.

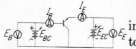

Fig. 11.23 Test circuit for the grounded collector configuration.

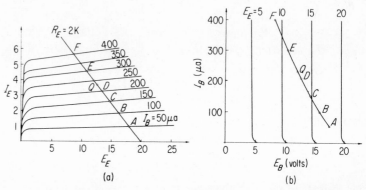

Fig. 11.24 Typical input and output characteristics for the grounded collector configuration.

The input characteristics E_B and I_B are shown in Fig. 11.24(b). They are quite different in appearance from what has been seen for the grounded emitter and grounded base configurations. The parameter kept constant for each curve in the family is the output voltage E_E. As can be seen, $E_E > E_B$ at all points in the graph. These curves are strikingly linear also. The fact that both input and output characteristics are linear produces an interesting result in the graphical construction of the grounded-collector transfer curve.

To show how the grounded collector behaves as a voltage amplifier, the input-voltage versus output-current transfer curve is obtained in much the same manner as was done for the other configurations. The simplified circuit that will be used to explain the grounded collector amplifier is shown in Fig. 11.25. This circuit

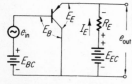

Fig. 11.25 A simplified grounded collector amplifier circuit.

utilizes two-battery bias and is not a practical circuit as such, but it is sufficient to show the graphical construction. Practical circuits will be dealt with in Sec. 11.10. The output voltage is taken from emitter to ground. The first step in obtaining the desired transfer curve is to draw the load line on the output characteristics. A load line for $R_E = 2$ K is shown in Fig. 11.24(a) for a supply voltage of 20 volts. From this, the points A, B, C, $\cdots$, F can be obtained, marking the intersection of the load line with $I_B = $ const curves. These points are plotted on the input characteristics, as shown in Fig. 11.24(b).

From these two curves (the load line and input curve), the points A, B, C, $\cdots$, F are transposed to an I_E and E_B coordinate system, as shown in Fig. 11.26. The first thing one notices is the linearity of the resultant transfer curve. The input voltage can vary from about 9 to 18 volts and produce a resultant current I_E free of significant distortion. This is contrary to the results found

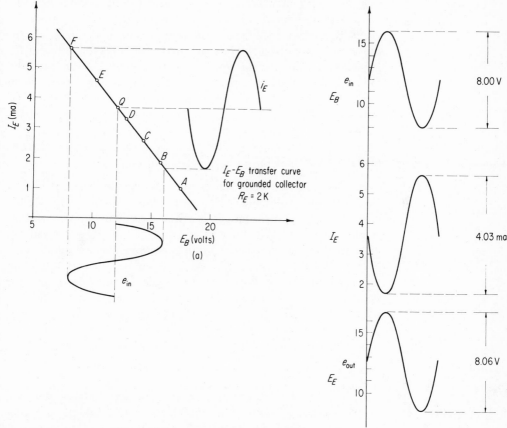

Fig. 11.26 (*Above*) I_E–E_B transfer curve. (*Right*) Voltage and current waveforms for the grounded collector amplifier.

in the other two configurations, where the input voltage variation had to be kept down to several hundredths of a volt. The second result obtained from the graphical construction is seen more easily when the input and output voltages are compared. Assume an input voltage of 8 volts pp, as shown in Fig. 11.26. The resultant current variation is about 4.03 ma pp. Since $e_{out} = E_{EC} - I_E R_E$, Fig. 11.25, the output voltage variation is about 8.06 volts pp. The resulting wave forms are shown in Fig. 11.26(b). There is no phase change between input and output voltages, and the voltage gain $A_v \approx 1.00$. Actually, the ac voltage-equivalent circuit calculations will show that the gain for a grounded collector approaches 1, but cannot exceed 1. The error in this graphical solution is within reason. At least it shows that the gain is of the order of 1.

In all three characteristics just mentioned, the grounded collector configuration resembles the grounded-plate, or cathode-follower, tube circuit. The resemblance does not end here. The analytical approach of the next chapter will confirm that the input impedance is high and the output impedance very low for the grounded collector. Thus the grounded collector is useful for impedance matching in much the same way as the cathode-follower vacuum-tube circuit. One interesting application is to use this configuration as the output stage in an audio amplifier. Since the output impedance can be made very low, and power transistors can easily conduct current of the order of 1 amp, a loudspeaker can be connected directly into the emitter circuit, obviating the need for an output transformer.

In the grounded collector circuit, the emitter voltage follows the base voltage. In Fig. 11.25, as the base voltage is raised, the current in the emitter circuit is reduced and the emitter voltage raises. As the base voltage is reduced, the emitter current increases and the emitter to ground voltage also increases. The net result is that the difference between the emitter-to-ground voltage and the base-to-ground voltage is the base-to-emitter voltage. It is this voltage that determines the current flow in the emitter circuit. To show how closely the emitter voltage follows the base voltage, look at the data shown in Example 11.10. It was obtained experimentally for a 2N94 transistor.

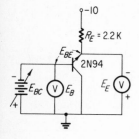

Example 11.10 A 2N94 transistor of the NPN type was connected as shown. E_B was varied and the corresponding values of E_E were measured. The tabulated values follow.

1. Tabulated values of E_B, E_E, and the difference $E_E - E_B$ are as follows:

E_B, volts	E_E, volts	$E_E - E_B$, volts
−0.85	−1.06	−0.21
−1.80	−1.99	−0.19
−2.60	−2.79	−0.19
−3.60	−3.78	−0.18
−4.60	−4.78	−0.18
−5.00	−5.17	−0.17
−6.40	−6.55	−0.15
−7.00	−7.14	−0.14
−7.60	−7.74	−0.14
−8.40	−8.53	−0.13
−9.20	−9.32	−0.12
−9.20	−9.70	−0.10
−10.00	−9.97	−0.03 ← cutoff

2. Since the 2N94 is an NPN transistor, the more negative the emitter with respect to the base, the greater the emitter current, and consequently the lower the emitter-to-ground voltage. This is shown in the tabulation given above. As the bias voltage $E_E - E_B$ approaches 0 volt, the transistor approaches cutoff and the emitter-to-ground voltage approaches the supply voltage, 10 volts.

3. Thus, as the base voltage is varied, the emitter voltage follows, and the difference between base and emitter voltages shows where the operating point of the transistor lies on the input curve.

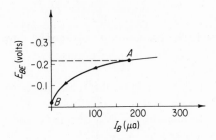

As the base voltage varies from −0.85 volts to −10.00 volts, the voltage E_{BE} varies from point A to point B, where

$$E_{BE} = E_E - E_B.$$

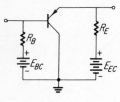

Two-battery bias

(a)

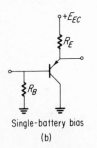

Single-battery bias

(b)

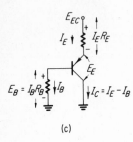

(c)

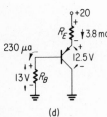

(d)

Fig. 11.27 Biasing schemes for the grounded collector.

11.10   Biasing the Grounded Collector

The elementary two-battery circuit of Fig. 11.27(a) can be converted into the single-battery circuit of Fig. 11.27(b) by biasing the base with the voltage drop across R_B. This is possible because E_B and E_E are both positive with respect to ground. The dc currents are indicated in Fig. 11.27(c). Positive current is being used. The voltage drop across R_B is given by $I_B R_B$, where I_B is the base current at the Q point selected. R_B is selected from this relationship because both I_B and E_B are determined by the load line. R_B is given by

$$R_B = \frac{E_B}{I_B} \tag{11.25}$$

where both E_B and I_B are determined from the Q point selected on the load line.

Example 11.11   Using the load line and transistor characteristics shown in Fig. 11.26, find the proper resistors to bias the transistor to the Q point, $E_B = 12$ volts.

1.   The load resistor R_E is given in the example as $R_E = 2$ K. Choosing the Q point $E_B = 12$ volts gives the following dc values for I_E, E_E, and I_B as obtained from the characteristics.

$$I_E = 3.8 \text{ ma}$$

$$E_E = 12.5 \text{ volts}$$

$$I_B = 230 \ \mu a$$

2.   Using Eq. (11.25),

$$R_B = \frac{12}{230 \times 10^{-6}} \cong 52 \text{ K}$$

3.   The biased transistor circuit becomes as shown below.

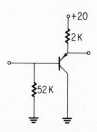

To illustrate the linearity of the grounded collector configuration, Fig. 11.28 shows the experimental results obtained for a 2N94 transistor. The audio-oscillator output voltage is a 1000-cps sine

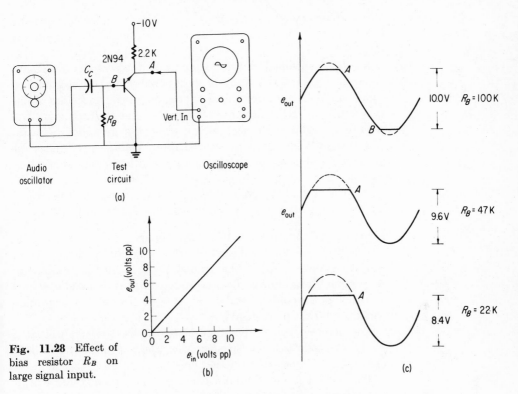

Fig. 11.28 Effect of bias resistor R_B on large signal input.

wave. Its peak-to-peak value is varied from 0 to 12 volts. The output and input voltages are measured with the oscilloscope by alternately connecting the lead labeled "Vert. In" to points A and B, respectively. As the curve in Fig. 11.28(b) shows, the input voltage can be varied from 0 to about 10 volts peak to peak before distortion occurs. The distortion is produced when the input signal has become large enough to drive the transistor to saturation and cutoff.

Typical examples of results for various base bias resistors are shown in Fig. 11.28(c). For all three cases, E_{in} was 12 volts peak to peak, enough to produce distortion. For $R_B = 100$ K, the transistor is driven to cutoff at point A and to saturation at point B. It is a characteristic of transistors, when driven to saturation, to have a very small voltage drop across them. In this case, the voltage drop was so small that it was not possible to measure it with the oscilloscope, and therefore the peak-to-peak voltage output measured 10 volts. For $R_B = 47$ K, the Q point is obviously moved closer to the cutoff point, since the output wave form shows severe cutoff clipping for a 12-volt pp input voltage, but no saturation clipping. For $R_B = 22$ K, the Q point moves still closer to the cutoff point and more of the signal is clipped.

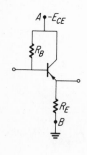

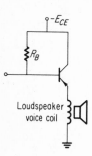

Of the three choices for R_B, $R_B = 100$ K seems to be the optimum one because it biases the transistor in the middle of its operating range.

As was mentioned in the beginning of the section, the circuit shown in Fig. 11.27(b) is not the most popular form of the grounded collector. The reason for this is that if one uses transistors of one type, such as NPN, then the grounded emitter amplifiers will require a negative supply voltage while the grounded collector amplifier will require a positive supply voltage. To make the grounded collector compatible with the other circuits (that is, use the same supply voltage), the circuit shown in Fig. 11.27(b) is changed so that the point $+E_{EC}$ is made ground and the ground points are put at a negative potential. (See Example 11.12.)

Example 11.12   In order to change the basic grounded-collector circuit into one that will operate from a negative rather than a positive supply voltage, the ground point is moved from point A to point B, as shown. The resulting circuit is at middle, left.

Although the collector is not at dc ground now, it is at ac ground, which is all that is necessary.

Another important advantage of this revised circuit is that if R_E is replaced by a loudspeaker, then the loudspeaker will have one side at ground potential, as shown at bottom, left.

This circuit is not the final one. As can be seen, there will be a dc current through the loudspeaker, which will cause an average extension or depression of the cone. This could produce distortion because of the unequal swings of the cone that will result when a signal is present. This can be overcome in one of two ways. First, and most simply, the speaker can be coupled to the output of the grounded collector by means of a large capacitor. The circuit is shown below.

As can be seen, the coupling capacitor has to be very large so as to have a small reactance compared with that of the speaker, so that the ac current flow through the speaker will be a maximum. R_E is of the order of 1000 ohms. Its resistance is large compared with the parallel circuit composed of C and the loudspeaker. R_E determines the dc Q point, as before.

The second method used to get rid of the dc current in the

loudspeaker voice coil is to put the loudspeaker in the cross arm of a balanced bridge circuit, where the arms of the bridge are transistors. This circuit is shown in Chapter 13.

11.11 Temperature Effects

Because the material used in transistors sustains intrinsic conduction as well as extrinsic conduction, the number of charged carriers is temperature-dependent. The extrinsic conduction is fixed by the doping and is not significantly temperature-dependent at normal temperatures (10 °C to 100 °C). The intrinsic conduction depends upon thermal agitation for the formation of electron-hole pair production. This pair production is an exponential function of temperature. If the temperature gets high enough, the increase in the intrinsic conduction can become appreciable, compared with the extrinsic conduction, and can alter the operating point and characteristics of the transistor used.

For example, in Fig. 11.29 two sets of characteristics for a grounded emitter stage are shown. The data for Fig. 11.29(a) were obtained while keeping the ambient temperature at 25 °C, and Fig. 11.29(b) was obtained for $T = 100$ °C. A 2-K load line is drawn for each case. If the Q point is selected as $I_b = 200$ μa for $T = 25$ °C at $T = 100$ °C, the Q point will have moved up to the point marked Q'. This may be enough to place the transistor in the nonlinear portion of the I_C and E_B transfer curve, producing severe distortion. The base current will stay constant because of the biasing method used. This means that the collector current will change. This is shown in Fig. 11.30 by taking the load-

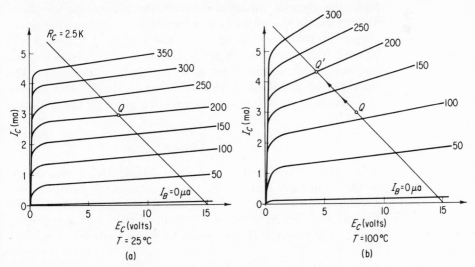

Fig. 11.29 Shift of operating point due to temperature changes in collector characteristics.

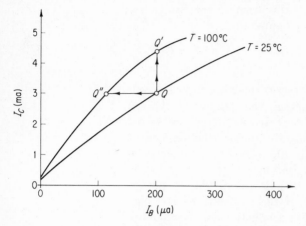

Fig. 11.30 Bias current reduced to compensate for temperature changes.

line points of Fig. 11.29(a) and (b) and constructing the corresponding I_C and I_B curves for the two temperatures. In this particular case, the collector current varies from 3 ma to 25 °C to 4.4 ma at 100 °C. In order to bring the collector current at 100 °C down to its previous value of 3 ma at 25 °C, the base current must be reduced to about 112 μa, that is, to point Q'', Fig. 11.30. This reduction in base current can be accomplished in several ways. One way is to reduce the base-to-emitter bias voltage as the temperature goes up. This is usually done by making the biasing resistor temperature-dependent. Another way to reduce temperature effects is to use degenerative feedback, such as the current and voltage feedback circuits discussed previously.

If it is possible to produce enough compensation by using feedback, then special voltage-divider circuits will not be necessary. To investigate the efficiency of a feedback circuit in reducing the change in collector current, a simple analytical approach will be used. This approach will not be used to produce design criteria, but only to show the effect of major circuit components. The *current stability factor* indicates the effect on the collector current of a change in the saturation current. It is given by the definition

$$S_i = \left(\frac{\Delta I_C}{\Delta I_{CO}}\right)_{EBE} \tag{11.26}$$

where ΔI_C is the change in collector current; ΔI_{CO} is the change in the saturation current, due to an increase in intrinsic conduction; and the ratio is evaluated at a constant supply voltage. The ratio has no units, being a ratio of current, and is defined in such a way that the factor should be numerically equal to zero, ideally.

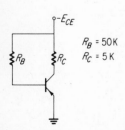

$R_B = 50\text{K}$
$R_C = 5\text{K}$

Example 11.13 Find the stability factor for the circuit at left. Assume the following parameters for the transistor $\alpha = 0.97$, and $\Delta I_{CO} = 20$ μa for a change of 50 °C. Find ΔI_C.

1. Redrawing the circuit, inserting current directions, and indicating voltage drops, and the accompanying new circuit results.

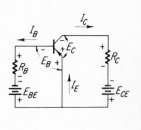

2. The circuit equations become

$$I_E = I_B + I_C \tag{1}$$

$$-E_{BE} + I_B R_B + E_B = 0 \tag{2}$$

$$I_C = I_{CO} + \alpha I_E \tag{3}$$

Eliminating I_B between Eqs. (1) and (2),

$$I_E R_B - I_C R_B = E_{BE} \tag{4}$$

where $E_B \ll E_{BE}$ and has been neglected. Choosing I_E and I_C as independent variables, Eqs. (3) and (4) can be put into the form useful for determinant solution:

$$I_E R_B - I_C R_B = E_{BE}$$

$$-I_E \alpha + I_C = I_{CO}$$

Solving for I_C

$$I_C = I_{CO} \left(\frac{1}{1 - \alpha} \right) + E_{BE} \frac{1}{R_B} \left(\frac{\alpha}{1 - \alpha} \right) \tag{5}$$

3. Applying the definition for S_i, given by Eq. (11.26),

$$S_i = \left(\frac{\Delta I_C}{\Delta I_{CO}} \right)_{E_{BE}} = \frac{1}{1 - \alpha}$$

$$= 33 \tag{6}$$

4. For the change in saturation current given, that is, $\Delta I_{CO} = 20\ \mu a$, the corresponding change in I_C becomes

$$\Delta I_C = S_i\, \Delta I_{CO}$$

$$= 0.67\ \text{ma}$$

Example 11.14 For the circuit shown, utilizing an emitter "swamping" resistor, find S_i for a change of $\Delta I_{CO} = 20\ \mu a$ in 50 °C, calculate ΔI_C. Use $\alpha = 0.97$.

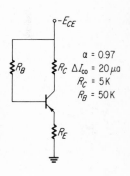

$\alpha = 0.97$
$\Delta I_{co} = 20\ \mu a$
$R_C = 5\text{K}$
$R_B = 50\text{K}$

1. Redrawing the circuit, inserting current directions, and indicating voltage drops results in the accompanying new circuit.

2. The circuit equations become

$$I_E = I_C + I_B \tag{1}$$

$$I_C = I_{CO} + \alpha I_E \tag{2}$$

$$-E_{BE} + I_B R_B + E_B + I_E R_E = 0 \tag{3}$$

$$-E_{CE} + I_C R_C + E_C + I_E R_E = 0 \tag{4}$$

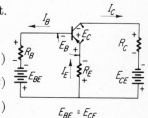

$E_{BE} = E_{CE}$

3. Eliminating I_B between Eqs. (1) and (3) and neglecting E_B with respect to E_{BE},

$$I_E(R_B + R_E) - I_C R_B = E_{BE} \qquad (5)$$

Rewriting Eq. (2),

$$-I_E \alpha + I_C = I_{CO} \qquad (6)$$

Solving Eqs. (5) and (6) simultaneously for I_C,

$$I_C = I_{CO}\left[\frac{R_B + R_E}{R_B + R_E - \alpha R_B}\right] + E_{BE}\left[\frac{\alpha}{R_B + R_E - \alpha R_B}\right]$$

4. Applying the definition for S_i, Eq. (11.26),

$$S_i = \left(\frac{\Delta I_C}{\Delta I_{CO}}\right)_{EBE} = \frac{1}{1 - \alpha(R_B/(R_B + R_E))]} \qquad (7)$$

5. Equation (7) shows that S_i is reduced if R_E increases or if R_B decreases.

6. For the values given in the statement of the problem, and letting $R_E = 1$ K, S_i can be found by substituting into Eq. (7):

$$S_i = 20$$

This is a reduction from the preceding example, where the amplifier was uncompensated. The corresponding value for ΔI_C becomes

$$\Delta I_C = S_i = \Delta I_{CO}$$

$$= 0.4 \text{ ma}$$

7. To show how S_i depends upon R_E, a graph of S_i versus R_E is shown for the given transistor.

As can be seen, the stability factor goes down rapidly with increase in R_E. After R_E reaches a value of about 5 K, there is very little change in S_i. For this transistor, then, an optimum value of R_E would then be $R_E = 5$ K.

The two examples show the difference between an uncompensated, amplifier and an amplifier utilizing current feedback. Example 11.14 shows that S_i is reduced very rapidly with an increase in R_E. For many applications, a resistor in the emitter lead is adequate temperature compensation. It should be noted that Eq. (7) in Example 11.14 shows that S_i will decrease if R_E is increased, but it also shows that as $R_B \to 0$, $S_i \to 1$. R_B can be made equal to zero by using interstage transformers and separate battery supplies for base and collector. The temperature stability can be greatly improved.

Equation (7) in Example 11.14 does not tell the entire story. It shows that if $R_B \to 0$, then the value of R_E is not important. The stability of I_C versus temperature change is always increased

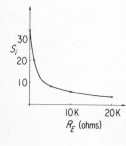

S_i

30

20

10

10K   20K

R_E (ohms)

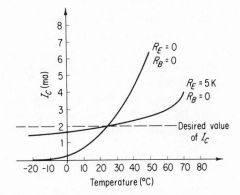

Fig. 11.31 Effect of R_E on temperature stability of I_C.

with the inclusion of an emitter swamping resistor. Figure 11.31 shows qualitatively how the inclusion of a swamping resistor R_E increases the collector current stability. In both cases, R_B is zero.

In essence, as the temperature increases, the base current increases. This is equivalent to saying that the effective base-emitter junction resistance R_{BE} decreases with temperature. A large value of resistance in the emitter lead (the swamping resistor R_E) will make R_{BE} a smaller fraction of the total resistance in this circuit. Therefore, since R_E does not change much with temperature, R_{BE} produces a small change in the total resistance. For more precise temperature compensation, thermistors or diodes are used to lower the base bias voltage as the temperature rises, thereby tending to keep the collector current at the original value.

Figure 11.32 shows two stabilizing circuits employed with transformer coupling. In each case, R_B and D_B are components whose resistance decreases with temperature, thus lowering the base bias voltage. In Fig. 11.32(a), R_B is a thermistor. In Fig. 11.32(b), D_B is a diode. In each case, C_B is a bypass capacitor used to make the bottom of T_1 an ac ground. Of the two circuits, the diode stabilization is the more successful. Essentially, the base-emitter junction of the transistor is a forward-biased diode, as is D_B.

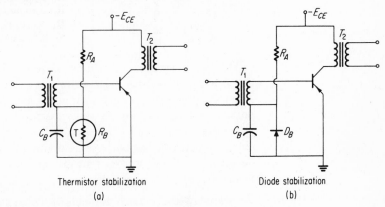

Fig. 11.32 Temperature stabilization in transformer coupling.

Thermistor stabilization
(a)

Diode stabilization
(b)

If D_B is selected to have characteristics similar to the base-emitter junction, the diode D_B will track over a large range of temperature ($\Delta T = 50\ °C$); that is, with increase in temperature, the base voltage will drop in such a way as to keep the collector current at the Q point.

In summing up, then, for RC-coupled circuits a swamping resistor R_E is used to reduce temperature shifts of I_C. For transformer-coupled circuits, diode stabilization is preferred.

11.12 Remote Temperature Indicator*

The fact that the collector current of a transistor increases with temperature is utilized to good effect in a temperature-measuring circuit designed by Sylvania Electric Products Inc. The circuit and parts list is shown in Fig. 11.33. This is a bridge circuit, with R_1 and Q_1 in one leg and R_3 and Q_2 in the other leg. The meter M_1 shows the unbalance in current due to a difference in temperature between Q_1 and Q_2. Since Q_1 is the sensing transistor and Q_2 is the control transistor, meter M_1 is placed in the circuit with polarity as shown so that as Q_1 increases in temperature, its collector voltage drops and the current through M_1 increases. A device like this can be calibrated by placing the control transistor Q_2 in a constant-temperature bath, such as a melting ice bath, and then raising the temperature of Q_1 in known steps. At each temperature, the meter face is appropriately marked.

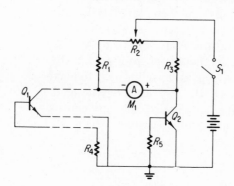

Fig. 11.33 Simple temperature measuring circuit.

B_1: 22.5-volt battery, Burgess 5156SC or equivalent
M_1: Meter, 0–25 ma (dc)
Q_1, Q_2: Sylvania 2N1038 transistors
R_1, R_3: 1 K, $\frac{1}{2}$-watt resistor
R_2: 250-ohm potentiometer
R_4, R_5: 10 K, $\frac{1}{2}$-watt resistor
S_1: SPST toggle switch

The potentiometer R_2 is used to zero the meter by balancing the bridge when both transistors are at the same temperature. The reproducibility of such a circuit is excellent. Also, the sensing portion is small in volume (it is essentially transistor Q_1 and its associated wires) and can be placed at large distances from the circuit for monitoring purposes. Circuits utilizing the temperature-dependence of collector current are found in various temperature-measuring and fire alarm devices.

* Sylvania Products Inc., "Performance Tested Transistor Circuits," 1958.

Problems

11.1 Draw a schematic of the grounded emitter circuit using a 2N647 transistor (NPN type). Use single-battery series-resistance biasing.

11.2 Repeat Problem 11.1, using a PNP transistor.

11.3 Plot β as a function of collector current at $E_C = 4$ volts for the 2N647.

11.4 Find β for a high-gain transistor such as the 2N406 and for a high-power transistor such as the 2N1070. Use a point essentially in the middle of each set of characteristics.

11.5 Using the E_B–I_C transfer curve and the I_B–I_C transfer curve given for the 2N647 transistor, reconstruct the input (E_B–I_B) characteristics.

11.6 Using the input characteristics found in Problem 11.5, determine the dc-input resistance of the 2N647 as a function of base-to-emitter voltage. Express your answer in the form of a graph.

11.7 Find the voltage gain for the 2N647 in the grounded emitter configuration, using the approximate method for the Q point $I_B = 1.0$ ma, $E_C = 2.5$ volts, and $R_C = 100$.

11.8 Find A_v by the simplified method for the 2N647 in the grounded base configuration for a load resistance of 500 ohms and a Q point of $I_B = 0.8$ ma, $E_C = 4$ volts.

11.9 Find A_v by the simplified method for the 2N647 in the grounded collector configuration for a load resistor of 500 ohms and a Q point of $I_B = 0.8$ ma, $E_C = 4$ volts.

11.10 For the circuit shown, determine the following:
(a) R_1 for a base current of 0.7 ma.
(b) R_B.
(c) E_{CE} for an R_C value of 1 K.

11.11 Using the 2N406 in the grounded emitter configuration and a Q point of $I_b = 40$ μa, $E_C = 6$ volts, determine (a) the biasing resistance R_1, (b) the value of E_{CE} for $R_C = 5$ K, and (c) the current gain.

11.12 Do Problem 11.11 over again for the case where a swamping resistor $R_E = 1000$ ohms is used in conjunction with a bypass capacitor C_E.

11.13 Find S_i for the accompanying circuit.

11.14 Find S_i, the current stability factor, for the circuit.

11.15 Find S_i for the accompanying circuit.

11.16 Find S_i for the accompanying circuit.

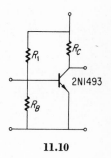

11.10

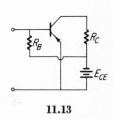

11.13

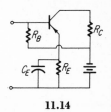

11.14

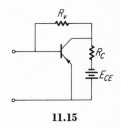

11.15

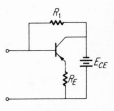

11.16

12.1 Four-Terminal Networks

12

Transistors— Analytical

The analysis of transistor circuits has been systematized into a four-terminal network description. Based on experience gained in vacuum-tube circuit analysis, the transistor is treated as a formal four-terminal network, and an equivalent circuit is derived from the analysis of this four-terminal network. This will allow one to use all the techniques of an analytical approach.

The four-terminal network applied to a transistor in the grounded emitter configuration is illustrated in Fig. 12.1(a). The input terminals are the base and emitter leads, and the output terminals are the collector and emitter leads. We shall try analytically to replace the transistor by equivalent generators and resistors and perform our calculations on the resultant equivalent circuit. This is what has been done to the vacuum tube. The ac voltage-equivalent circuit for the vacuum tube is shown in Fig. 12.1(b), for reference. Because grid current was considered zero for vacuum-tube circuits, the input circuit contains no generator. The output circuit has a generator that represents the amplified voltage μe_g.

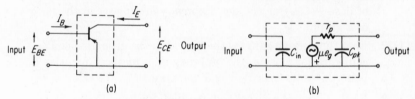

Fig. 12.1 Four terminal networks.

In the transistor there are four variables, two voltages, and the resultant currents. In Fig. 12.1(a), the four variables are E_{BE}, E_{CE}, I_B, and I_E. Positive current is used throughout the analytical development. In obtaining the ac equivalent circuit, two of the variables are called *dependent*. Any two can be called the *dependent variables*. Each choice of dependent variables will produce a different ac equivalent circuit. However, each circuit will give us the same answers as to voltage gain, current gain, input and output impedance, etc. If the voltages are chosen as the dependent variables, the ac parameters developed are called *open-circuit parameters:*

$$\left. \begin{array}{l} E_{BE} = f_1(I_B,\, I_C) \\[2mm] E_{CE} = f_2(I_B,\, I_C) \end{array} \right\} \quad \text{Open-circuit parameters}$$

Short-circuit parameters are obtained when the currents are chosen as the dependent variables:

$$\left. \begin{array}{l} I_B = f_1(E_{BE},\, E_{CE}) \\[2mm] I_C = f_2(E_{BE},\, E_{CE}) \end{array} \right\} \quad \text{Short-circuit parameters}$$

Finally, a third set of parameters can be obtained by choosing one voltage and one current as dependent variables. This is usually done because these particular parameters can be easily measured experimentally. The resulting parameters are called *hybrid,* or *h, parameters*:

$$\left. \begin{array}{l} E_{BE} = f_1(I_B, E_{CE}) \\[2mm] I_C = f_2(I_B, E_{CE}) \end{array} \right\} \quad \text{Hybrid parameters}$$

Another set of parameters is the one that will be used in this text. They are called the T parameters. These parameters can be derived from the open-circuit parameters by a simple algebraic manipulation. They are useful because of the simplicity of the resulting ac equivalent circuit. Since manufacturers usually state the parameters for a transistor, these must be converted into equivalent T types. The conversion is simple, and a table of conversions will be derived

12.2 AC Equivalent Circuits

In this section, the various equivalent circuit descriptions will be investigated. The initial effort will be an application of elementary four-terminal network theory. It will illustrate how the choice of dependent variables determines the ac equivalent circuit. A convention must be followed in setting up the initial parameters. The accepted method is to use the grounded base configuration, as shown in Fig. 12.2.

Positive currents will be used in the calculations. Since the current will be used as an operator to help obtain other results, the choice of current direction makes no difference.

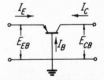

Fig. 12.2 Transistor parameters in the grounded base configuration.

OPEN-CIRCUIT PARAMETERS

If the voltages E_{EB} and E_{CB} of Fig. 12.2 are considered the dependent variables, then functionally,

$$\begin{array}{l} E_{EB} = f_1(I_E, I_C) \\[2mm] E_{CB} = f_2(I_E, I_C) \end{array} \tag{12.1}$$

Taking the total derivative of each function,

$$dE_{EB} = \left(\frac{\partial f_1}{\partial I_E}\right)_{I_C} dI_E + \left(\frac{\partial f_1}{\partial I_C}\right)_{I_E} dI_C$$

$$\tag{12.2}$$

$$dE_{CB} = \left(\frac{\partial f_2}{\partial I_E}\right)_{I_C} dI_E + \left(\frac{\partial f_2}{\partial I_C}\right)_{I_E} dI_C$$

Since $f_1 = E_{EB}$ and $f_2 = E_{CB}$,

$$dE_{EB} = \left(\frac{\partial E_{EB}}{\partial I_E}\right)_{I_C} dI_E + \left(\frac{\partial E_{EB}}{\partial I_C}\right)_{I_E} dI_C$$

$$dE_{CB} = \left(\frac{\partial E_{CB}}{\partial I_E}\right)_{I_C} dI_E + \left(\frac{\partial E_{CB}}{\partial I_C}\right)_{I_E} dI_C$$

(12.3)

In the small signal approximation, the differentials are replaced by the ac signals:

$$dE_{EB} = e_{EB} \qquad dI_E = i_E$$

$$dE_{CB} = e_{CB} \qquad dI_C = i_C$$

(12.4)

and the partial derivative ratios are relabeled:

$$r_{11} = \left(\frac{\partial E_{EB}}{\partial I_E}\right)_{I_C} \qquad r_{12} = \left(\frac{\partial E_{EB}}{\partial I_C}\right)_{I_E}$$

$$r_{21} = \left(\frac{\partial E_{CB}}{\partial I_E}\right)_{I_C} \qquad r_{22} = \left(\frac{\partial E_{CB}}{\partial I_C}\right)_{I_E}$$

(12.5)

Substituting Eqs. (12.4) and (12.5) in Eq. (12.3),

$$e_{EB} = r_{11}i_E + r_{12}i_C$$

$$e_{CB} = r_{21}i_E + r_{22}i_C$$

(12.6)

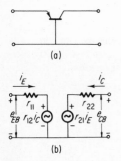

(a)

(b)

Fig. 12.3 Transistor and equivalent open circuit parameters.

It should be noted that the quantities defined in Eq. (12.5) have the units of volts/amp = ohms. The quantities r_{11} and r_{22} can be thought of as ac resistances, since the numerator stands for the voltage across the terminals in question and the denominator stands for the resultant current. The other two quantities are cross ratios. For example, for r_{12}, the numerator refers to the emitter circuit and the denominator refers to the collector circuit. For this reason, the cross terms $r_{12}i_C$ and $r_{21}i_E$ are put into the equivalent circuit as voltage generators. They produce the interaction between input and output circuits. Equations (12.6) produce the equivalent circuit shown in Fig. 12.3(b). Inspection of Fig. 12.3(b) will show that Eqs. (12.6) will result from a simple summation of voltage drops. Thus the open-circuit parameters look like two equivalent circuits of Thévenin's theorem. Essentially, what has been derived is the fact that the transistor circuit of Fig. 12.3(a) can be replaced by the circuit of Fig. 12.3(b).

The parameters r_{11}, r_{12}, r_{21}, and r_{22} are designated open-circuit parameters because of the way they are obtained experimentally. For example, looking at Eqs. (12.6), if

$$i_C = 0$$

(12.7)

Then

$$e_{EB} = r_{11}i_e$$

$$e_{CB} = r_{21}i_e$$

(12.8)

or

$$r_{11} = \left(\frac{e_{EB}}{i_E}\right)_{i_C=0}$$

(12.9)

$$r_{21} = \left(\frac{e_{CB}}{i_E}\right)_{i_C=0}$$

(12.10)

Equation (12.9) suggests that the circuit shown in Fig. 12.4(a) can be used to find r_{11} and r_{21}. If the collector circuit is open-circuited while a generator of voltage e_{EB} is applied to the input terminals, the measurement of I_E, e_{CB}, and e_{EB} will be enough to calculate r_{11} and r_{21}, using Eqs. (12.9) and (12.10). The quantities r_{12} and r_{22} can be obtained from measurements indicated in Fig. 12.4(b). The emitter circuit is open-circuited and a generator e_{CB} is applied across the output terminals. In this case,

$$i_E = 0$$

(12.11)

and Eqs. (12.6) become

$$r_{22} = \left(\frac{e_{CB}}{i_C}\right)_{i_E=0}$$

(12.12)

$$r_{12} = \left(\frac{e_{EB}}{i_C}\right)_{i_E=0}$$

(12.13)

The measurements of e_{EB}, e_{CB}, and i_C in Fig. 12.4(b) are enough, with the use of Eqs. (12.12) and (12.13), to determine r_{22} and r_{12}.

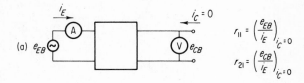

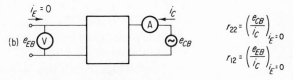

Fig. 12.4 Test methods for obtaining open circuit parameters.

As can be seen, these parameters are obtained by alternately opening the input and the output circuits, and making the appropriate measurements. In practice, the circuits cannot be physically opened, since this would disrupt the dc biasing. What one does is place a large value of inductance alternately in series with the input and output circuits. The inductor will make i_C and i_E approach zero but not disturb the dc voltages and currents.

SHORT-CIRCUIT PARAMETERS

To show how the choice of dependent variables determines the ac equivalent circuit, assume that the currents I_C and I_E of Fig. 12.2 are the dependent parameters. The functional dependence between the four parameters becomes

$$I_E = f_1(E_{EB}, E_{CB})$$
$$I_C = f_2(E_{EB}, E_{CB})$$

(12.14)

Taking the total derivative of Eqs. (12.14),

$$dI_E = \left(\frac{\partial I_E}{\partial E_{EB}}\right)_{E_{CB}} dE_{EB} + \left(\frac{\partial I_E}{\partial E_{CB}}\right)_{E_{EB}} dE_{CB}$$

$$dI_C = \left(\frac{\partial I_E}{\partial E_{EB}}\right)_{E_{CB}} dE_{EB} + \left(\frac{\partial I_E}{\partial E_{CB}}\right)_{E_{EB}} dE_{CB}$$

(12.15)

Using the small signal approximations for the increments,

$$dI_E = i_E \qquad dE_{EB} = e_{EB}$$
$$dI_C = i_C \qquad dE_{CB} = e_{CB}$$

(12.16)

with the definitions

$$g_{11} = \left(\frac{\partial I_E}{\partial E_{EB}}\right)_{E_{CB}} \qquad g_{12} = \left(\frac{\partial I_E}{\partial E_{CB}}\right)_{E_{EB}}$$

$$g_{21} = \left(\frac{\partial I_C}{\partial E_{EB}}\right)_{E_{CB}} \qquad g_{22} = \left(\frac{\partial I_C}{\partial E_{CB}}\right)_{E_{EB}}$$

(12.17)

Equations (12.15) become

$$i_E = g_{11}e_{EB} + g_{12}e_{CB}$$

(12.18)

$$i_C = g_{21}e_{EB} + g_{22}e_{CB}$$

(12.19)

The units for the g_{11} are amp/volts = mho. They are like transconductances. The parameters g_{11} and g_{22} are then equivalent to reciprocals of true resistance, since both denominator and numerator of each refer to a given circuit. For example, the numerator and denominator of g_{11} refer to the input circuit. For this reason, the parameters g_{11} and g_{22} are placed into the equivalent circuit as equivalent resistances in parallel with the terms $g_{12}e_{CB}$ and $g_{21}e_{EB}$, respectively. These latter terms have the units of current and appear in the equivalent circuit as current generators. Figure

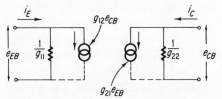

Fig. 12.5 Short circuit parameters and equivalent circuit.

12.5 shows the equivalent circuit resulting from short-circuit parameters. Summing up the currents at branch points in Fig. 12.5 will produce Eqs. (12.18) and (12.19). Thus the transistor of Fig. 12.2 has been replaced by the ac equivalent circuit of Fig. 12.5.

The reason for designating g_{11}, g_{12}, g_{21}, and g_{22} as short-circuit parameters is due to the method used to determine them experimentally. Looking at Eqs. (12.18) and (12.19), if $e_{CB} = 0$, then

$$g_{11} = \left(\frac{i_E}{e_{EB}}\right)_{e_{CB}=0} \tag{12.20}$$

$$g_{21} = \left(\frac{i_C}{e_{EB}}\right)_{e_{CB}=0} \tag{12.21}$$

The measurements indicated by Eqs. (12.20) and (12.21) are shown in Fig. 12.6(a). The output is short-circuited, a generator e_{EB} is connected to the input terminals, and i_E and i_C are measured. This is enough, using Eqs. (12.20) and (12.21), to determine g_{11} and g_{21}. Similarly, if $e_{EB} = 0$ in Eqs. (12.18) and (12.19), then

$$g_{12} = \left(\frac{i_E}{e_{CB}}\right)_{e_{EB}=0} \tag{12.22}$$

$$g_{22} = \left(\frac{i_C}{e_{CB}}\right)_{e_{EB}=0} \tag{12.23}$$

The appropriate measuring configuration for these parameters is indicated in Fig. 12.6(b), where the input is short-circuited, a generator e_{CB} is applied across the output terminals, and i_E and i_C are measured. Application of Eqs. (12.22) and (12.23) will then yield g_{12} and g_{22}.

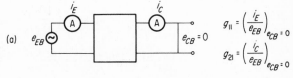

Fig. 12.6 Test methods for obtaining short circuit parameters.

Actually, the transistor is not short-circuited, for this would alter the dc biases. In practice, the short circuit for ac voltages is accomplished by placing a large capacitor alternately across the input and the output circuits. This procedure does not affect the dc voltages and currents. The capacitor will place the emitter at ac ground when across the input terminals and will place the collector at ac ground when placed across the output terminals.

12.3   Hybrid Parameters

In the open-circuit parameters, the measurement of r_{11} and r_{12} by making $i_C = 0$ is made difficult because of the low frequency used for the external generator, 1000 cps. This means that the inductor placed in series with the output circuit (to make $i_C = 0$) must have a larger reactance than the 50-K to 2-meg collector circuit resistance. The inductor would therefore have to be very large physically. Similarly, for the short-circuit parameters, in the measurement of g_{21} and g_{22}, e_{EB} must be made equal to zero. This is accomplished by shorting the input with a capacitor. Since the input resistance of a transistor is of the order of 50 to 200 ohms, this means that the capacitor must be very large in order to make its reactance low compared with the input resistance at 1000 cps, the test frequency. This had led to the definition of a hybrid set of variables. This set of parameters allows for easy measurement, since it will require $i_E = 0$ and $e_{CB} = 0$, both easily accomplished with reasonably sized components.

To obtain the hybrid set of parameters, E_{EB} and I_C are chosen as the dependent variables.

$$E_{EB} = f_1(E_{CB}, I_E)$$
$$I_C = f_2(E_{CB}, I_E) \tag{12.24}$$

Taking the total derivative of each variable,

$$dE_{EB} = \left(\frac{\partial E_{EB}}{\partial I_E}\right)_{E_{CB}} dI_E + \left(\frac{\partial E_{EB}}{\partial E_{CB}}\right)_{I_E} dE_{CB}$$

$$dI_C = \left(\frac{\partial I_C}{\partial I_E}\right)_{E_{CB}} dI_E + \left(\frac{\partial I_C}{\partial E_{CB}}\right) dE_{CB} \tag{12.25}$$

Defining the partial derivatives as

$$h_{11} = \left(\frac{\partial E_{EB}}{\partial I_E}\right)_{E_{CB}} \qquad h_{12} = \left(\frac{\partial E_{EB}}{\partial E_{CB}}\right)_{I_E}$$

$$h_{21} = \left(\frac{\partial I_C}{\partial I_E}\right)_{E_{CB}} \qquad h_{22} = \left(\frac{\partial I_C}{\partial E_{CB}}\right)_{I_E} \tag{12.26}$$

and using the small signal approximations for the differentials, Eqs. (12.25) become

$$e_{EB} = h_{11}i_E + h_{12}e_{CB} \qquad (12.27)$$

$$i_C = h_{21}i_E + h_{22}e_{CB} \qquad (12.28)$$

The ac equivalent circuit suggested by Eqs. (12.27) and (12.28) is shown in Fig. 12.7. The quantity h_{12} is a ratio of voltage and is unitless, and the quantity h_{21} is a ratio of currents and is unitless.

Fig. 12.7 Hybrid parameters and the equivalent circuit.

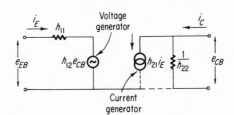

The parameter h_{11} is essentially the input-circuit resistance and the ratio $1/h_{22}$ comprises the output-circuit resistance. The ac equivalent circuit has turned out to be a composite of the open-circuit and short-circuit equivalent circuits. The voltage generator $h_{12}e_{CB}$ shows the interaction between output and input circuits, and the current generator $h_{21}I_E$ shows the interaction between input and output circuits. The parameter h_{21} should be recognized as the definition for α, given in the preceding chapter.

To obtain the parameters experimentally, i_E can be made equal to zero with a moderate value inductance in series with the input circuit, and e_{CB} can be made equal to zero by shunting the output with a relatively small capacitor. Another measurement advantage of the hybrid parameters is the fact that the quantities h_{11}, h_{12}, h_{21}, h_{22} can be obtained graphically from the dc input and output characteristics of the transistor. A set of typical characteristics for a transistor is shown in Fig. 12.8. Figure 12.8(a) and (b) are the conventional collector and emitter characteristics in the grounded base configuration.

PARAMETER h_{11}

The parameter h_{11} can be obtained from Fig. 12.8(b). The definition of h_{11} is given by

$$h_{11} = \left(\frac{\partial E_{EB}}{\partial I_E}\right)_{E_{CB}} \qquad (12.29)$$

This is seen to be the slope of the E_{CB} = const curve of Fig. 12.8(b).

From previous experience with graphical constructions of transfer

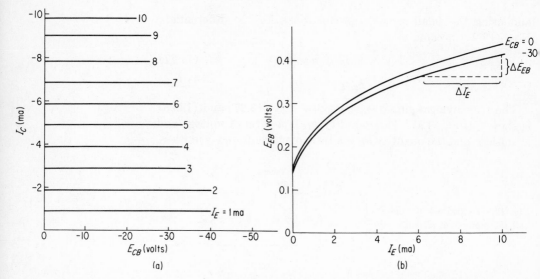

Fig. 12.8 Typical grounded base dc characteristics.

curves in Chapter 11, it is known that the transistor will be operated in the linear portion of the input characteristics. This is in the region $E_{EB} = 0.350$ volt to $E_{EB} = 0.450$ volt. There is no great difference in the slope of the $E_{CB} = 0$ and the $E_{CB} = -30$-volt curves, shown in Fig. 12.8(b). Therefore, arbitrarily, the $E_{CB} = -30$-volt curve will be used. For $I_E = 8.0$ ma, the slope of the E_{EB} and I_E curve can be approximated by the triangle constructed in Fig. 12.8(b):

$$\Delta E_{EB} = (0.413 - 0.363) \text{ volts} = 0.05 \text{ volt}$$

$$\Delta I_E = (10 - 6) \times 10^{-3} \text{ amp} = 4 \times 10^{-3} \text{ amp}$$

$$h_{11} \cong \left(\frac{\Delta E_{EB}}{\Delta I_E}\right)_{E_{CB}=-30} = \frac{0.05}{4 \times 10^{-3}} = 12.5 \text{ ohms}$$

PARAMETER h_{12}

For h_{12}, the approximation to be used is

$$h_{12} = \left(\frac{\Delta E_{EB}}{\Delta E_{CB}}\right)_{I_E=6 \text{ ma}}$$

This is also evaluated from the input characteristics.

For the line $I_E = 6$ ma, the respective changes in E_{EB} and E_{CB} are

$$E_{EB} = (0.363 - 0.391) \text{ volt}$$

$$E_{CB} = (30 - 0) \text{ volts}$$

$$h_{12} = 930 \times 10^{-6}$$

PARAMETER h_{21}

For h_{21}, the output characteristics must be referred to because of the definition for h_{21}:

$$h_{21} = \left(\frac{\Delta I_C}{\Delta I_E}\right)_{ECB}$$

For the case $E_{CB} = -10$ volts, $I_C = -6$ ma, the respective changes in I_C and I_E become

$$\Delta I_C = (-4.97 + 5.90) \times 10^{-3} \text{ amp} \Big]$$

$$\Delta I_E = (4 - 5) \times 10^{-3} \text{ amp} \qquad \Big\}_{ECB=-10}$$

$$h_{21} = -0.97$$

PARAMETER h_{22}

For h_{22}, the output characteristics are again referred to. Since h_{22} is defined approximately by

$$h_{22} = \left(\frac{\Delta I_C}{\Delta E_{CB}}\right)_{I_E}$$

it can be seen, referring to Fig. 12.8(a), that h_{22} is given by the slope of the $I_E = $ const curves. Using $I_E = 6$ ma, the respective changes in I_C and E_{CB} are found to be

$$\Delta I_C = (-5.95 + 5.88) \times 10^{-3} \text{ amp} \Big]$$

$$\Delta E_{CB} = (-30 + 0) \text{ volts} \qquad \Big\}_{I_B=6 \text{ ma}}$$

$$h_{22} = 2.3 \times 10^{-6} \text{ ohm}$$

The equivalent circuit for the transistor illustrated in Fig. 12.8 is shown in Fig. 12.9 with the appropriate h parameters. The polarity of e_1 is as shown because h_{12} was positive. The values thus found show that in the grounded base configuration, the transistor has a small ac input resistance, h_{11}, and a relatively high output resistance, $1/h_{22}$.

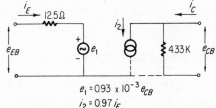

Fig. 12.9 Typical hybrid parameter values.

$$e_1 = 0.93 \times 10^{-3}\, e_{CB}$$
$$i_2 = 0.97\, i_E$$

12.4 *T* Type of Equivalent Circuit

It has been shown how the choice of dependent variables determines the ac equivalent circuit. Also, experimentally, it is difficult to measure some of these derived parameters. Then the hybrid parameters were introduced. It was shown that they can be measured easily and that they can be obtained graphically, if need be. Manufacturers have almost uniformly adopted the specification of hybrid parameters in their data sheets. It is becoming increasingly popular, however, to use another ac equivalent circuit for the transistor, one that is different from the three discussed so far. This circuit is most like the open-circuit parameters, being obtained from the open-circuit equivalent circuit by an algebraic manipulation. This ac equivalent circuit is called the *T* type of equivalent circuit, and is shown in Fig. 12.10.

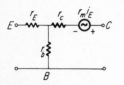

Fig. 12.10 The *T*-type equivalent circuit.

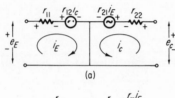

(a)

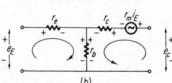

(b)

Fig. 12.11 Converting from open circuit to *T* circuit parameters.

To obtain the parameters r_e, r_b, r_c, and r_m from the open-circuit parameter, consider Fig. 12.11. Figure 12.11(a) is the open-circuit ac equivalent circuit, redrawn, and Fig. 12.11(b) is the *T* type of equivalent circuit. Writing down the equations for both circuits gives the open-circuit equations

$$e_E = i_E r_{11} + i_C r_{12} \tag{12.30}$$

$$e_C = i_E r_{21} + i_C r_{22} \tag{12.31}$$

and the *T* type of equations

$$e_E = i_E(r_e + r_b) + i_C r_b \tag{12.32}$$

$$e_C = i_E(r_m + r_b) + i_C(r_c + r_b) \tag{12.33}$$

Comparing Eqs. (12.30) and (12.32),

$$r_{11} = r_e + r_b$$
$$r_{12} = r_b \tag{12.34}$$

Comparing Eqs. (12.31) and (12.33),

$$r_{21} = r_m + r_b$$
$$r_{22} = r_c + r_b$$

(12.35)

Solving Eqs. (12.34) and (12.35) for the T type of parameter,

$$r_e = r_{11} - r_{12}$$
$$r_b = r_{12}$$
$$r_c = r_{22} - r_{12}$$
$$r_m = r_{21} - r_{12}$$

(12.36)

What has been shown is that the T type of equivalent circuit is a legitimate ac equivalent circuit, since it can be derived from the open-circuit parameters. In doing this, the transformation equations between the grounded base open-circuit parameters and the T-type parameters have been derived. One of the advantages of the T type of ac equivalent circuit is the ease in using it to derive analytical results. Also, the circuit intuitively appears to be what one would expect for a transistor, based on experience gained from work on graphical constructions.

12.5 Transformation Equations

More and more manufacturers are supplying the T-type parameters, along with the hybrid parameters, in their specification sheets. In the other cases, one has to convert from the hybrid parameters (which are easy to measure) to the T-type parameters (which are easy to use in calculations). It is the purpose of this section to show how easy it is to go from one set of parameters to another, by deriving the necessary transformation equations.

To start the derivation, both circuits are drawn. In this case they are the hybrid and T-type circuits, as shown in Fig. 12.12. Then the circuital equations are written down for the two cases. Next the two sets of equations obtained are put into the same form and the transformations are read off by comparing factors.

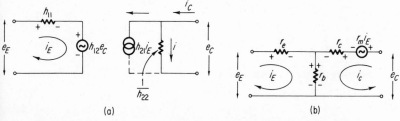

(a)   (b)

Fig. 12.12 Converting from hybrid circuit to T circuit parameters.

For the T type of circuit:

$$e_E = i_E(r_e + r_b) + i_C r_b \tag{12.37}$$

$$e_C = i_E(r_m + r_b) + i_C(r_c + r_b) \tag{12.38}$$

For the hybrid circuit:

$$e_E = i_E h_{11} + e_C h_{12} \tag{12.39}$$

$$i_C = i_E h_{21} + i \tag{12.40}$$

where

$$i = e_C h_{22} \tag{12.41}$$

Eliminating i between Eqs. (12.40) and (12.41),

$$e_C h_{22} = -i_E h_{21} + i_C \tag{12.42}$$

In order to be able to read off the transformation set, Eqs. (12.39) and (12.42) must be put in terms of i_E and i_C so that they will look like Eqs. (12.37) and (12.38).

Eliminating e_C from Eqs. (12.39) and (12.42),

$$e_E = i_E \left(h_{11} - \frac{h_{12} h_{21}}{h_{22}} \right) + i_C \frac{h_{12}}{h_{22}} \tag{12.43}$$

and solving Eq. (12.42) for e_C,

$$e_C = -i_E \frac{h_{21}}{h_{22}} + i_C \frac{1}{h_{22}} \tag{12.44}$$

Equations (12.43) and (12.44) for the hybrid circuit are now in the form of Eqs. (12.37) and (12.38) for the T type of circuit. Comparing Eqs. (12.37) and (12.43),

$$r_e + r_b = h_{11} - \frac{h_{12} h_{21}}{h_{22}}$$

$$r_b = \frac{h_{12}}{h_{22}} \tag{12.45}$$

Comparing Eqs. (12.38) and (12.44),

$$r_m + r_b = -\frac{h_{21}}{h_{22}}$$

$$r_b + r_c = \frac{1}{h_{22}} \tag{12.46}$$

Solving Eqs. (12.45) and (12.46) for the T-type parameters,

$$r_e = h_{11} - \frac{h_{12}}{h_{22}} (h_{21} + 1) \tag{12.47}$$

$$r_b = \frac{h_{12}}{h_{22}} \tag{12.48}$$

$$r_c = \frac{1 - h_{12}}{h_{22}} \tag{12.49}$$

$$r_m = -\frac{h_{21} + h_{12}}{h_{22}} \tag{12.50}$$

Using the values for the h parameters, as found graphically in Sec. 12.4, the values of the T parameters can be calculated from the equations derived above. The results are shown in Table 12.1.

TABLE 12.1

h Parameters	T Parameters, ohms
$h_{11} = 12.5$ ohms	$r_e = 5.2$
$h_{12} = 560 \times 10^{-6}$	$r_b = 243$
$h_{21} = -0.97$	$r_c = 435$ K
$h_{22} = 2.3 \times 10^{-6}$ mho	$r_{\text{in}} = 422$ K

Example 12.1 Using the specification sheet for a 2N76 PNP transistor (see Appendix), find the T parameters for this transistor.

1.   From the specifications labeled "Design Center," the following data are obtained:

Output admittance (input open circuit), h_{22} 1.0 μmhos

Current amplification (output short circuit), h_{21} -0.95

Input impedance (output short circuit), h_{11} 32 ohms

Voltage feedback ratio (input open circuit), h_{12} 3×10^{-4}

Remember that the specifications are given with respect to the common base configuration. From the above data, the values for the h parameters are

$$h_{11} = 32 \text{ ohms}$$

$$h_{12} = 3 \times 10^{-4}$$

$$h_{21} = -0.95$$

$$h_{22} = 1 \times 10^{-6} \text{ mho}$$

2. To obtain r_e, use Eq. (12.47):

$$r_e = h_{11} - \frac{h_{12}}{h_{22}} (h_{21} + 1)$$

$$= 32 - \frac{3 \times 10^{-4}}{1 \times 10^{-6}} (-0.95 + 1)$$

$$= 17 \text{ ohms}$$

3. For r_b, use Eq. (12.48).

$$r_b = \frac{h_{12}}{h_{22}}$$

$$= \frac{3 \times 10^{-4}}{1 \times 10^{-6}}$$

$$= 300 \text{ ohms}$$

4. For r_c, use Eq. (12.49):

$$r_c = \frac{1 - h_{12}}{h_{22}}$$

$$= \frac{1 - 3 \times 10^{-4}}{1 \times 10^{-6}}$$

$$= 1 \text{ meg}$$

5. For r_m, use Eq. (12.50):

$$r_m = -\frac{h_{12} + h_{21}}{h_{22}}$$

$$= \frac{0.95 - 3 \times 10^{-4}}{1 \times 10^{-6}}$$

$$= 950 \text{ K}$$

12.6 AC Equivalent-Circuit Calculations

Transistors can furnish not only a voltage gain, but also a significant current gain. This makes the transistor useful in servo systems, where most mechanical devices are actuated by large current. Voltage gain and current gain will be calculated for the transistor configurations. Another simple step is to combine these two gains to form the power gain for a stage. In addition, for impedance matching in both voltage- and current-gain applications, the input and output impedances will be calculated for each configuration. Using Fig. 12.13(a) as a general example, the voltage gain A_v is given by

$$A_v = \frac{e_{\text{out}}}{e_{\text{in}}} \qquad (12.51)$$

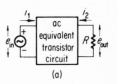

(a)

The current gain A_i is given by

$$A_i = \frac{i_2}{i_1} \qquad (12.52)$$

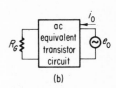

(b)

The input resistance can also be obtained from Fig. 12.13(a):

$$r_{\text{in}} = \frac{e_{\text{in}}}{i_1} \qquad (12.53)$$

Fig. 12.13 (a) Configuration for input impedance calculations. (b) Configuration for output impedance calculations.

The output resistance is a function of the resistance of the preceding stage and the transistor under discussion. The load resistance is excluded from this calculation. It is always in parallel with the output resistance thus found. Therefore it can be taken into account in finding the total resistance, as seen by the next stage. The output resistance is found by shunting the input of the transistor with the output resistance of the preceding stage, removing the load resistor, and placing a generator e_0 across the load terminals. The resultant current drawn from the generator e_0 is i_0. The output resistance is then defined as

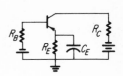

$$r_{\text{out}} = \frac{e_0}{i_0} \qquad (12.54)$$

Sometimes the power gain for a circuit is needed; since $A_v = e_{\text{out}}/e_{\text{in}}$ and $A_i = i_2/i_1$, the product will give the power gain

$$A_p = A_v A_i \qquad (12.55)$$

In the analytical work to follow (that is, converting a transistor circuit into its ac equivalent circuit), the transistor is replaced by the T-parameter equivalent circuit; then the other circuit components having appreciable ac voltage drops are drawn in, and the calculations carried out.

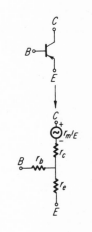

Example 12.2 Given the circuit shown, top, right, draw the ac equivalent circuit.

1. The transistor is replaced by its equivalent T type of circuit.

2. Now the remainder of the circuit having an appreciable ac voltage drop is drawn in, bottom, right.

3. Note that the ac equivalent circuit was oriented with the transistor leads E, B, and C. R_E and C_E do not appear because C_E shunts R_E for ac voltages and can be represented by a short circuit for ac voltages and currents. The batteries also appear as

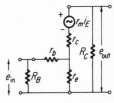

short circuits, since the ac voltage drops across them are essentially zero. This leaves only R_B and R_C in the external circuit.

For this grounded emitter circuit, the signal input would be across R_B, and the amplified output voltage would appear across R_C. Also, a glance at the final circuit will show that current will flow in the input as well as the output circuit.

12.7 Grounded Emitter Configuration

A typical circuit for a grounded-emitter amplifier is shown in Fig. 12.14(a). R_1 and R_2 form the biasing network, R_C is the load resistor, and the R_E and C_E combination is the current feedback, or the swamping circuit. R_E is bypassed by C_E so that R_E is taken into account only in the dc-biasing calculations. In Fig. 12.14(b), a generator having an emf of e_g and an internal resistance R_G is placed across the input terminals. To convert this circuit into an ac equivalent circuit, the transistor is replaced by its ac equivalent T-type circuit between the points E, B, and C. Then the remainder of the circuit is drawn in. Components having negligibly small ac voltage drops across them are replaced by short circuits. The resultant ac equivalent circuit is shown in Fig. 12.14(c). R_1 and R_2 are in parallel across the input of the transistor. The voltage e_{in} is the fraction of e_g that appears across the input terminals of the transistor and is amplified. Usually the network consisting of R_G, R_1, and R_2 is treated separately, and the calculation of A_v is given as defined by Eq. (12.51):

$$A_v = \frac{e_{out}}{e_{in}}$$

where e_{in} can always be obtained in terms of e_g by the relation

$$e_{in} = e_g \left[\frac{R_1 R_2/(R_1 + R_2)}{R_G + [R_1 R_2/(R_1 + R_2)]} \right] \qquad (12.56)$$

Therefore the calculations are made on the circuit shown in Fig. 12.15.

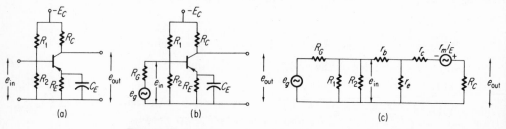

Fig. 12.14 Reduction of an amplifier circuit into a T-type ac equivalent circuit.

CURRENT GAIN

Drawing in loop current i_1 and i_2, and indicating the voltage-drop polarities, the following equations are obtained from Fig. 12.15:

$$i_E = i_1 + i_2$$

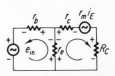

Fig. 12.15 Simplified ac equivalent circuit for the grounded emitter.

Note that the positive value for i_e is given by i_e entering the emitter terminal. This was established when the ac equivalent circuit was derived; it also fixes the polarity of the voltage generator $r_m i_E$.

$$e_{in} + i_1(r_b + r_e) + i_2 r_e = 0$$
$$r_m(i_1 + i_2) - i_2(R_C + r_c + r_e) - i_1 r_e = 0 \qquad (12.57)$$

Putting the equations in ordered form,

$$i_1(r_b + r_e) + i_2 r_e = -e_{in}$$
$$i_1(r_m - r_e) - i_2(R_C + r_c - r_m + r_e) = 0 \qquad (12.58)$$

Using the definitions given by Eqs. (12.52) and (12.58), i_1 and i_2 can be solved for (and a ratio formed)

$$i_1 = \frac{\begin{vmatrix} -e_{in} & r_e \\ 0 & -(R_C + r_c - r_m + r_e) \end{vmatrix}}{|D|} = \frac{e_{in}(R_C + r_c - r_m + r_e)}{|D|}$$

$$i_2 = \frac{\begin{vmatrix} r_b + r_e & -e_{in} \\ r_m - r_e & 0 \end{vmatrix}}{|D|} = \frac{e_{in}(r_m - r_e)}{|D|}$$

The denominator is not worked out because it will cancel in the ratio formed to obtain A_i:

$$A_i = \frac{i_2}{i_1} = \frac{r_m - r_e}{R_C + r_c - r_m + r_e}$$

Again a simplification can be made by realizing that r_e is small compared with the other terms. A_i becomes

$$A_i \cong \frac{r_m}{R_C + r_c - r_m} \qquad (12.59)$$

Using the parameters for a 2N76, and $R_C = 10$ K, A_i becomes

$$A_i = 15.8$$

VOLTAGE GAIN

The voltage gain is given by

$$A_v = \frac{e_{out}}{e_{in}} = \frac{i_2 R_C}{e_{in}} \qquad (12.60)$$

The current i_2 must be known. Solving Eqs. (12.58) for i_2,

$$i_2 = \frac{\begin{vmatrix} r_b + r_e & -e_{in} \\ r_m - r_e & 0 \end{vmatrix}}{\begin{vmatrix} r_b + r_e & r_e \\ r_m - r_e & -(R_C + r_c - r_m + r_e) \end{vmatrix}}$$

$$= \frac{-e_{in}(r_m - r_e)}{(r_b + r_e)(R_C + r_c - r_m + r_e) + r_e(r_m - r_e)}$$

At this point several simplifications can be made because of the relative orders of magnitude of the T parameters. For example, $r_e \ll r_m$, and $r_e \ll (r_c - r_m)$. Therefore i_2 can be rewritten as

$$i_2 = \frac{-e_m r_m}{r_b(R_C + r_c - r_m) + r_e(R_C + r_c)}$$

$$= -\left(\frac{r_m}{R_C + r_c - r_m}\right)\left[\frac{e_{in}}{r_b + r_e[(R_C + r_c)/(R_C + r_c - r_m)]}\right]$$

$$\text{(12.61)}$$

Using Eq. (12.59),

$$i_2 = -\frac{A_i e_{in}}{r_b + r_e[(R_C + r_c)/(R_C + r_c - r_m)]} \qquad (12.62)$$

Putting this in the equation for A_v, Eq. (12.60),

$$A_v = -\frac{A_i R_C}{r_b + r_e[(R_C + r_c)/(R_C + r_c - r_m)]} \qquad (12.63)$$

With the parameters for a 2N76 as given in Example 12.1, and $R_C = 10$ K,

$$A_v = -268$$

The voltage gain for a grounded emitter is high, similar to a pentode. The negative sign shows that there is a 180-deg phase reversal between input and output signals. It should be mentioned that the gain derived is for the simplified transistor circuit, Fig. 12.15, and that the gain in an actual circuit will be lower than this value when the bias and coupling networks are taken into account.

INPUT RESISTANCE

One needs i_1 and e_{in} to form the ratio $r_{in} = e_{in}/i_1$; using Eq. (12.53) and the circuit Eqs. (12.58), i_1 is found to be

$$i_1 = \frac{e_{in}(R_C + r_c - r_m + r_e)}{|D|}$$

Evaluating $|D|$,

$$i_1 = \frac{-e_{in}(R_C + r_c - r_m + r_e)}{(r_b + r_e)(R_C + r_c - r_m + r_e) + r_e(r_m - r_e)} \quad (12.64)$$

Since i_1 turns out to be negative, this means that in Fig. 12.15, i_1 should be going in the other direction. This can be taken care of by noting this and writing r_{in} as

$$r_{in} = \frac{e_{in}}{|i_1|}$$

$$= \frac{(r_b + r_e)(R_C + r_c - r_m + r_e) + r_e(r_m - r_e)}{R_C + r_c - r_m + r_e} \quad (12.65)$$

As can be seen, the equations are not "neat." Some of them are quite complicated, but simplifying assumptions help reduce them in many cases. In Eq. (12.65) using $r_e \ll r_m$ and $r_e \ll (r_c - r_m)$, r_{in} can be written as

$$r_{in} = r_b + r_e + \frac{r_e r_m}{R_C + r_c - r_m} \quad (12.66)$$

Grouping terms common in r_e,

$$r_{in} = r_b + r_e\left(1 + \frac{r_m}{R_C + r_c - r_m}\right)$$

$$= r_b + r_e(1 + A_i) \quad (12.67)$$

The input resistance is a function of the load resistor R_C. To get a feel for the order of magnitude of r_{in}, it is calculated using $R_C = 10$ K and the parameters for a 2N76 transistor:

$$r_{in} = 300 + 17(1 + 15.8)$$

$$= 586 \text{ ohms}$$

Compared to vacuum tube circuits, this is extremely low, but it also shows that the parallel combination of R_1 and R_2 in Fig. 12.14(c). can be fairly small before it begins to shunt the input resistance.

OUTPUT RESISTANCE

To calculate the output resistance of the amplifier under consideration, the definition indicated in Fig. 12.13(b) must be used. Redrawing Fig. 12.15 to conform to this definition, one obtains Fig. 16(a). R_G represents the output resistance of the previous stage or generator, R_1 and R_2 are the biasing resistors of Fig. 12.14(a), and R_{in} is the parallel combination of R_G, R_1, and R_2. Representing these three resistors by R_{in} produces the diagram in Fig. 12.16(b). The calculations are performed on this circuit.

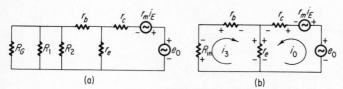

Fig. 12.16 AC equivalent circuit for output impedance calculations on the grounded emitter.

Drawing loop currents i_0 and i_3, and showing voltage-drop polarities, the circuit equations can be written

$$i_e = -i_0 - i_3$$

$$e_0 - r_m(-i_0 - i_3) - i_0(r_c + r_e) - i_3 r_e = 0 \quad (12.68)$$

$$i_3(R_{in} + r_b + r_e) - i_0 r_e = 0$$

Rearranging the last two equations for solution by determinants,

$$\begin{aligned} i_0(r_m - r_c - r_e) + i_3(r_m - r_e) &= -e_0 \\ -i_0 r_e \qquad\qquad + i_3(R_{in} + r_b + r_e) &= 0 \end{aligned} \quad (12.69)$$

Solving for i_0,

$$i_0 = \frac{\begin{vmatrix} -e_0 & (r_m - r_e) \\ 0 & (R_{in} + r_b + r_e) \end{vmatrix}}{\begin{vmatrix} (r_m - r_c - r_e) & (r_m - r_e) \\ -r_e & (R_{in} + r_b + r_e) \end{vmatrix}}$$

$$= \frac{-e_0(R_{in} + r_b + r_e)}{(r_m - r_c - r_e)(R_{in} + r_b + r_e) + r_e(r_m - r_e)} \quad (12.70)$$

Forming the ratio e_0/i_0 in Eq. (12.70),

$$r_{out} = \frac{e_0}{i_0} = r_e + r_c - r_m + \frac{r_e^2 - r_e r_m}{R_{in} + r_b + r_e} \quad (12.71)$$

Again, for all practical purposes, the r_e and r_e^2 terms can be neglected; r_{out} becomes

$$r_{out} = r_c - r_m - \frac{r_e r_m}{R_{in} + r_b + r_e} \quad (12.72)$$

For the 2N76 transistor, assuming a value $R_{in} = 5$ K,

$$r_{out} \cong 47 \text{ K}$$

The total resistance as seen by the next stage will be r_{out} and R_C in parallel. Using $R_C = 10$ K, the total output resistance R_{out} becomes

$$R_{out} = \frac{R_C r_{out}}{R_C + r_{out}} \cong 8.25 \text{ K}$$

The effect of R_C on A_v, A_i, and r_{in} should be shown for the grounded emitter configuration. This is done in the following examples.

Example 12.3 To show how A_i varies with R_C for a 2N76 transistor.

1. With the use of Eq. (12.66),

$$A_i = \frac{r_m}{R_C + r_c - r_m}$$

The following tabulated values are obtained.
2.

R_C, ohms	A_i
500	18.8
1000	18.6
5000	17.3
10000	15.8
50000	9.5

3. Obviously, the greater R_C, the smaller the current through R_C, and hence the smaller A_i. For $R_C \rightarrow 0$, one obtains the conditions for the definition of β, Eq. (11.2). Equation (12.66) reduces to

$$\lim_{R_C \rightarrow 0} A_i = \beta = \frac{r_m}{r_c - r_m}$$

$$\beta = \frac{r_m/r_c}{1 - (r_m/r_c)}$$

For the transistor, the dc currents are related by

$$I_C + I_B = I_E$$

$$I_C = \alpha I_E$$

$$I_C = \beta I_B$$

Putting the first equation in terms of I_B by using the other two equations, one obtains

$$\beta I_B + I_B = \frac{\beta I_B}{\alpha}$$

Solving for β,

$$\beta = \frac{\alpha}{1 - \alpha}$$

Comparing this equation with the one obtained using ac measurements, α can be determined from the ratio

$$\alpha = \frac{r_m}{r_c}$$

Thus

$$\beta = \frac{r_m}{r_c - r_m}$$

and

$$\alpha = \frac{r_m}{r_c}$$

For the 2N76,

$$\alpha = 0.97$$

$$\beta = 32.3$$

Example 12.4   Show the effect of R_C on the voltage gain for the grounded emitter. Use the 2N76 as an example.

1.   Equation (12.65), derived by neglecting r_e, is used to obtain A_v;

$$A_v = -\frac{R_C}{r_b} \frac{1}{\{[(R_C + r_c)/r_m] + 1\}}$$

2.   The resulting calculations are tabulated below for several values of R_C:

R_C, ohms	A_v
500	−14.8
1,000	−29.4
5,000	−142
10,000	−268
50,000	−993

3.   As can be seen, the gain increases almost linearly with R_C. The transistor can produce very large voltage gains per stage.

Example 12.5 r_{in} for the 2N76 can be calculated using Eq. (12.67).

1.

$$r_{in} = r_b + r_e + \frac{r_e r_m}{R_C + r_c - r_m}$$

2. The resultant values of r_{in} for several values of R_C are tabulated below:

R_C, ohms	r_{in}, ohms
500	636
1,000	634
5,000	610
10,000	586
50,000	478

3. As can be seen, r_{in} does not vary too much with a change in R_C.

For the grounded emitter, the analytical solutions show that there is a substantial voltage gain and a moderate current gain. The input resistance is low and the output resistance is moderately high. This information will be useful when multistage calculations are made later.

12.8 The Grounded Base Configuration

A typical circuit diagram for the grounded base configuration is shown in Fig. 12.17(a). R_E is the emitter return resistor used to establish the bias between emitter and base, R_C is the load resistor, and R_1 and R_2 form the primary bias circuit. A generator is placed across the input terminals for ac calculations, Fig. 12.17(b). R_G represents the internal resistance of the generator, and e_g its emf.

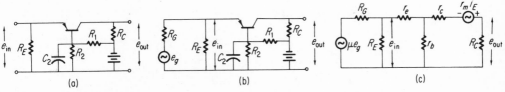

Fig. 12.17 Reduction of the grounded base circuit into its ac equivalent circuit.

To obtain the ac equivalent circuit, the transistor is replaced by its equivalent circuit, and all components having negligible ac-voltage drops are short-circuited. The result is the diagram in Fig. 12.17(c). The calculations, however, are usually made in relation to e_{in}. Since e_{in} can be obtained from the relation

$$e_{in} = e_g \left(\frac{R_E}{R_G + R_E} \right) \qquad (12.73)$$

there is no problem in using e_{in} for the analytical approach. It should be kept in mind that the calculations will give the voltage and current gain, as well as the input resistance for the transistor circuit, independent of the coupling circuit, that is, R_G and R_E. These resistors will be taken care of when multistage circuits are discussed.

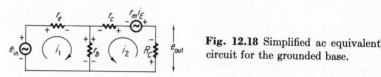

Fig. 12.18 Simplified ac equivalent circuit for the grounded base.

The circuit on which the calculations will be performed is shown in Fig. 12.18. Drawing in loop currents i_1 and i_2, and indicating the individual voltage-drop polarities, the following equations can now be written down:

$$i_e = i_1 \tag{12.74}$$

$$e_{in} - i_1(r_e + r_b) - i_2 r_b = 0 \tag{12.75}$$

$$i_1 r_m + i_2(R_C + r_b + r_c) + i_1 r_b = 0 \tag{12.76}$$

$$e_{out} = -i_2 R_C \tag{12.77}$$

Rewriting Eqs. (12.75) and (12.76) for determinantal solution:

$$\begin{aligned} i_1(r_e + r_b) + i_2 r_b &= e_{in} \\ i_1(r_m + r_b) + i_2(R_C + r_c + r_b) &= 0 \end{aligned} \tag{12.78}$$

Voltage Gain

Solving for i_2 in Eqs. (12.78),

$$i_2 = \frac{\begin{vmatrix} r_e + r_b & e_{in} \\ r_m + r_b & 0 \end{vmatrix}}{\begin{vmatrix} r_e + r_b & r_b \\ r_m + r_b & R_C + r_c + r_b \end{vmatrix}}$$

$$= -\frac{e_{in}(r_m + r_b)}{(r_e + r_b)(R_C + r_c + r_b) - r_b(r_m + r_b)} \tag{12.79}$$

$$A_v = \frac{e_{out}}{e_{in}} = -\frac{i_2 R_C}{e_{in}} = \frac{(r_m + r_b)R_C}{(r_e + r_b)(R_C + r_c + r_b) - r_b(r_m + r_b)} \tag{12.80}$$

Neglecting r_e in the denominator and removing parentheses,

$$A_v \cong \frac{R_C}{r_b}\left(\frac{r_m + r_b}{R_C + r_c - r_m}\right) \tag{12.81}$$

Now $r_b \ll r_m$, so that Eq. (12.81) can be written as

$$A_v = \frac{R_C}{r_b}\left(\frac{r_m}{R_C + r_c - r_m}\right) \tag{12.82}$$

Comparing this result to the grounded emitter, one sees that there is no phase reversal in the grounded-base case and that the voltage gain is much higher. For the case where $R_C = 10$ K, $A_v = 526.7$, or about twice the voltage gain for the grounded emitter. There are other differences, too, as calculations on r_{in} and r_{out} show.

CURRENT GAIN

Using Eqs. (12.78) and solving for i_1,

$$i_1 = \frac{\begin{vmatrix} e_{\text{in}} & r_b \\ 0 & R_C + r_c + r_b \end{vmatrix}}{|D|} = \frac{e_{\text{in}}(R_C + r_c + r_b)}{|D|} \tag{12.83}$$

Dividing i_2 by i_1 gives the current gain:

$$A_i = \frac{i_2}{i_1} = \frac{-(r_m + r_b)}{R_C + r_c + r_b} \tag{12.84}$$

Since $r_c > r_m$, A_i is less than unity. In fact, as R_C approaches zero, the ratio approaches

$$\lim_{R_C \to 0} A_i = -\frac{r_m + r_b}{r_c + r_b} \tag{12.85}$$

Since $r_b \ll r_m$ and $r_b \ll r_c$, the equation can be written as

$$A_i = -\frac{r_m}{r_c} \tag{12.86}$$

or

$$i_2 = -\frac{r_m}{r_c} i_1 \tag{12.87}$$

The negative sign indicates that the current i_2 should be flowing in the other direction in Fig. 12.18. From this diagram, $i_1 = i_e$ and $-i_2 = i_C$; therefore Eq. (12.87) can be written as

$$i_e = \left(\frac{r_m}{r_c}\right) i_C \tag{12.88}$$

But one also knows that i_e and i_C are related by the equation

$$i_e = \alpha i_C \qquad (12.89)$$

Therefore the quantity α can be measured by using ac parameters. Comparing Eqs. (12.88) and (12.89),

$$\alpha = \frac{r_m}{r_c} \qquad (12.90)$$

In the grounded base configuration, α is the maximum current gain that one can obtain, and this occurs for $R_C = 0$.

INPUT RESISTANCE

Using Eqs. (12.78) and solving for i_1 completely,

$$i_1 = \frac{\begin{vmatrix} e_{in} & r_b \\ 0 & R_C + r_c + r_b \end{vmatrix}}{\begin{vmatrix} r_e + r_b & r_b \\ r_m + r_b & R_C + r_c + r_b \end{vmatrix}}$$

$$= \frac{e_{in}(R_C + r_c + r_b)}{(r_e + r_b)(R_C + r_c + r_b) - r_b(r_m + r_b)} \qquad (12.91)$$

Forming the ratio e_{in}/i_1 gives one r_{in}:

$$r_{in} = r_e + r_b - \frac{r_b(r_m + r_b)}{R_C + r_c + r_b}$$

Neglecting r_b in the numerator,

$$r_{in} = r_e + r_b - \frac{r_b r_m}{R_C + r_c + r_b} \qquad (12.92)$$

Since A_i is given by the ratio

$$A_i = \frac{r_m}{R_C + r_c + r_b}$$

and this ratio appears in Eq. (12.92), the equation can be written in simplified form as

$$r_{in} = r_e + r_b(1 - A_i) \qquad (12.93)$$

To show the magnitude of r_{in} for the grounded base configuration, use the parameters for the 2N76 transistor and $R_C = 10$ K.

$$r_{in} = 35 \text{ ohms}$$

This is very low and relegates the grounded base configuration to special circuits.

OUTPUT RESISTANCE

In order to calculate r_{out}, a different circuit from that shown in Fig. 12.18 is used. The correct circuit is shown in Fig. 12.19. The

Fig. 12.19 AC equivalent circuit for output impedance calculations on the grounded base.

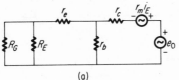

(a)

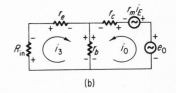

(b)

parallel combination R_G and R_E is relabeled R_{in}, and the final circuit is shown in Fig. 12.19(b). The resulting circuit equations are

$$i_e = i_3$$

$$i_3(R_{in} + r_b + r_e) + i_0 r_b = 0 \qquad (12.94)$$

$$e_0 - i_3 r_m - i_0(r_c + r_b) - i_3 r_b = 0$$

Rewriting the last two equations,

$$i_3(R_{in} + r_b + r_e) + i_0 r_b = 0$$
$$i_3(r_m + r_b) + i_0(r_c + r_b) = e_0 \qquad (12.95)$$

Solving for i_0

$$i_0 = \frac{e_0(R_{in} + r_b + r_e)}{(r_c + r_b)(R_{in} + r_b + r_e) - r_b(r_m + r_b)} \qquad (12.96)$$

Solving this for the ratio $r_{out} = e_0/i_0$,

$$r_{out} = r_c - r_b\left(\frac{r_m - R_{in}}{r_b + R_{in}}\right) \qquad (12.97)$$

An order-of-magnitude calculation, using the 2N76 and assuming $R_{in} = 5$ K, gives

$$r_{out} = 946 \text{ K}$$

This is rather high. The grounded-base circuit thus has a very low input resistance and a very high output resistance. Its voltage gain is similar to that of the grounded emitter and the current gain is less than unity. Since this resistance is in parallel with R_C, Fig. 12.17(c), the parallel combination is essentially equal to R_C.

Example 12.6 Using the parameters for a 2N76 transistor and selected values of R_C, the values of A_v, A_i, r_{in}, and r_{out} will be calculated.

1. For A_v, using Eq. (12.82),

$$A_v = \frac{R_C}{r_b}\left(\frac{r_m}{R_C + r_c - r_m}\right)$$

The following values are tabulated for a range of R_C values:

R_C	A_v	R_C	A_v
500	31.3	10,000	527
1,000	62.0	50,000	1,580
5,000	288		

2. For A_i, the equation is given as

$$A_i = \frac{r_m + r_b}{R_C + r_c + r_b}$$

once it is seen that the negative sign simply signified that the original current direction should have been reversed. It is obvious that as the emitter current increases, so will the collector current; hence the rationalization for using the positive sign for A_i. Values of A_i versus R_C are tabulated below.

R_C, ohms	A_i	R_C, ohms	A_i
500	0.949	10,000	0.940
1,000	0.949	50,000	0.904
5,000	0.945		

The current gain is rather insensitive to changes in R_C. As $R_C \rightarrow 0$, $A_i \rightarrow 0.95$, which is the value of α for the 2N76.

3. The equation for r_{in}, as given by Eq. (12.93), is

$$r_{\text{in}} = r_e + r_b(1 - A_i)$$

The values of r_{in} versus R_C are tabulated below for several values of R_C.

R_C, ohms	r_{in}, ohms	R_C, ohms	r_{in}, ohms
500	32.3	10,000	35.0
1,000	32.3	50,000	45.8
5,000	33.5		

Here, again, the parameter is rather independent of R_C because it depends upon A_i.

4.   Using Eq. (12.97), r_{out} versus R_{in} is tabulated below.

R_{in}, ohms	r_{out}, K	R_{in}, ohms	r_{out}, K
500	644	10,000	973
1,000	781	50,000	995
5,000	947		

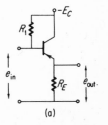

(a)

R_{out} rapidly approaches r_c as R_C increases. The value of r_{out} remains high even for small values of R_C.

12.9   The Grounded Collector Configuration

A typical grounded collector circuit is shown in Fig. 12.20(a). The input is between base and ground, and the output is between emitter and ground. R_1 is a biasing resistor and R_E is the load resistor. The ac equivalent circuit for this circuit is shown in Fig. 12.20(b). The generator, consisting of e_g and R_G, is also shown placed across the input. As was done before, the effect of the voltage divider R_G and R_1 is put off until multistage circuits are discussed. One simply considers the simplified circuit shown in Fig. 12.20(c).

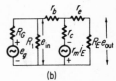

(b)

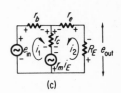

(c)

VOLTAGE GAIN

Drawing in loop currents i_1 and i_2 and indicating the respective voltage drops, the following equations result from Fig. 12.20(c):

$$i_e = i_2 \tag{12.98}$$

$$e_{in} - i_1(r_b + r_c) - i_2 r_c + r_m i_2 = 0 \tag{12.99}$$

$$e_{in} - i_1 r_b + i_2(r_e + R_E) = 0 \tag{12.100}$$

$$e_{out} = -i_2 R_E \tag{12.101}$$

Rearranging Eqs. (12.99) and (12.100) for determinantal use,

$$i_1(r_c + r_b) + i_2(r_c - r_m) = e_{in}$$

$$i_1 r_b - i_2(R_E + r_e) = e_{in} \tag{12.102}$$

Solving for i_2

$$i_2 = -\frac{e_{in} r_c}{(r_c + r_b)(R_E + r_e) + r_b(r_c - r_m)} \tag{12.103}$$

The voltage gain A_v is found by using Eq. (12.103) in Eq. (12.101):

$$A_v = \frac{R_E r_c}{(r_c + r_b)(R_E + r_e) + r_b(r_c - r_m)} \tag{12.104}$$

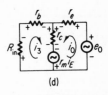

(d)

Fig. 12.20 The reduction of the grounded collector circuit into its ac equivalent circuit.

This can be simplified by realizing that $r_b \ll r_c$ and $r_e \ll R_E$. Equation (12.104) becomes

$$A_v = \frac{R_E r_c}{R_E r_c + r_b(r_c - r_m)} \qquad (12.105)$$

Dividing through numerator and denominator by $R_E r_c$

$$A_v = \frac{1}{1 + (r_b/R_E)[1 - (r_m/r_c)]} \qquad (12.106)$$

Using the definition given in Eq. (12.90),

$$A_v = \frac{1}{1 + (r_b/R_E)(1 - \alpha)} \qquad (12.107)$$

It is evident the $A_v < 1$ for all R_E. However, the quantity $(1 - \alpha)$ is so small that

$$\frac{r_b}{R_E}(1 - \alpha) < 1 \qquad (12.108)$$

and for practically all values of R_E, $A_v = 1$ to an excellent approximation.

CURRENT GAIN

Calculating i_1 from Eqs. (12.102),

$$i_1 = -\frac{(R_E + r_e + r_c - r_m)e_{\text{in}}}{|D|} \qquad (12.109)$$

Using Eqs. (12.103) and (12.109), A_i can be found:

$$A_i = \frac{i_2}{i_1} = \frac{-r_c}{(r_c - r_m) - (R_E + r_e)} \qquad (12.110)$$

Again, looking at Fig. 20(c), one can see that i_e must split up into $i_b + i_c$; therefore the minus sign in Eq. (12.110) simply signifies that i_1 should be reversed. Dividing numerator and denominator by r_c and using Eq. (12.90), A_i becomes

$$A_i = \frac{1}{1 - \alpha - [(R_E + r_e)/r_c]} \qquad (12.111)$$

The inequality $(R_E + r_e) < r_c$ holds for a large range of R_E; therefore A_i can be reduced to

$$A_i = \frac{1}{1 - \alpha} \qquad (12.112)$$

For a 2N76, for example, $\alpha = 0.95$ and $A_i \cong 20$. Thus the grounded collector gives an appreciable current gain even though the voltage gain is unity.

INPUT RESISTANCE

Using Eq. (12.109) and solving for the ratio e_{in}/i_1,

$$r_{in} = \frac{(r_c + r_b)(R_E + r_e) + r_b(r_c - r_m)}{R_E + r_e + r_c - r_m} \qquad (12.113)$$

Simplifications are in order; in the numerator, $r_b < r_c$ and $r_e < R_E$; and in the denominator, $r_e < (R_E + r_c - r_m)$:

$$r_{in} \cong \frac{r_c R_E + r_b(r_c - r_m)}{R_E + r_c - r_m} \qquad (12.114)$$

Dividing numerator and denominator by the product $r_c R_E$ and using Eq. (12.90),

$$r_{in} = \frac{1 + (r_b/R_E)(1 - \alpha)}{(1/r_c) + (1/R_E)(1 - \alpha)} \qquad (12.115)$$

The term in the numerator containing $(1 - \alpha)$ is less than unity for a large range of R_E. For example, for $R_E = 5$ K and the parameters for a 2N76,

$$\frac{r_b}{R_E}(1 - \alpha) = 0.003$$

Therefore Eq. (12.115) reduces to

$$r_{in} = \frac{r_c}{1 + (r_c/R_E)(1 - \alpha)} \qquad (12.116)$$

The input resistance of the grounded collector is high and is dependent on the external circuit.

OUTPUT RESISTANCE

Altering Fig. 12.20(b) in order to calculate r_{out}, one obtains Fig. 12.20(d), where R_{in} consists of R_G and R_1 in parallel; also, R_E has been removed. Drawing in loop currents i_0 and i_3, and designating the voltage drops, one obtains the following equations:

$$i_e = i_0$$

$$i_0 r_e - i_3(R_{in} + r_b) = e_0 \qquad (12.117)$$

$$i_0(r_e + r_c - r_m) + i_3 r_c = e_0$$

Solving for i_0

$$i_0 = \frac{e_0(r_c + R_{in} + r_b)}{r_e r_c + (R_{in} + r_b)(r_e + r_c - r_m)} \qquad (12.118)$$

This can be solved for r_{out} by forming the ratio e_0/i_0.

$$r_{out} = \frac{r_e r_c + (R_{in} + r_b)(r_e + r_c - r_m)}{r_c + R_{in} + r_b} \qquad (12.119)$$

This can be simplified by realizing that in the numerator $r_e < (r_c - r_m)$ and in the denominator $(R_{in} + r_b) < r_c$ for a large range of R_{in}. With the aid of Eq. (12.90), Eq. (12.119) can then be put in the form

$$r_{out} = r_e + (R_{in} + r_b)(1 - \alpha) \qquad (12.120)$$

The output resistance depends strongly on R_{in}. It is also rather low in value. As can be seen on reflection, the characteristics of the grounded collector closely resemble those of the vacuum-tube cathode follower.

Example 12.7 The order of magnitude of the characteristics of the grounded collector are investigated by using 2N76 parameters and varying R_C and R_{in}.

1. For A_v, for all practical purposes, the voltage gain of the grounded collector is given by unity:

$$A_v = 1$$

It is independent of R_E.

2. For A_i, the current gain is also independent of the external circuit for all practical purposes, and is given by

$$A_i = \frac{1}{1 - \alpha}$$

For the 2N76, $\alpha = 0.95$ and $A_i = 20$.

3. For r_{in}, the input resistance is highly dependent on R_E. A tabulation of r_{in} for various values of R_E, using 2N76 parameters,

R_E, ohms	r_{in}, K
500	9.9
1,000	19.6
5,000	91.0
10,000	167
50,000	500

will point this out. It increases to very high values rapidly with increase in R_E, and r_{in} approaches r_c as R_E increases.

4.   The r_{out} parameter depends upon the external circuit, namely, on R_{in}. R_{in} is made up of R_G and R_1 in parallel. R_1 is of the order of 100 K, while R_G depends upon the resistance of the previous stage and can be either large or small, depending upon the configuration. To show how r_{out} depends upon R_{in}, a few values of r_{out} have been tabulated, using the 2N76 parameters.

R_{in}, ohms	r_{out}, ohms
500	57
1,000	82
5,000	282
10,000	532
50,000	2,532

As can be seen, r_{out} does depend on the value of R_{in}. However, since r_{out} is in parallel with R_E, the net output resistance is reduced somewhat.

12.10   Effect of Degenerative Feedback

To show the effect of degenerative feedback on the action of a typical circuit, consider the modified ground-emitter stage with an unbypassed emitter resistor R_E. Feedback occurs not only at very low frequencies, to reduce temperature effects, but also at the signal frequencies. The circuit of Fig. 12.21 (a) can be reduced to the ac equivalent circuit shown in Fig. 12.21 (b), where e_{in} is the fraction of the applied generator voltage that appears from base to ground. To show the effect of R_E on the circuit, A_v and r_{in} will be calculated.

From the diagram in Fig. 12.21 (b), the following equations can be written down:

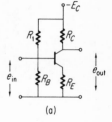

(a)

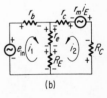

(b)

Fig. 12.21 The unbypassed emitter resistor and equivalent circuit.

$$i_e = -i_1 - i_2$$

$$i_{in} - i_1(r_b + r_e + R_E) - i_2(r_e + R_E) = 0$$

$$e_{in} - i_1 r_b + i_2(R_C + r_c) + r_m(-i_1 - i_2) = 0$$

$$(12.121)$$

$$e_{out} = -i_2 R_C$$

Rewriting the two loop equations for determinantal use,

$$i_1(r_b + R_E + r_e) + i_2(R_E + r_e) = e_{in} \qquad (12.122)$$

$$i_1(r_b + r_m) - i_2(R_C + r_c - r_m) = e_{in} \qquad (12.123)$$

Calculation of A_v

To obtain A_v, one needs i_2, since

$$A_v = \frac{e_{\text{out}}}{e_{\text{in}}} = -\frac{i_2 R_C}{e_{\text{in}}} \qquad (12.124)$$

Solving Eqs. (12.122) and (12.123) for i_2,

$$i_2 = \frac{\begin{vmatrix} (r_b + R_E + r_e) & e_{\text{in}} \\ (r_b + r_m) & e_{\text{in}} \end{vmatrix}}{\begin{vmatrix} (r_b + R_E + r_e) & (R_E + r_e) \\ (r_b + r_m) & -(R_C + r_c - r_m) \end{vmatrix}}$$

$$= \frac{-e_{\text{in}}(R_E + r_e - r_m)}{(r_b + R_E + r_e)(R_C + r_c - r_m) + (R_E + r_e)(r_b + r_m)} \qquad (12.125)$$

Neglecting r_e in the numerator and regrouping the denominator, we have

$$i_2 = \frac{-e_{\text{in}}(R_E - r_m)}{r_b(R_C + r_c - r_m + r_e) + R_E(R_C + r_c + r_b) + r_e r_m} \qquad (12.126)$$

In the quantity $(R_C + r_c + r_b)$, $r_b < (R_C + r_c)$ and can be neglected, and also r_e can be neglected in the first term in the denominator.

$$i_2 = \frac{-e_{\text{in}}(R_E - r_m)}{r_b(R_C + r_c - r_m) + R_E(R_C + r_c) + r_e r_m} \qquad (12.127)$$

Putting Eq. (12.127) into Eq. (12.124),

$$A_v = -\frac{R_C(r_m - R_E)}{r_b(R_C + r_c - r_m) + R_E(R_C + r_c) + r_e r_m} \qquad (12.128)$$

Calculation of r_{in}

The input resistance r_{in} is given by

$$r_{\text{in}} = \frac{e_{\text{in}}}{i_1}$$

Solving for i_1 in Eqs. (12.122) and (12.123),

$$i_1 = \frac{\begin{vmatrix} e_{in} & (R_E + r_e) \\ e_{in} & -(R_C + r_c - r_m) \end{vmatrix}}{\begin{vmatrix} (r_b + R_E + r_e) & (R_E + r_e) \\ (r_b + r_m) & -(R_C + r_c - r_m) \end{vmatrix}}$$

$$= \frac{e_{in}(R_C + r_c - r_m + R_E + r_e)}{(r_b + R_E + r_e)(R_C + r_c - r_m) + (R_E + r_e)(r_b + r_m)}$$

(12.129)

Using the approximations made in obtaining Eq. (12.127), Eq. (12.129) is solved for $e_{in}/i_1 = r_{in}$;

$$r_{in} = \frac{r_b(R_C + r_c - r_m) + R_E(R_C + r_c) + r_e r_m}{R_C + R_E + r_c - r_m} \quad (12.130)$$

ORDER OF MAGNITUDE

The quantitative values for A_v and r_{in} for a 2N76 transistor are shown in the table below. The calculations involve Eqs. (12.128) and (12.130). An R_C value of 10 K was used.

R_E, ohms	A_v	r_{in}, K
0	−268	0.68
50	−112	1.69
100	−70	2.7
200	−40	4.7
500	−18	10.8
1000	−9.1	20.9
5000	−1.9	102

As might be expected from the previous vacuum tube work, r_{in} increases with increase in feedback. Something is sacrificed, and this is the voltage gain. The voltage gain is reduced drastically as R_E increases. Compare these values with those for the case where $R_E = 0$, that is, the uncompensated amplifier. Inclusion of an unbypassed emitter resistor is sometimes a useful trick when the characteristics of a grounded emitter and a moderately high input impedance are needed.

TABLE 12.2

	Grounded Emitter	Grounded Base	Grounded Collector
A_v	$-\dfrac{A_i R_C}{r_b + r_e[(R_C + r_c)/(R_C + r_c - r_m)]}$	$\left(\dfrac{R_C}{r_b}\right)\left(\dfrac{r_m}{R_C + r_c - r_m}\right)$	1
A_i	$\dfrac{r_m}{R_C + r_c - r_m}$	$\dfrac{r_m}{R_C + r_c + r_b}$	$\dfrac{1}{1 - \alpha}$
r_{in}	$r_b + r_e(1 + A_i)$	$r_e + r_b(1 - A_i)$	$\dfrac{r_c}{1 + (r_c/R_B)(1 - \alpha)}$
r_{out}	$r_c - r_m - \dfrac{r_e r_m}{R_{in} + r_b}$	$r_c - r_b\left(\dfrac{r_m - R_{in}}{r_b + R_{in}}\right)$	$r_e + (R_{in} + r_b)(1 - \alpha)$

12.11 Summary of Equations

Table 12.2 is a convenient summary of the equations developed here for the values of A_v, A_i, r_{in}, and r_{out}.

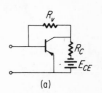

(a)

Problems

12.1 Given that the hybrid parameters for the common base configurations are

$h_{11} = 12$ ohms $\qquad h_{21} = -0.97$

$h_{12} = 560 \times 10^{-6} \qquad h_{22} = 2.2 \times 10^{-6}$ mho

Find the short-circuit parameters for the common base, using the conversion tables in the appendix.

(b)

12.2 Derive the conversion formulas to go from hybrid to open-circuit parameters in the common base configuration.

12.3 Given that the hybrid parameters for a given transistor in the common base configuration are $h_{11} = 39$ ohms, $h_{12} = 38 \times 10^{-6}$, $h_{21} = -0.98$, and $h_{22} = 0.49$ μmho; find the open-circuit parameters r_{11}, r_{12}, r_{21}, and r_{22}.

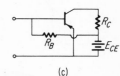

(c)

12.6

12.4 Using the hybrid parameters given in Problem 12.3, determine the T parameters for the transistor.

12.5 Using the hybrid parameters given for the 2N76 listed under "Design Center," find the corresponding T parameters.

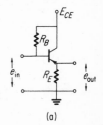

(a)

12.6 Determine the input impedance of the accompanying circuits in the midfrequency range.

12.7 Determine the current gain for each of the circuits.

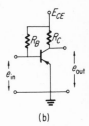

(b)

12.8 A transistor with T parameters of $r_e = 10$ ohms, $r_b = 200$ ohms, $r_c = 1$ meg, and $\alpha = 0.98$ is used in the common emitter configuration. If the source impedance is 500 ohms and the load is 20 K, calculate the overall voltage gain.

12.7

12.9 For the common emitter circuit shown, determine the following: (a) A_i, (b) A_v, (c) input resistance.

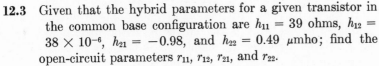

$$R_B = 300 \text{ K} \qquad r_e = 20$$
$$R_C = 10 \text{ K} \qquad r_b = 400$$
$$r_c = 1 \text{ meg}$$
$$r_m = 0.95 \text{ meg}$$

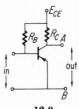

12.9

12.10 For the circuit shown in Problem 12.9, determine the Thévenin's equivalent circuit that can be put between terminals A and B. Assume $R_{in} = 10$ K.

13.1 Impedance Matching

13

Multistage and Power-Amplifier Circuits

For a stage of amplification to be useful, one must have an input generator and connect the output of the stage to either another stage for further amplification or to a transducing device. The stage can be used either for voltage amplification or for power amplification. In order to use a stage efficiently, one tries to secure maximum gain. This means that care must be used in connecting one stage to another, since gain can actually be lost if stages are mismatched. The conditions for matching voltage amplifiers and power amplifiers are not the same.

Voltage Transfer

Figure 13.1 shows a stage of amplification with a generator connected to it. The generator can represent either a previous stage, or a transducing device such as a crystal cartridge in a phonograph system, or a microphone, etc. The emf of the input device or stage is represented by e_g and the internal resistance by R_G.

In the transistor stage, R_{in} represents the input resistance as seen by the generator. It is usually composed of the input resistance r_{in} in parallel with biasing resistors. The voltage actually amplified by the stage is e_{in}. Therefore, if the stage is to be used as a voltage amplifier, e_{in} should be as large as possible. In the voltage divider of Fig. 13.1, e_{in} is given by

$$e_{in} = e_g \left(\frac{r_{in}}{R_G + r_{in}} \right) \tag{13.1}$$

A little reflection will convince one that as $r_{in} \to \infty$, the ratio of resistances approaches unity:

$$\lim_{r_{in} \to \infty} \left(\frac{r_{in}}{R_G + r_{in}} \right) = 1 \tag{13.2}$$

This makes $e_{in} = e_g$, the maximum value. Obviously r_{in} cannot approach infinity in practice, but the result shows what must be done for maximum voltage transfer:

$$r_{in} > R_G \tag{13.3}$$

This inequality ensures $e_{in} \to e_g$. Recollecting the characteristics of the three configurations as given in Chapter 12, the grounded base configuration has a very low input resistance, and there would be difficulty satisfying Eq. (13.3); on the other hand, the grounded collector does have a large input resistance but low gain, $A_v = 1$. Therefore the grounded emitter seems to be the choice, although it has only a moderately high input resistance. For cases where the internal resistance of the voltage generator is very high, special circuits must be used to increase the input resistance of the amplifier stage. For example, one could use a grounded col-

428

lector, for its very high input resistance, and connect this to a grounded emitter to secure the necessary amplification. There is excellent voltage transfer throughout the chain.

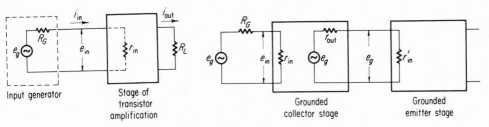

Fig. 13.1 Input generator connected to a transistor amplifier can result in impedance mismatches.

Fig. 13.2 Use of a grounded collector to help match impedances.

Figure 13.2 illustrates this in block diagram form. Usually r_{in} for the grounded collector is very large, for most situations, $r_{in} > R_G$. This ensures that $e_{in} \cong e_g$. The gain for the grounded collector is essentially unity for a large range of load resistance; thus the emf of the output generator for the grounded collector is $e_{out} = e_{in} \cong e_g$, as shown. Now r_{out} for the grounded collector can be made rather small, and connecting this stage to a grounded emitter ensures that $r'_{in} > r_{out}$. Therefore the input voltage to the grounded emitter is, for all practical purposes, e_g, the input generator emf. So, at the expense of an impedance matching stage (the grounded collector), the grounded emitter stage can be used very efficiently. In retrospect, for correct voltage transfer, Eq. (13.3) must be satisfied.

POWER TRANSFER

In applications where power amplification is important, the condition derived in Sec. 1.15 is used. For Fig. 13.1, this means that the following condition must be met:

$$R_G = r_{in} \tag{13.4}$$

where R_G is the internal resistance of the generator and r_{in} is the input resistance of the power-amplifier stage. The input power for the circuit of Fig. 13.1 is given by

$$P_{in} = \tfrac{1}{2}I_{in}^2 r_{in} \tag{13.5}$$

The input current is given by

$$I_{in} = \frac{E_g}{R_G + r_{in}} \tag{13.6}$$

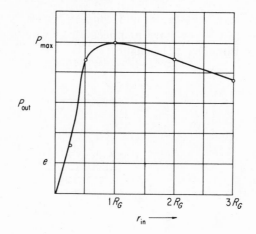

Fig. 13.3 Power transfer from generator to transistor as a function of r_{in}.

Putting this into Eq. (13.5),

$$P_{in} = \frac{1}{2}\left(\frac{E_g}{R_G + r_{in}}\right)^2 r_{in} \qquad (13.7)$$

This equation shows how P_{in} depends upon r_{in}. Power transfer requires that Eq. (13.4) be satisfied, but as is often the case, this cannot always be done. A graph of Eq. (13.7) in Fig. 13.3 shows that if mismatch must be tolerated, it should be on the side $r_{in} > R_G$. For $r_{in} < R_G$, the output power drops off rapidly with a change in r_{in}. For the case $r_{in} > R_G$, the dropoff is not so great for a similar change in r_{in}. Therefore, if there must be a mismatch, it should be made on the high side, that is, $r_{in} > R_G$. This will keep the mismatch small up to $r_{in} = 2R_G$.

13.2 Calculation of Overall Gain—Single Stage

The effect of the biasing resistors and the internal resistance of the input generator will now be taken into account in calculating the gain of a stage of amplification. This is a prelude to calculations for multistage circuits, which will use the same method of attack.

Consider the circuit for a typical grounded emitter, shown in Fig. 13.4(a). Resistors R_1 and R_B form the biasing network, R_E is the swamping resistor bypassed by C_E, R_C is the load resistor, and R_G is the internal resistance of the generator. In Fig. 13.4(b), the same circuit is redrawn in the ac equivalent form. R_I is the parallel combination of R_1 and R_B:

$$R_I = \frac{R_1 R_B}{R_1 + R_B} \qquad (13.8)$$

and r_{in} is the input resistance of the grounded emitter stage. The transistor is represented by the four-terminal block. It is not necessary to insert the ac equivalent circuit because all the equations for this block have been calculated in Chapter 12.

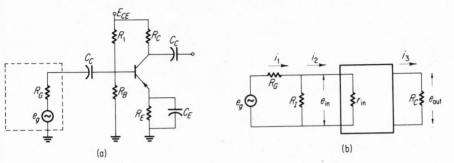

Fig. 13.4 Reduction of generator-transistor circuit for calculations.

The overall current gain for the stage is given by Eq. (12.66):

$$A_i = \frac{r_m}{R_C + r_c - r_m} \qquad (13.9)$$

Also, from Fig. 13.4(b),

$$A_i = \frac{I_3}{I_2} \qquad (13.10)$$

where I_2 and I_3 are maximum values of i_2 and i_3.

The power gain is given by

$$A_P = \frac{\frac{1}{2}I_3{}^2 R_C}{\frac{1}{2}I_2{}^2 r_{in}} \qquad (13.11)$$

where the numerator is the average power developed in the output resistor, and the denominator is the average power developed in the input resistance of the transistor stage. Now the ratio of currents in Eq. (13.11) is nothing more than the overall current gain, that is,

$$A_i = \frac{I_3}{I_2} \qquad (13.12)$$

so that Eq. (13.11) can be written as

$$A_P = A_i{}^2 \left(\frac{R_C}{r_{in}}\right) \qquad (13.13)$$

The power gain is also given by

$$A_P = A_i A_v \qquad (13.14)$$

Solving this for A_v,

$$A_v = \frac{A_P}{A_i} \qquad (13.15)$$

Eliminating A_P between Eqs. (13.13) and (13.15),

$$A_v = A_i \left(\frac{R_C}{r_{in}}\right) \qquad (13.16)$$

Thus the current gain is used as an intermediary in calculating A_v, the voltage gain. This, however, is not the final form. The voltage divider R_G, R_I, and r_{in} has not been taken into account. Equation (13.16) gives the *available voltage gain* A_v:

$$A_v = \frac{e_{out}}{e_{in}} \qquad (13.17)$$

where e_{out} and e_{in} are given in Fig. 13.4. To secure the *overall voltage gain*, one needs

$$A_{ov} = \frac{e_{out}}{e_g} \qquad (13.18)$$

In Eq. (13.18), e_g can be replaced by the equation connecting e_{in} and e_g:

$$e_{in} = e_g \left(\frac{R_{in}}{R_G + R_{in}}\right) \qquad (13.19)$$

where R_{in} is the parallel combination of R_I and r_{in}. Equation (13.8) becomes

$$A_{ov} = \left(\frac{R_{in}}{R_G + R_{in}}\right)\left(\frac{e_{out}}{e_{in}}\right) \qquad (13.20)$$

Replacing the second term on the right, e_{out}/e_{in}, by Eq. (13.16), the overall voltage gain becomes

$$A_{ov} = A_i \left(\frac{R_C}{r_{in}}\right)\left(\frac{R_{in}}{R_G + R_{in}}\right) \qquad (13.21)$$

TABLE 13.1   Voltage Gain

Available voltage gain	$A_v = A_i \left(\dfrac{R_c}{r_{in}}\right)$
Overall voltage gain	$A_{ov} = A_v \left(\dfrac{R_{in}}{R_G + R_{in}}\right)$

It should be noted that Eqs. (13.16) and (13.21) give only the absolute values when using this method. If one wishes to take into account the input generator resistance, Eq. (13.21) is used. Otherwise, one uses Eq. (13.16), which gives the available voltage gain, before the input generator is applied. See Table 13.1.

Example 13.1 Find the overall gain for the circuit below. Use the characteristics for the 2N76, developed in Chapter 12. Compare this with the available voltage one would have before the input generator is applied.

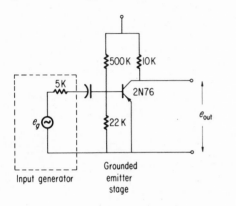

$r_e = 17$ ohms
$r_b = 300$ ohms
$r_c = 1$ meg
$r_m = 950$ K

1. The ac equivalent circuit for the given circuit is now shown.

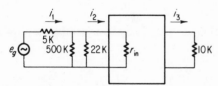

Here the coupling capacitors are assumed to have negligible reactance for this calculation.

2. The circuit can be further modified by combining the parallel resistors to form R_I.

$$R_I = \frac{500 \text{ K} \times 22 \text{ K}}{500 \text{ K} + 22 \text{ K}} = 21.2 \text{ K}$$

Then r_{in} is calculated, using Eq. (12.67),

$$r_{\text{in}} = r_b + r_e + \frac{r_e r_m}{R_C + r_c - r_m}$$

$$= 300 + 17 + \frac{17 \times 950 \text{ K}}{10 \text{ K} + 1000 \text{ K} - 950 \text{ K}}$$

$$= 586 \text{ ohms}$$

and the circuit values become as shown at right.

For later use, R_{in} is calculated:

$$R_{in} = \frac{21.2 \text{ K} \times 586}{21.2 \text{ K} + 586}$$

$$= 570 \text{ ohms}$$

3. The available voltage gain, before the generator is applied, is given by Eq. (13.16) as

$$A_v = A_i \left(\frac{R_C}{r_{in}}\right)$$

Calculating A_i with the aid of Eq. (12.66),

$$A_i = \frac{r_m}{R_C + r_C - r_m}$$

$$= \frac{950 \text{ K}}{10 \text{ K} + 1000 \text{ K} - 950 \text{ K}}$$

$$= 15.8$$

and the available voltage gain becomes

$$A_v = 15.8 \left(\frac{10 \text{ K}}{586}\right)$$

$$= 270$$

4. This can be compared to the overall gain obtained when the generator is applied, Eq. (13.21):

$$A_{ov} = A_i \left(\frac{R_C}{r_{in}}\right)\left(\frac{R_{in}}{R_G + R_{in}}\right)$$

$$= 15.8 \left(\frac{10 \text{ K}}{586}\right)\left(\frac{570}{5 \text{ K} + 570}\right)$$

$$= 27.6$$

5. A large reduction from the theoretical gain occurs with this amplifier because of the poor match between input generator and the amplifier. Equation (13.3) is certainly not satisfied.

Example 13.2 To show how the overall gain can be increased with a better matching of stages, recalculate the voltage gain for the circuit of Example 13.1 by using a 2N525 transistor. The characteristics are shown below for 2N525:

$$r_e = 12.5 \text{ ohms} \qquad r_c = 1.67 \text{ meg}$$

$$r_b = 840 \text{ ohms} \qquad r_m = 1.63 \text{ meg}$$

1. The input resistance r_{in} is given by

$$r_{in} = r_b + r_e + \frac{r_e r_m}{R_C + r_c - r_{in}}$$

$$= 1.26 \text{ K}$$

This larger value of r_{in} will produce better matching between generator and amplifier.

R_{in} is given by

$$R_{in} = \frac{R_I r_{in}}{R_I + r_{in}}$$

$$= 1.19 \text{ K}$$

2. The overall current gain can now be calculated:

$$A_i = \frac{I_3}{I_2} = \frac{r_m}{R_C + r_c - r_m}$$

$$= 32.6$$

3. The maximum or available voltage gain, given by Eq. (13.20), becomes

$$A_v = A_i \left(\frac{R_C}{r_{in}}\right) = 259$$

This is close to the voltage gain calculated for Example 13.1.

4. The overall gain, taking into account R_G, becomes

$$A_{ov} = A_i \left(\frac{R_C}{r_{in}}\right)\left(\frac{R_{in}}{R_G + R_{in}}\right)$$

$$= 259 \times \frac{1.19 \text{ K}}{(5 + 1.19) \text{ K}}$$

$$= 49.8$$

As can be seen, although the available gain is about the same for the two examples, the overall gain is practically doubled because a transistor with a higher input resistance was used.

13.3   Overall Gain—Multistage Amplifiers

As an exercise in calculation of the overall gain of several stages of amplification, two 2-stage amplifiers will be used as examples. In the first example, two grounded-emitter stages will be used.

The second example will involve a more complicated circuit, comprising a grounded collector and grounded emitter and utilizing a special dc connection to stabilize the collector current against temperature variations.

FIRST EXAMPLE

Referring to Fig. 13.5(a), the first example consists of two grounded-emitter stages in cascade. Both stages have bypassed swamping resistors, so that these components do not enter into the ac equivalent circuit. The input generator is assumed to have an internal resistance of 800 ohms. The reactance of the coupling capacitors is considered negligible in these calculations.

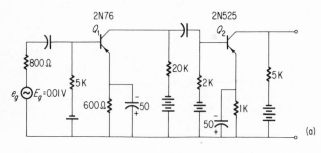

(a)

Fig. 13.5 Two-stage amplifier and equivalent circuit.

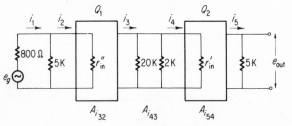

(b)

In calculating the overall voltage gain, the procedure given in Sec. 13.2 will be used; that is, Eq. (13.21) will be used, where the current gain A_i is the overall current gain between the input of the first stage and the output of the second stage, taking into account the interstage current gains. Figure 13.5(b) shows the ac equivalent circuit for the schematic shown in Fig. 13.5(a). The transistors are represented by block diagrams. It is not necessary to insert the ac equivalent circuits for the transistors for these calculations because the items of interest, such as A_i per stage, have already been calculated in Chapter 12. Those results will now be used here.

The current gain must first be calculated. The overall gain is given by the product of the individual current gains:

$$A_i = A_{i_{32}} A_{i_{43}} A_{i_{54}} \tag{13.22}$$

where

$$A_{i_{54}} = \text{current gain for } Q_2 \text{ stage}$$

$$A_{i_{43}} = \text{interstage current gain}$$

$$A_{i_{32}} = \text{current gain for } Q_1 \text{ stage}$$

The T parameters for the 2N76 and 2N525 are given below:

	2N76	2N525
	r_e = 17 ohms	r_e = 12.5 ohms
	r_b = 300 ohms	r_b = 840 ohms
	r_c = 1 meg	r_c = 1.67 meg
	r_m = 950 K	r_m = 1.63 meg

Calculating $A_{i_{54}}$ for Q_2,

$$A_{i_{54}} = \frac{r_m}{R_C + r_c - r_m}$$

$$= \frac{1.63 \times 10^6}{5 \times 10^3 + (1.67 - 1.63) \times 10^6}$$

$$= 36.2 \qquad\qquad (13.23)$$

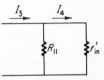

Fig. 13.6 Simplification of input circuit of Q_2.

In calculating $A_{i_{43}}$, the interstage current gain, the interstage circuit is simplified by combining the 20-K and 2-K resistors, since they are in parallel as shown in Fig. 13.6, and the current gain is given by

$$A_{i_{43}} = \frac{I_4}{I_3} = \frac{R_{||}}{R_{||} + r'_{in}} \qquad\qquad (13.24)$$

where

$$R_{||} = \frac{20 \text{ K} \times 2 \text{ K}}{20 \text{ K} + 2 \text{ K}} = 1.82 \text{ K}$$

and

$$r'_{in} = r_e + r_b + \frac{r_e r_m}{R_C + r_c - r_m} \qquad\qquad (13.25)$$

$$= 12.5 + 840 + \frac{12.5 \times 1.63 \times 10^6}{5 \times 10^3 + (1.67 - 1.63) \times 10^6}$$

$$= 1.26 \text{ K}$$

Now

$$A_{i_{43}} = \frac{1.82 \text{ K}}{1.82 \text{ K} + 1.26 \text{ K}} = 0.592$$

Calculating $A_{i_{32}}$ for Q_1,

$$A_{i_{32}} = \frac{r_m}{R_C + r_c - r_m}$$

$$= \frac{950 \text{ K}}{745 + (1000 - 950) \text{ K}}$$

$$= 18.7 \tag{13.26}$$

Note that the value for R_C used here must be the 20-K and 2-K resistors, and r'_{in} in parallel. The overall current gain becomes

$$A_i = 18.7 \times 0.592 \times 36.2 = 306$$

The power gain is given by

$$A_P = A_i^2 \left(\frac{R_C}{r''_{in}}\right) \tag{13.27}$$

where

$$R_C = \text{load resistance for } Q_2 \text{ stage}$$

$$r''_{in} = \text{input resistance of } Q_1 \text{ stage}$$

$$r''_{in} = r_e + r_b + \frac{r_e r_m}{R_C + r_c - r_m} \tag{13.28}$$

Here, again, R_C is given by 20-K, 2-K and 1.26-K resistors in parallel.

$$r''_{in} = 618 \text{ ohms}$$

Using A_i and A_P to obtain A_v,

$$A_v = \frac{A_P}{A_i} = A_i \left(\frac{R_C}{r''_{in}}\right)$$

$$= 306 \times \frac{5 \times 10^3}{618} = 2470 \tag{13.29}$$

Remember that this is the available voltage gain before the input generator is applied. The overall gain, with R_G taken into account, is given by

$$A_{ov} = A_i \left(\frac{R_C}{r''_{in}}\right)\left(\frac{R_{in}}{R_G + R_{in}}\right) \tag{13.30}$$

where R_{in} is the parallel combination of the 5-K resistor and r''_{in}.

$$R_{\text{in}} = \frac{5 \times 10^3 \times 618}{5 \times 10^3 + 618} = 550 \text{ ohms}$$

The overall gain becomes

$$A_{\text{ov}} = 2470 \times \left(\frac{550}{800 + 550}\right) = 1000$$

SECOND EXAMPLE

In the second example, a word concerning the circuit is in order. The circuit under consideration is given in Fig. 13.7(a). C_1 and C_2 are coupling capacitors. Their reactances will be considered negligible for the calculations. Q_1 is in the grounded collector configuration and Q_2 is in the grounded emitter configuration. C_3 is a bypass capacitor which places both the collector of Q_1 and the emitter of Q_2 at ac ground. R_2 is the load resistor for Q_1, and the output of this stage is coupled through C_2 to the base of Q_2. The Q_1 circuit is temperature-stabilized by making R_1 small and using R_2 as a swamping resistor. Figure 13.7(b) shows that the dc current flowing through Q_2 also passes through Q_1 and R_2. Q_1 and R_2 form the swamping resistor for Q_2. Since the current in the stabilizing circuit is common to the two stages, Q_1 stabilizes the changing current due to the changing Q_1 and Q_2 base-emitter junction resistances.

Even though the dc paths may be complicated in the ac equivalent circuit, Fig. 13.7(a) reduces to a grounded collector stage coupled to a common emitter stage. Using the component values listed below, a calculation will be made on the voltage gain for the circuit.

$R_1 = 500$ ohms	$R_4 = 10$ K	$R_G = 1000$ ohms
$R_2 = 1000$ ohms	$Q_1 = 2\text{N}76$	$e_g = 0.01$ volt
$R_3 = 10$ K	$Q_2 = 2\text{N}525$	

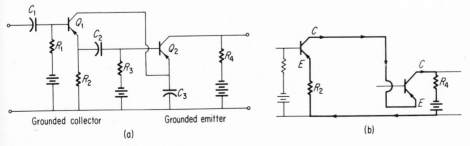

Grounded collector   Grounded emitter

(a)   (b)

Fig. 13.7 Two-stage amplifier utilizing common current stabilizing circuit.

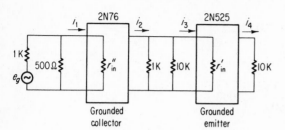

Fig. 13.8 Reduction of the two-stage amplifier in Fig. 3.7(a).

The ac equivalent circuit is shown in Fig. 13.8.
The input resistance for the 2N525 is given by

$$r'_{in} = r_e + r_b + \frac{r_e r_m}{R_C + r_c - r_m}$$

$$= 12.5 + 840 + \frac{12.5 \times 1.63 \times 10^6}{10 \times 10^3 + (1.67 - 1.63) \times 10^6}$$

$$= 1247 \text{ ohms} \tag{13.31}$$

The input resistance for the 2N76 is given by

$$r''_{in} = \frac{r_c}{1 + (r_c/R_E)(1 - \alpha)} \tag{13.32}$$

where the value used for R_E must be the parallel combination of R_2, R_3, and r'_{in}, and α is given by $\alpha = r_m/r_c$.

$$R_E = \boxed{\;\;1\text{ K}\quad 10\text{ K}\quad 1.247\text{ K}\;\;} = 526 \text{ ohms}$$

$$\alpha = \frac{r_m}{r_c} = \frac{1.63 \times 10^6}{1.67 \times 10^6} = 0.976$$

Using these values in Eq. (13.32),

$$r''_{in} = \frac{1.67 \times 10^6}{1 + [(1.67 \times 10^6)/526](1 - 0.976)}$$

$$= 21.7 \text{ K}$$

The current gain for each stage can also be calculated. For the 2N525 stage,

$$A_{i_{43}} = \frac{r_m}{R_C + r_c - r_m}$$

$$= \frac{1.63 \times 10^6}{10 \times 10^3 + (1.67 - 1.63) \times 10^6}$$

$$= 32.6$$

For the 2N76 stage, a grounded collector,

$$A_{i_{21}} = \frac{1}{1 - \alpha}$$

$$= \frac{1}{1 - 0.976}$$

$$= 41.7$$

Note in Eq. (12.111) that the approximation is made $(R_E + r_e) < r_c$ to obtain Eq. (13.34). This is certainly the case here because the value for R_E must be taken as the parallel combination of R_2, R_3, and r'_{in}, which is small. One more step has to be performed before the power gain can be calculated. The interstage current gain is given by

$$A_{i_{22}} = \frac{I_3}{I_2}$$

The parallel combination of R_2 and R_3 is shown in Fig. 13.9 as the 9.1-K resistor. The current gain is now given by

$$A_{i_{22}} = \frac{I_3}{I_2} = \frac{9.1 \times 10^3}{(9.1 + 1.247) \times 10^3} = 0.88$$

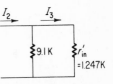

Fig. 13.9 Equivalent input circuit for the 2N525 in Fig. 13.8.

The available voltage gain can now be calculated. First,

$$A_i = A_{i_{21}} A_{i_{22}} A_{i_{23}}$$

$$= 41.7 \times 0.88 \times 32.6 = 1127$$

and

$$A_P = A_i^2 \left(\frac{R_4}{r''_{in}}\right)$$

From this one obtains A_v:

$$A_v = \frac{A_P}{A_i} = A_i \left(\frac{R_4}{r''_{in}}\right) \qquad (13.33)$$

Using the calculated values obtained thus far,

$$A_v = 1127 \times \left(\frac{10 \cdot \times 10^3}{21.7 \times 10^3}\right) = 520$$

Remember, this is the *available* voltage gain. To find the *overall* voltage gain—that is, with the generator connected—the following calculation can be made

$$A_{ov} = A_i \left(\frac{R_4}{r''_{in}}\right)\left(\frac{R_{in}}{R_G + R_{in}}\right) \qquad (13.34)$$

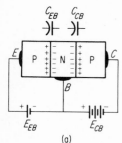

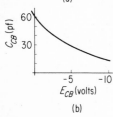

(a)

(b)

Fig. 13.10 Effect of collector voltage on C_{CB}.

Fig. 13.11 Interelectrode capacitance in a transistor.

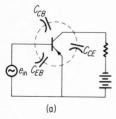

(a)

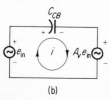

(b)

Fig. 13.12 Simplified calculation on imput impedance in the grounded emitter configuration.

where R_{in} is the parallel combination of R_1 and r''_{in}. Since $R_1 < r''_{in}$, $R_{in} \cong R_1$.

$$A_{ov} = 520 \times \left(\frac{500}{1000 + 500}\right) = 173$$

13.4 Interelectrode Capacitance

The major part of shunting capacitance in a transistor is due to the junctions. A transistor has two junctions. The base collector junction is reverse-biased. Wherever there is a separation of charge, one has a capacitor. This is what happens at the junctions. In the base-collector junction, for instance, there is a depletion layer, so-called because this region has few extrinsic conductors. The width of this layer increases with increase in reverse bias. The effect is exactly like that of an effective parallel-plate capacitor, labeled C_{CB} in Fig. 13.10(a). The way this effective capacitance varies with applied voltage E_{CB} is shown in Fig. 13.10(b). The order of magnitude for C_{CB} is 50 pf.

In a dynamic circuit, there is appreciable current flow through the base-collector junction due to the extrinsic carriers that have drifted over from the base-emitter junction. This introduction of charges increases the effective capacitance C_{CB}. Therefore C_{CB} is also dependent upon I_E. Since the base-emitter junction is forward, biased, the depletion layer is extremely small, and therefore the charges have a small effective separation. This means that C_{EB} is much larger than C_{CB}. It is from 10 to 50 times as large as C_{CB}. Figure 13.11 shows the capacitors in schematic form. The measured collector-to-emitter capacitance C_{CE} is about 5 to 10 times as large as C_{CB}.

In using these capacitances in ac equivalent circuits, the capacitor connecting input and output circuits can be algebraically placed in the input circuits. The procedure is shown for a grounded emitter stage in Fig. 13.12. For the calculation, the capacitor C_{EB} is seen to be in parallel with e_{in}, while C_{CB} couples the input and output circuits. The simplified circuit in Fig. 13.12(a) is replaced by the one shown in Fig. 13.12(b). The transistor is replaced by an equivalent voltage generator, $A_v e_{in}$. The difference in currents through C_{CB}, due to the two generators e_{in} and $A_v e_{in}$, gives the equivalent capacitance reactance as seen by generator e_{in}. Z_{in} is defined as

$$Z_{in} = \frac{e_{in}}{i} \qquad (13.35)$$

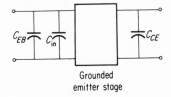

Fig. 13.13 Equivalent shunting capacitances for the grounded emitter.

Grounded
emitter stage

In the equivalent circuit,

$$e_{\text{in}} - (-jX_{C_{CB}})i - A_v e_{\text{in}} = 0 \qquad (13.36)$$

Solving for e_{in}/i in Eq. (13.36),

$$Z_{\text{in}} = \frac{e_{\text{in}}}{i} = -j\frac{X_{C_{CB}}}{1 - A_v}$$

$$= -j\frac{1}{2\pi f C_{CB}(1 - A_v)} \qquad (13.37)$$

As can be seen, the input impedance is an effective capacitor, given by

$$C_{\text{in}} = C_{CB}(1 - A_v) \qquad (13.38)$$

and for ac equivalent circuits, the capacitor C_{CB}, instead of coupling the input and output circuits, can be put into the form given by Eq. (13.38) and placed across the input. This is shown in Fig. 13.13. The input is shunted by C_{EB} and C_{in} in parallel. This can amount to several hundred picofarads. This is not too serious because r_{in} is rather small, so that the upper cutoff frequency is still high.

13.5 Frequency Response

Because the input and output circuits are strongly interrelated for a transistor, a slightly different mode of attack will be used in producing, and using, the ac equivalent circuit. As in the vacuum tube calculations, use will be made of the low-frequency, mid-frequency, and high-frequency approximations. The calculations will differ in that the output circuit is replaced by an equivalent ac-voltage generator and series resistance. The series resistance will be r_{out}, calculated in Chapter 12. An example is shown first for a vacuum tube circuit, since this should· be rather familiar. Consider the standard ac equivalent circuit for the plate circuit of a tube, shown in Fig. 13.14(a). The voltage gain for this circuit is

$$A_v \text{ (mid)} = -g_m R_{||} \qquad (13.39)$$

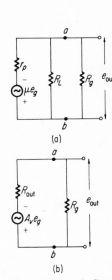

Fig. 13.14 (a) Conventional ac equivalent circuit for a vacuum tube. (b) Equivalent circuit utilizing the definition of R_{out}.

where $R_\parallel$ is the parallel combination of r_p, R_L, and R_g. In our new nomenclature this will be called the *overall voltage gain*, since it includes the effect of R_g. This circuit can be converted into the Thévenin equivalent circuit shown in Fig. 13.14(b) and then solved, where

$$A_v = \frac{-\mu R_L}{r_p + R_L} \tag{13.40}$$

and

$$R_{\text{out}} = \frac{r_p R_L}{r_p + R_L} \tag{13.41}$$

These two quantities are recognized as the voltage gain and output resistance for the circuit before R_g is applied. The voltage gain given in Eq. (13.40) would now be called the *available voltage gain*, found by calculations on the simple circuit.

Calculation of A_v by using Fig. 13.14(b) will also result in Eq. (13.39); that is, the two circuits are equivalent. The procedure shown in Fig. 13.14(b) is the method that will be used to produce our first-order calculations on the frequency response of a transistor stage.

The calculations will be performed for the grounded emitter.

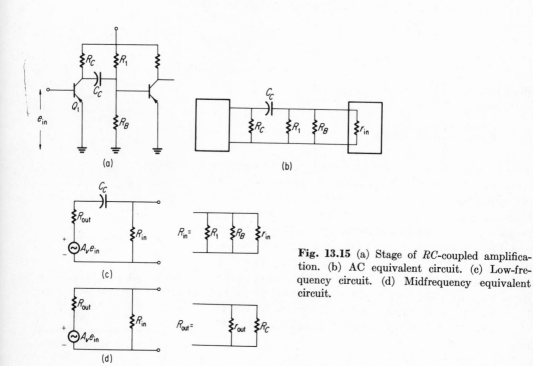

Fig. 13.15 (a) Stage of RC-coupled amplification. (b) AC equivalent circuit. (c) Low-frequency circuit. (d) Midfrequency equivalent circuit.

Two stages of grounded-emitter amplification are shown in Fig. 13.15(a). The dc equivalent circuit for the interstage section is shown in Fig. 13.15(b). Each preceding and succeeding stage will have the same form. Therefore calculations on a general case will suffice for all examples. The circuit of Fig. 13.15(b) is modified so that the definition of R_{out} can be used in Fig. 13.15(c), where R_{out} is the parallel combination of r_{out} and R_C. Also, R_{in} has been defined as the parallel combination of R_1, R_B, and r_{in}. The interelectrode capacitances are not included, but the coupling capacitor remains. This should be recognized as the definition for the low-frequency equivalent circuit.

For the midfrequency case, the reactance of C_C is considered small, and C_C is replaced by a short circuit. This circuit is shown in Fig. 13.15(d).

MIDFREQUENCY

In the midfrequency circuit, the output is taken across R_{in}, Fig. 13.15(d). The output voltage can be written as

$$e_{out} = A_v e_{in} \left(\frac{R_{in}}{R_{out} + R_{in}} \right) \tag{13.42}$$

Multiplying numerator and denominator by R_{out},

$$e_{out} = \frac{A_v e_{in}}{R_{out}} \left(\frac{R_{out} R_{in}}{R_{out} + R_{in}} \right)$$

By defining $R_{||}$ as

$$R_{||} = \frac{R_{out} R_{in}}{R_{out} + R_{in}} \tag{13.43}$$

e_{out} can be written as

$$e_{out} = A_v e_{in} \left(\frac{R_{||}}{R_{out}} \right)$$

Solving for A_{ov},

$$A_{ov} \text{ (mid)} = \frac{e_{out}}{e_{in}} = A_v \left(\frac{R_{||}}{R_{out}} \right) \tag{13.44}$$

LOW FREQUENCY

From Fig. 13.15(c), the output voltage can be written down as

$$e_{out} = A_v e_{in} \frac{R_{in}}{R_{in} + R_{out} - jX_{C_C}} \tag{13.45}$$

Multiplying numerator and denominator by R_{out} and factoring out $(R_{\text{out}} + R_{\text{in}})$ in the denominator,

$$e_{\text{out}} = \left(\frac{A_v e_{\text{in}}}{R_{\text{out}}}\right)\left(\frac{R_{\text{out}} R_{\text{in}}}{R_{\text{out}} + R_{\text{in}}}\right)\left(\frac{1}{1 - j[X_{Cc}/(R_{\text{out}} + R_{\text{in}})]}\right) \quad (13.46)$$

Using the definition given by Eq. (13.43),

$$e_{\text{out}} = A_v e_{\text{in}} \left(\frac{R_{||}}{R_{\text{out}}}\right)\left(\frac{1}{1 - j[X_{Cc}/(R_{\text{out}} + R_{\text{in}})]}\right)$$

Solving for the overall voltage gain, $e_{\text{out}}/e_{\text{in}}$,

$$A_{\text{ov}} = \frac{e_{\text{out}}}{e_{\text{in}}} = A_v \left(\frac{R_{||}}{R_{\text{out}}}\right)\left(\frac{1}{1 - j[X_{Cc}/(R_{\text{out}} + R_{\text{in}})]}\right) \quad (13.47)$$

Using Eq. (13.44), Eq. (13.47) can be written as

$$A_{\text{ov}}\,(\text{low}) = A_{\text{ov}}\,(\text{mid}) \left(\frac{1}{1 - j[X_{Cc}/(R_{\text{out}} + R_{\text{in}})]}\right) \quad (13.48)$$

The absolute value of Eq. (13.48) is

$$|\,A_{\text{ov}}\,(\text{low})\,| = \frac{A_{\text{ov}}\,(\text{mid})}{\sqrt{1 + [X_{Cc}/(R_{\text{out}} + R_{\text{in}})]^2}} \quad (13.49)$$

Using the half-power definition, the condition on C_C becomes

$$f_1 = \frac{1}{2\pi(R_{\text{in}} + R_{\text{out}})C_C} \quad (13.50)$$

Equation (13.49) can now be stated in terms of f_1, Eq. (13.50),

$$|\,A_{\text{ov}}\,(\text{low})\,| = \frac{A_{\text{ov}}\,(\text{mid})}{\sqrt{1 + (f_1/f)^2}} \quad (13.51)$$

At the low-frequency end, the gain falls off in exactly the same manner as for a vacuum tube RC-coupling circuit, and for the same reason.

High Frequency

For the high-frequency case, there are two reasons for the gain's falling off. First, the shunting effect of the interelectrode capacitances becomes important; second, there is a physical limitation due to the diffusion transit time of the carriers, which is usually expressed as a decrease in α with frequency. The shunting effect of the interelectrode capacitances will be dealt with first.

Figure 13.16(a) shows the assorted resistors and capacitors that enter into the high-frequency equivalent circuit. This has been converted, in Fig. 13.16(b), into an equivalent circuit

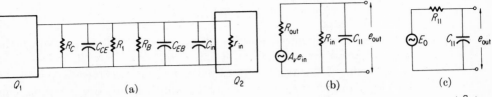

Fig. 13.16 (a) High-frequency equivalent circuit. (b) Reduction by use of R_{out}. (c) Further reduction by Thévenin's theorem.

$$E_0 = A_v e_{in}\left(\frac{R_{||}}{R_{out}}\right)$$

utilizing R_{out}. R_{in} is as defined before, and $C_{||}$ is composed of C_{CE}, C_{EB}, and C_{in} in parallel. The equivalent circuit can be further reduced by taking $C_{||}$ as the load and combining R_{out} and R_{in}. The circuit so formed is shown in Fig. 13.16(c) where

$$E_0 = A_v e_{in}\left(\frac{R_{||}}{R_{out}}\right)$$

$$R_{||} = \frac{R_{out}R_{in}}{R_{out} + R_{in}}$$

The output voltage can be written down as

$$e_{out} = A_v e_{in}\left(\frac{R_{||}}{R_{out}}\right)\left(\frac{-jX_{C||}}{R_{||} - jX_{C||}}\right) \tag{13.52}$$

Forming the ratio e_{out}/e_{in} and dividing numerator and denominator by $-JX_{C||}$,

$$A_{ov}\text{ (high)} = A_v\left(\frac{R_{||}}{R_{out}}\right)\left(\frac{1}{1 + j(R_{||}/X_{C||})}\right) \tag{13.53}$$

$$= \frac{A_{ov}\text{ (mid)}}{1 + [j(R_{||}/X_{C||})]} \tag{13.54}$$

The absolute value of A_{ov} (high) becomes

$$|A_{ov}\text{ (high)}| = \frac{A_{ov}\text{ (mid)}}{\sqrt{1 + (R_{||}/X_{C||})^2}} \tag{13.55}$$

This is exactly the same form obtained in the vacuum tube case; so, from that example,

$$f_2 = \frac{1}{2\pi R_{||}C_{||}} \tag{13.56}$$

and

$$|A_{ov}\text{ (high)}| = \frac{A_{ov}\text{ (mid)}}{\sqrt{1 + (f/f_2)^2}} \tag{13.57}$$

ALPHA CUTOFF

Alpha cutoff is defined as that frequency at which the value of α has been reduced to 0.707 of its value at 1000 cps. Although the transistor may operate at this cutoff frequency, its gain will be reduced drastically. Usually a transistor is chosen to operate in frequency ranges well below the alpha cutoff frequency, f_{co}. For our purposes, alpha decreases as f increases because the transit time for charged carriers is finite in a transistor. As f increases, the transit time becomes an appreciable portion of the period of the applied frequency, and this is reflected in lower gain.

Rigorously, it can be shown that β and α are functions of the diffusion length of the charge carriers, which can be related to the transit time (with average velocities known) and hence depends upon the frequency of the applied signal. The most important parameter appears to be the base width. If the base width is made smaller, the cutoff frequency increases. If the base width is made too small, resistance r_b increases too much, and the transistor gain goes down again. The value of α does not change significantly until one approaches f_{co}. Therefore the effect is felt only in the high-frequency range. The effect can be taken care of by writing the high-frequency gain as

$$| A_{ov} \text{ (high)} | = \frac{A_{ov} \text{ (mid)}}{\sqrt{1 + (f/f_2)^2}} K(f) \qquad (13.58)$$

where $K(f)$ is an empirically derived function that approximates the α cutoff characteristics.

$$K(f) = \frac{1}{\sqrt{1 + (f/f_{co})^2}} \qquad (13.59)$$

Thus, incorporating Eq. (13.59), the complete high-frequency gain equation becomes

$$| A_{ov} \text{ (high)} | = \frac{A_{ov} \text{ (mid)}}{\sqrt{1 + (f/f_2)^2} \sqrt{1 + (f/f_{co})^2}} \qquad (13.60)$$

The cutoff frequency depends upon the configuration used. It is higher for the grounded base than for the grounded emitter. The empirical relationship between cutoff frequencies is

$$f_{coe} = f_{cob}(1 - \alpha) \qquad (13.61)$$

where

$f_{coe} = \alpha$ cutoff frequency for grounded emitter configuration

$f_{cob} = \alpha$ cutoff frequency for grounded base configuration

$\alpha = r_m/r_c$ at midfrequency range

So, one of the advantages of the grounded-base configuration is that it can operate as an amplifier at much higher frequencies than can the grounded emitter. The cutoff factor in Eq. (13.60) does not affect the shape of the curve if $f_2 < f_{co}$. It becomes important when f_2 is close to f_{co}. To show this effect, the upper half-power point for Eq. (13.60) will be derived. The upper half-power point is defined as the frequency at which $|A_{ov} \text{ (high)}|$ is reduced to 0.707 of its maximum value. This means that the following holds true at the half-power point:

$$\sqrt{1 + \left(\frac{f'_2}{f_2}\right)^2} \sqrt{1 + \left(\frac{f'_2}{f_{co}}\right)^2} = \sqrt{2} \qquad (13.62)$$

where f'_2 is the new upper half-power point. Squaring both sides and eliminating quantity signs,

$$(f'_2)^4 \frac{1}{f_2^2 f_{co}^2} + (f'_2)^2 \left(\frac{1}{f_2^2} + \frac{1}{f_{co}^2}\right) - 1 = 0$$

Multiplying through by $f_2^2 f_{co}^2$,

$$(f'_2)^4 + (f'_2)^2 (f_2^2 + f_{co}^2) - f_2^2 f_{co}^2 = 0 \qquad (13.63)$$

Solving for $(f'_2)^2$ by means of the quadratic formula, and retaining only the positive root,

$$(f'_2)^2 = \frac{f_2^2 + f_{co}^2}{2} \left[\sqrt{1 + 4 \frac{f_2^2 f_{co}^2}{(f_2^2 + f_{co}^2)^2}} - 1 \right] \qquad (13.64)$$

Taking the square root and using only the positive root,

$$f'_2 = \sqrt{\left(\frac{f_2^2 + f_{co}^2}{2}\right) \left(\sqrt{1 + 4 \frac{f^2 f_{2co}^2}{(f_2^2 + f_{co}^2)^2}} - 1\right)} \qquad (13.65)$$

This equation shows how the overall upper half-power point is reduced because of the effect of the alpha cutoff characteristics. For example, if $f_{co} = 5f_2$, then Eq. (13.65) shows that the new upper half-power point will be $f'_2 = 0.954f_2$. The band width is reduced slightly because of the alpha cutoff factor. The tabulation below indicates how f'_2 depends upon the ratio of f_2/f_{co}.

f_2/f_{co}	f'_2
0.01	1.000
0.1	0.999
0.2	0.966
0.5	0.835
1.0	0.643

Coupling Capacitors

The value of C_C is obtained by using Eq. (13.50) and choosing a value for f_1, the lower half-power point. For example, for a value of $R_{in} = 500$ ohms and $R_{out} = 5$ K and choosing $f_1 = 10$ cps, the minimum value of C_C becomes

$$C_C = \frac{1}{2\pi f_1 (R_{in} + R_{out})}$$

$$= \frac{1}{2\pi \times 10 \times (500 + 5000)} \cong 3 \ \mu f$$

In general, the coupling capacitors in transistor audio circuits are high, of the order of 5 μf. This is due to the small values of R_{out} and R_{in}. Normally, electrolytic capacitors are used as coupling capacitors in order to obtain large capacitance values in a small size container. Polarity must be observed; otherwise, excessive leakage currents will result.

Emitter Bypass Capacitor

The actual equation to be used to calculate C_E is rather complicated. However, a simpler method can be substituted because C_E is essentially a bypass capacitor, placed in the circuit to keep the emitter at ac ground. It therefore has a minimum value, and any value above this is satisfactory. For this reason, the method used in calculating C_K for vacuum tubes will suffice. In some applications the method will give a value of C_E larger than required. The approximate equation to be used is empirical in nature and is given by

$$X_{C_E} \leq \tfrac{1}{2} R_E$$

Solving for C_E,

$$C_E \geq \frac{2}{2\pi f_1 R_E} \tag{13.66}$$

For audio work, f_1 is usually taken as 10 cps. This will make Eq. (13.66) become

$$C_E \geq \frac{32}{R_E} \ \mu f$$

where R_E is given in kilohms (K). To show the order of magnitude of C_E for audio work, it will be calculated for $R_E = 1$ K.

$$C_E \geq 32 \ \mu f$$

To make things even simpler, one will find that a check of several audio amplifiers in tube manuals will show a value of $C_E = 50$ μf for R_E values that range from 330 ohms to 2.2 K. Thus, a rule of thumb is to use $C_E = 50$ μf for most applications.

13.6 Power Amplifiers

The design of a transistor power amplifier does not differ significantly from the vacuum tube in many points. For example, the single-ended amplifier must be operated class A for minimum distortion, the load line must be kept below the maximum power-dissipation curve, and the driving voltage must be kept below a critical amplitude so as to keep the second harmonic distortion below 5 percent. But a transistor is not a vacuum tube, and there are differences between them. A few of these differences are listed below.

1. The operating temperature of a transistor is critical, so care must be taken to keep it within bounds, heat sinks are used for this purpose.

2. The average power dissipated by the transistor is greatest when no signal is present, as in vacuum tubes, so that an emitter resistor must be used to help keep the transistor stabilized.

3. Distortion can arise due to nonequal spacing of $I_B = $ const curves, to the nonlinear input characteristics, to movement of the bias point as the temperature changes, and to clipping. Both saturation and cutoff are very sharp for a transistor.

4. Because of the construction of a power transistor, its input resistance is very low, so that it will require an appreciable input power to drive it. For this reason, a transistor power amplifier requires a driver stage, which is essentially operated as a low-power amplifier with an output in the range of 10 to 50 mv, carefully matched to the stage it is driving.

TRANSISTOR OUTLINES

The low-power amplifiers are encapsulated in small cases such as the TO-1 and TO-8, shown in Fig. 13.17. They depend upon radiation and convection transfer to keep them at ambient temperature. For higher wattages, radiating fins and conduction contact of the collector junction with the metal chassis are used. Two common high-wattage models are the TO-3 and TO-36 outlines, shown in Fig. 13.17.

In both types, the case is connected thermally to the collector junction so that the circuit can be grounded in such a way that the collector can be at chassis ground. In this way, the metal chassis (usually aluminum), helps conduct heat away from the

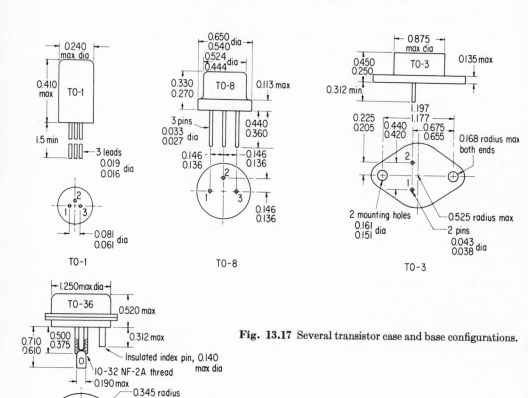

Fig. 13.17 Several transistor case and base configurations.

collector junction. In this case the flanges in the TO-3 are held in tight physical and electrical contact against the chassis by means of 4-40 machine screws. In the TO-36, the center threaded lug is used to make contact with the chassis.

Sometimes the circuit cannot be arranged conveniently to put the collector at chassis ground; then the collector must be insulated from the chassis. In that case, the TO-3 type, for example, is mounted as shown in Fig. 13.18. The flange is used as a radiating fin and the transistor must be used at reduced wattage.

THREE POWER TRANSISTORS

The following information concerns three power transistors, the 2N1068, 2N1070, and 2N1100. The data will be used in problems and examples in the subsequent sections. The 2N1068 and 2N1070 are medium-power transistors, and the 2N1100 is a high-power model. See Figs. 13.19, 13.20, 13.21. Note the ordinate and abscissa for both input and output characteristics for each transistor. In the 2N1068 input characteristics, the I_C and E_B transfer curve is given for $E_C = 4$ volts.

It should be noted at this time that the $E_C = $ const curves do not differ very much for the power amplifier, so that a typical

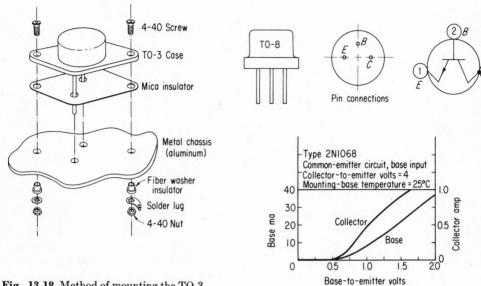

Fig. 13.18 Method of mounting the TO-3 type when the collector is not at chassis ground.

Fig. 13.19 Low-power transistor.

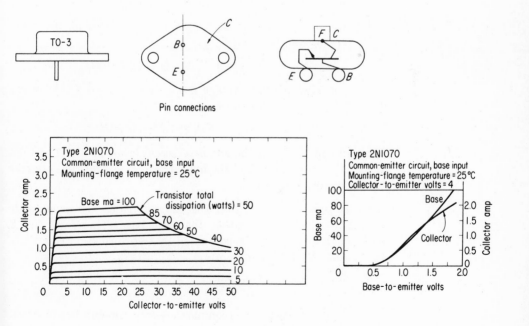

Fig. 13.20 Medium-power transistor.

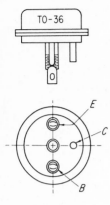

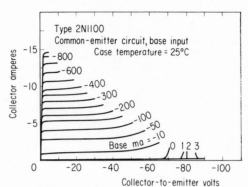

Fig. **13.21** High-power transistor.

curve such as shown is sufficient for graphical constructions. The slope of this curve gives h_{11}. For the linear portion, one obtains $h_{11} = 30$ ohms, a low value. Similarly, for the 2N1070 transistor, a higher-power transistor, $h_{11} = 10$ ohms. To obtain the order of magnitude of the average power dissipation under no-signal conditions, pick out a point in the output characteristics that is roughly in the center of the curves.

For the 2N1068, the point selected is $I_B = 8$ ma, $E_C = 15$ volts. The average power is given by

$$P_{av} = I_C E_C = 0.45 \times 15 = 6.75 \text{ watts}$$

For the 2N1070, the average power is calculated at the point $I_B = 50$ ma, $E_C = 15$ volts:

$$P_{av} = I_C E_C = 1.25 \times 15 = 18.7 \text{ watts}$$

For the 2N1100, the average power is calculated at the point $I_B = 300$ ma, $E_C = 10$ volts:

$$P_{av} = I_C E_C = 8.0 \times 10 = 80 \text{ watts}$$

Example 13.3 Using the characteristics shown in Fig. 13.19 for the 2N1068 transistor, find the hybrid parameters for the common base configuration.

1. It is to be noted that the characteristics are given with respect to the emitter rather than the base. Therefore, the h parameters found must be converted by transformation equations to the common base configuration.

2. For h_{11}, this is given by

$$h_{11} = \left(\frac{\Delta E_B}{\Delta I_B} \right)_{EC} \qquad \text{(common emitter configuration)}$$

Using the linear portion of the base characteristics, from about $I_B = 10$ ma to $I_B = 40$ ma, one can use the $E_C = $ const line itself.

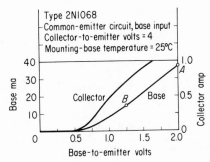

Points A and B in the above diagram are sufficient.

$$h_{11} = \frac{(2 - 1.25) \text{ volts}}{(37 - 15) \times 10^{-3} \text{ amp}} = 34 \text{ ohms}$$

3. The parameter h_{12} is also obtained from the input characteristics. However, this parameter is very small for a power amplifier, and it would be difficult to obtain it graphically. This is usually obtained by measurements. Assume $h_{12} = 75 \ \mu\text{mho}$ for this transistor.

4. The parameter h_{21} is obtained from the output characteristics. For the grounded emitter configuration, it is given by

$$h_{21} = \left(\frac{\Delta I_C}{\Delta I_B}\right)_{EC} \qquad \text{(common emitter configuration)}$$

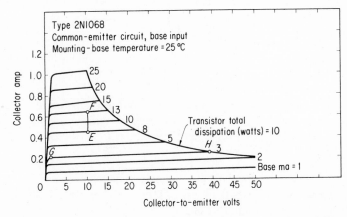

In the curves shown above, using an arbitrary value of $E_C = 10$ volts and points E and F,

$$h_{21} = \frac{(0.65 - 0.45) \text{ amp}}{(13 - 8) \times 10^{-3} \text{ amp}} = 40$$

In passing, it should be noted that this is the definition for β.

5. Using the output characteristics again, h_{22} for the common emitter configuration is given by

$$h_{22} = \left(\frac{\Delta I_C}{\Delta E_C}\right)_{I_B} \qquad \text{(common emitter configuration)}$$

All the curves for $I_B = $ const have about the same slope; so, using a "long" curve for accuracy (that is, $I_B = 3$ ma), one obtains

$$h_{22} = \frac{(0.25 - 0.20)\ \text{amp}}{(40 - 0)\ \text{volts}} = 1250\ \mu\text{mho}$$

from points G and H,

6. Using the conversion equations from Appendix II, convert from CE to CB:

$$h_{11}(\text{b}) = \frac{h_{11}(\text{e})}{1 + h_{21}(\text{e})} = \frac{34}{1 + 40} = 0.83$$

$$h_{12}(\text{b}) = \frac{h_{11}(\text{e})\,h_{22}(\text{e})}{1 + h_{21}(\text{e})} - h_{12}(\text{e})$$

$$= \frac{34 \times 1250 \times 10^{-6}}{1 + 40} - 75 \times 10^{-6}$$

$$= 962 \times 10^{-6}$$

$$h_{21}(\text{b}) = \frac{-h_{21}(\text{e})}{1 + h_{21}(\text{e})} = \frac{-40}{41} = -0.976$$

$$h_{22}(\text{b}) = \frac{h_{22}(\text{e})}{1 + h_{21}(\text{e})} = \frac{1250 \times 10^{-6}}{41} = 305\ \mu\text{mho}$$

Example 13.4 Transform the hybrid parameters found in Example 13.3 into T-type parameters and calculate the input resistance of the 2N1068 for a load resistance $R_C = 50$ ohms. Also, calculate r_{out}.

1. Using the transformation equations developed in Sec. 12.5, and noting that the hybrid parameters used must be those for the common base configuration,

$$r_e = h_{11} - \frac{h_{12}}{h_{22}}(h_{21} + 1) \qquad\qquad r_c = \frac{1 - h_{12}}{h_{22}}$$

$$r_b = \frac{h_{12}}{h_{22}} \qquad\qquad\qquad\qquad r_m = -\frac{h_{21} - h_{12}}{h_{22}}$$

2. Substituting:

$$r_e = 0.83 - \frac{962 \times 10^{-6}}{305 \times 10^{-6}} (-0.976 + 1) = 0.07 \text{ ohm}$$

$$r_b = \frac{965 \times 10^{-6}}{305 \times 10^{-6}} = 3.16 \text{ ohms}$$

$$r_c = \frac{1 - 965 \times 10^{-6}}{305 \times 10^{-6}} = 3280 \text{ ohms}$$

$$r_m = -\frac{-0.976 + 965 \times 10^{-6}}{305 \times 10^{-6}} = 3200 \text{ ohms}$$

Note should be taken of the extremely small values of r_e and r_b that appear above.

3. The input resistance is given by Eq. (12.67):

$$r_{\text{in}} = r_b + r_e(1 + A_i)$$

where

$$A_i = \frac{r_m}{R_C + r_c - r_m}$$

Finding A_i first,

$$A_i = \frac{3200}{50 + 3280 - 3200} = 24.6$$

Inserting this into the equation for r_{in},

$$r_{\text{in}} = 3.16 + 0.07(1 + 24.6) = 4.95$$

Although this is a very low value, it is characteristic of power amplifiers.

4. For r_{out}, one uses Eq. (12.72)

$$r_{\text{out}} = r_c - r_m - \frac{r_e r_m}{R_{\text{in}} + r_b + r_e}$$

In the third term, R_{in} is given by r_{in}, and the biasing resistors in parallel. In all practical cases, $r_{\text{in}} < R_B$, where R_B is the effective bias resistance. Therefore the value of $R_{\text{in}} \cong r_{\text{in}}$:

$$r_{\text{out}} = 3280 - 3200 - \frac{0.07 \times 3200}{4.95 + 3.23}$$

$$\cong 53 \text{ ohms}$$

13.7 Single-Ended Power Amplifier

Class A single-ended power amplifiers are shown in Fig. 13.22. One is transformer-coupled and the other is RC-coupled in the input circuit. Both outputs are transformer-coupled to the loudspeaker. In both amplifiers, R_1 and R_2 furnish the proper base bias to place the transistor in the middle of the linear portion of its I_C and E_B transfer curve. The emitter resistor R_E is inserted for stabilizing purposes. Since the emitter current is large for a power transistor, R_E is usually of the order of 5 to 50 ohms. This is too low a value to make C_E practicable (C_E would have to be >1000 μf), so C_E is usually omitted. The input transformer T_1 is used to match the driving stage to the power amplifier. A cheaper way is to use RC coupling. Usually there is significant mismatch this way, but the loss in gain due to this is accounted for in the overall design.

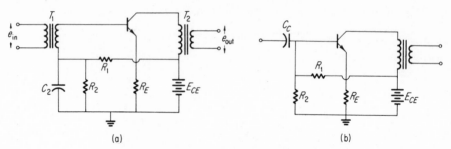

(a) (b)

Fig. 13.22 (a) Transformer coupled single-ended power amplifier. (b) RC-coupled single-ended power amplifier.

Figure 13.23 shows an idealized set of collector characteristics for a power transistor. A load line is drawn in. Since the $I_B = $ const curves are fairly straight almost to $E_C = 0$, point A is essentially at $E_C = 0$. It will be considered zero in this general discussion. Assuming the operation point to vary from A to B along the load line, as shown in Fig. 13.23(a), the resulting collector voltage and

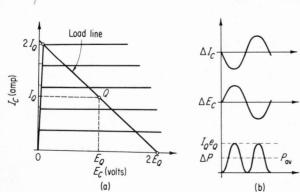

(a) (b)

Fig. 13.23 Linear representation of transistor characteristics.

current changes can be constructed. In terms of the Q-point coordinates, the dc power, with no signal present, is given by

$$P_{dc} = I_Q E_Q \qquad (13.67)$$

where E_Q is the supply voltage.

$$P_{ac} = I_{eff} E_{eff} = \frac{I_Q}{\sqrt{2}} \frac{E_Q}{\sqrt{2}} = \frac{1}{2} I_Q E_Q \qquad (13.68)$$

which is recognized as the average value of the instantaneous power curve developed in Fig. 13.23(b). The efficiency for this amplifier is given by

$$\text{Percent efficiency} = \frac{P_{ac}}{P_{dc}} \times 100 = \frac{\frac{1}{2} I_Q E_Q}{I_Q E_Q} \times 100 = 50 \text{ percent}$$

Because of the shape of the collector characteristics, the efficiency is higher than that for a triode tube, and in practice comes very close to this maximum value of 50 percent.

The ac power is transmitted to the loudspeaker by means of an impedance-matching transformer. The power dissipated in the transistor, with maximum signal, is given by the difference between the power supplied by the dc supply and the power dissipated by the loudspeaker:

$$P_{av} = P_{dc} - P_{ac}$$

Substituting the values found in Eqs. (13.67) and (13.68),

$$P_{av} = \tfrac{1}{2} I_Q E_Q \qquad \text{(max signal present)}$$

For no signal, the power dissipated in the transistor is given by Eq. (13.67). The design must take these facts into account.

Example 13.5 Using the characteristics for the 2N1068 transistor, Fig. 13.19, design a power amplifier that will be coupled to an 8-ohm loudspeaker.

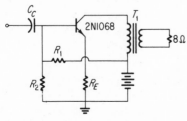

1. The T parameters for the 2N1068 have been calculated in Example 13.4. Also, on the basis of these parameters, it was shown that $r_{in} = 4.95$ ohms and $r_{out} = 53$ ohms. The r_{out} figure suggests that for maximum efficiency, the effective load resistance should be 53 ohms, as shown at right.

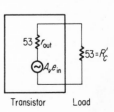

Transistor   Load

2.   The dc voltage at the collector will be essentially equal to the supply voltage E_{CE} because the dc-voltage drop across the primary of the output transformer is small. Thus the choice of E_{CE} will determine the Q point. What is usually done, however, is to construct the load line, choose the excursion along the load line, and then take the midpoint of this excursion as the Q point. Dropping a perpendicular from the Q point will give $E_Q = E_{CE}$.

Because of the interaction between the spacing of the I_B = const curves and the input characteristics, any load line chosen can be used over practically its entire length and not produce appreciable distortion. Therefore the only criterion to be met here is the power dissipation when no signal is present. To show the excellent linearity for large signals, three load lines, each for R_C = 53 ohms, have been drawn on the output characteristics.

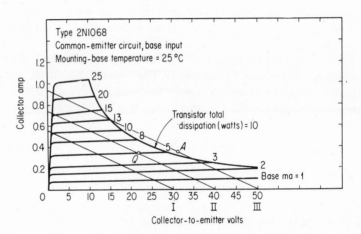

Load line I has an effective value of E'_{CE} = 30 volts, load line II has an effective value of E'_{CE} = 40 volts, and load line III has an effective value of E'_{CE} = 50 volts.

3.   From these load lines, the $I_C = E_B$ transfer curve is constructed by using the load line and input curve labeled "base." The results are shown on the next page.

The transfer curves can be considered as linear. Each shows saturation due to the fact that each load line is below the knee of the I_B = 25 ma curve. Obviously, transfer curve III will give the greatest ac power, but it should not be used because the Q point for load line III will fall well within the maximum power-dissipation curve. If one uses a change in I_C from I_C = 0.80 amp to I_C = 0.08 amp for transfer curve III, the median value for I_C will work out to I_C = 0.36 amp.

Going back to the output characteristics and the load line, the Q point is located as the point where I_C = 0.36 amp intersects

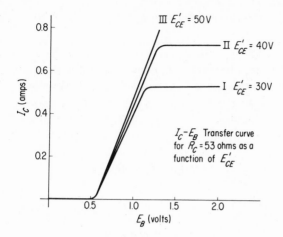

$I_C - E_B$ Transfer curve for $R_C = 53$ ohms as a function of E'_{CE}

load line III. This point is labeled A. As can be seen, it is well within the maximum dissipation curve.

4.   A good choice for the desired load line would be load line II. Looking at the I_C and E_B transfer curves developed in Part 3 of this example, and using a variation in I_C from $I_C = 0.68$ amp to $I_C = 0.20$ amp, one can see that the Q point will be located where the mean of these two values crosses load line II. This occurs for $I_C = I_Q = 0.44$ amp. This point is located on load line II and is labeled Q. The corresponding value of E_C is $E_C = E_Q = 22.0$ volts.

5.   Calculations will now be made on the maximum ac power, efficiency, R'_C, E_{CE}, and the turns ratio for the output transformer.

(a)   The dc power at the Q point is

$$I_Q = 0.44 \text{ amp}$$

$$E_Q = 22.0 \text{ volts}$$

$$P_{dc} = I_Q E_Q = 9.68 \text{ watts}$$

(b)   The maximum ac power is given by

$$P_{ac} = \tfrac{1}{8}(I_{max} - I_{min})(E_{max} - E_{min})$$

Along load line II,

$$I_{max} = 0.68 \text{ amp} \qquad E_{max} = 29 \text{ volts}$$

$$I_{min} = 0.20 \text{ amp} \qquad E_{min} = 4 \text{ volt}$$

$$P_{ac} = \tfrac{1}{8}(0.68 - 0.20)(29 - 4) = 1.50 \text{ watts}$$

(c)   The efficiency becomes

$$\text{Percent efficiency} = \frac{P_{ac}}{P_{dc}} \times 100 = \frac{1.50}{9.68} \times 100 = 15.5 \text{ percent}$$

(d) R'_C has been determined by the slope of the load line as

$$R'_C = 53 \text{ ohms}$$

(e) The supply voltage E_{CE} is found by the relation $E_{CE} = E_Q$ for transformer coupling. Therefore $E_{CE} = 17.5$ volts.

(f) It might be of interest to calculate the voltage gain of this power-amplifier stage also, to remind one that a power stage does have considerable gain:

$$A_v = \frac{\Delta E_C}{\Delta E_B} \quad \text{(along load line)}$$

$$\Delta E_C = R'_C \, \Delta I_C = 53 \times (0.68 - 0.20) = 25.4 \text{ volts}$$

$$\Delta E_B = (1.25 - 0.77) = 0.48 \text{ volt}$$

where ΔE_B was obtained from the I_C and E_B transfer curve. Thus

$$A_v = \frac{25.4}{0.48} = 53.0$$

which is pretty good for a power stage.

(g) The turns ratio for the output transfer is given by

$$\frac{N_1}{N_2} = \sqrt{\frac{R'_C}{R_{LS}}} = \sqrt{\frac{53}{8}}$$

$$\cong \frac{2.6}{1}$$

6. The dc components can be calculated, that is, R_1 and R_2, the biasing resistors.

(a) R_2 is independent of R_1 in this biasing method. The criterion on R_2 is simply

$$R_2 \geq 10 r_{\text{in}}$$

$$r_{\text{in}} = r_b + r_e(1 + A_i)$$

$$A_i = \frac{r_m}{R_C + r_c - r_m}$$

Using the values obtained in Example 13.4 for the T parameters,

$$r_e = 0.07 \qquad r_c = 3280$$

$$r_b = 4.95 \qquad r_m = 3200$$

$$A_i = \frac{3200}{53 + 3280 - 3200} = 24.0$$

and

$$r_{\text{in}} = 4.95 + 0.07(1 + 24.0) = 6.7 \text{ ohms}$$

Thus $R_2 \geq 67$ ohms is satisfactory. We shall choose $R_2 = 100$ ohms.

(b) R_1 is chosen by knowing the base current required at the Q point. From the load line and collector characteristics, this is established as $I_B = 5.0$ ma. The bias network is as shown at the right.

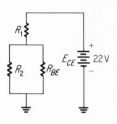

$$R_{BE} < R_2 \quad \text{and} \quad R_{BE} \ll R_1$$

Therefore

$$R_1 = \frac{E_{CE}}{I_B} = \frac{17.5 \text{ volts}}{5 \times 10^{-3} \text{ amp}} = 4.4 \text{ K}$$

(c) Because of the small base resistor value, $R_2 = 100$ ohms, the stabilization will be good for this circuit. If a swamping resistor is used here, it would be of the order of 3 to 6 ohms. In this case, R_1 would have to be recalculated, since the emitter voltage would be about 1 volt and this would subtract from E_{CE}. Also, R_E is so small it would require an inordinately large value of C_E for bypass purposes. For this reason no bypass capacitor is used. This will lower the gain slightly.

7. The complete schematic for the power amplifier is now shown.

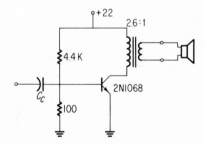

8. The value of C_C depends on R_{out} of the previous stage as well as R_{in} of this stage, and therefore is not calculated here. Also, second-harmonic distortion was not considered in this example because of the very good linearity of the I_C and E_B transfer curve.

13.8 Push-Pull Power Amplifiers

The single-ended power amplifier conducts current at all times. As was indicated in the preceding section, under no signal conditions, the power dissipated by the transistor is a maximum and may approach the maximum permissible power-dissipation curve. For this reason, as well as for efficiency, the push-pull power amplifier is very popular in transistor work.

In push-pull, the amplifiers are operated near class B, that is, near cutoff. Remember that cutoff bias is at zero volts for a transistor. Therefore an unbiased transistor is at cutoff. Also, under no signal conditions, the transistor is operating at close to zero power dissipation. These factors, coupled with the fact that two devices in push-pull produce more power than the devices operating in parallel in class A, make the push-pull amplifier desirable.

The simplicity of a push-pull amplifier operating at zero bias, that is, operating at cutoff, is shown in Fig. 13.24(a). Transformer T_1, the input transformer, acts as a phase splitter, producing two signals that are of equal amplitude but out of phase with each other by 180 deg. The output transformer T_2 is arranged as a summing device, that is, the secondary output is equal to the sum of the collector currents due to Q_1 and Q_2. The battery E_{CE} furnishes the dc supply voltage for both of the transistors. The circuit has good temperature stability because $R_B = 0$, being due only to the dc resistance of one-half of the secondary winding in T_1.

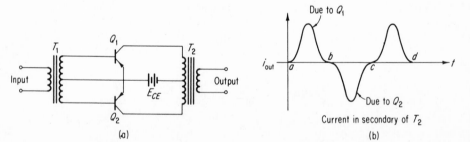

Fig. 13.24 Cross-over distortion is produced in the unbiased push-pull amplifier.

The circuit operates as follows: When the base of Q_1 is positive with respect to the center tap of T_1, the base Q_2 is negative. Since PNP-type transistors are used, Q_1 and Q_2 will conduct only if the base is negative with respect to the emitter (center tap). Therefore, under these conditions, Q_2 conducts and Q_1 is cut off. During the next half-cycle, the transistors switch roles and Q_1 conducts and Q_2 is cut off. The resultant currents are summed up by T_2 and are shown in Fig. 13.24(b). As can be seen, there is distortion at points a, b, c, d $\cdots$. This is cross-over distortion. It occurs when the emitter-base junction has a low voltage applied (remember that a semiconductor diode does not produce appreciable current until the impressed voltage exceeds about 0.5 volts). Cross-over distortion is especially bad in transistors. This distortion can be overcome in the same way it was overcome in vacuum tubes; that is, operate the transistors with a bias voltage slightly above cutoff. This is shown in Fig. 13.25.

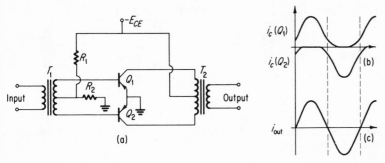

Fig. 13.25 Elimination of cross-over distortion by proper bias on the transistors.

The circuit in Fig. 13.25 is similar to that in Fig. 13.24(a) except for the addition of the biasing circuit R_1 and R_2. R_1 produces a small base current so that Q_1 and Q_2 operate slightly above cutoff, and R_2 is in the range of 30 to 100 ohms, to make R_B small for stabilization purposes. R_2 is not bypassed by a capacitor so that the center tap will be ac ground. The reason for this is that a capacitor across R_2 would tend to charge to the peak value of the signal voltage drop across R_2, since the emitter-base junctions act like diodes, with R_2 as a resistive load. This would disrupt the dc bias produced by R_1. The current in each transistor flows for slightly more than one half-cycle, as shown in Fig. 13.25(b). The graphical sum of i_c (Q_1) and i_c (Q_2) will produce an excellent sine curve, as shown in Fig. 13.25(c), if the proper choice of biasing current is made. This choice can be made graphically in the same manner as was done for the vacuum tube circuit, that is, by placing one of the I_C and E_B transfer curves upside down and then moving it about, with respect to the one that is right side up, until a resultant linear-transfer curve is found.

Figures 13.24 and 13.25 show transformer coupling to the input of the push-pull amplifier. If one wished to use RC coupling, the fact that the push-pull stage is operating at essentially class B necessitates a special coupling circuit. RC coupling has the advantage of better frequency response, so the special circuit is well worth it. When one considers dollars and cents, RC coupling, even with the special circuitry, is cheaper than a good input transformer T_1.

Figure 13.26 illustrates two RC-coupled push-pull amplifier circuits. Figure 13.26(a) is used to explain why it is *not* used in class B operation. During the half-cycle that Q_1 is conducting,

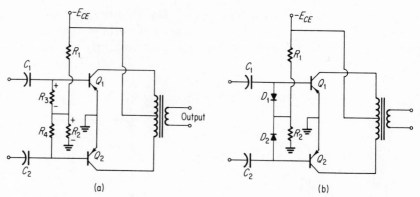

Fig. 13.26 (a) Incorrect RC-coupling of power amplifier. (b) Correct RC-coupling circuit.

capacitor C_1 charges up. During the second half-cycle, when Q_1 is cut off, C_1 discharges through R_3 and R_2, producing the voltage-drop polarities shown. These voltage drops are opposed to the bias voltage present, which is negative with respect to ground. Thus, on the half-cycle that Q_1 is cut off, R_3 has an appreciable voltage drop, which can actually alter the bias on the base because the time constant is large and C_1 discharges slowly. The reverse bias may last into the next half-cycle, when Q_1 should be conducting. C_2 and R_4 produce the same action on Q_2. The voltage drops across R_2 are neglected because R_2 is so small (10 to 50 ohms). To avoid this back bias, R_3 and R_4 are replaced by diodes, placed as shown in Fig. 13.26(b). The diodes are reversed with respect to the emitter-base junctions of Q_1 and Q_2. Thus, on the half-cycle that Q_1 is conducting, D_1 is reverse-biased and presents a resistance greater than r_{in}; during the second half-cycle, when Q_1 is cut off, D_1 is forward-biased, and presents a very small resistance to the discharge of C_1, thus reducing the time constant so that C_1 is discharged before the next half-cycle occurs.

13.9 Phase Splitters

The simplest phase splitter is the center-tapped transformer secondary, as used in Fig. 13.25(a). This has the advantage of low base-circuit resistance and therefore good stability. But the interstage gain is low, sometimes less than unity, and a transformer with good response is expensive.

Another type of phase splitter, an exact counterpart of a vacuum tube circuit, is shown in Fig. 13.27. The outputs e_1 and e_2 will be 180 deg out of phase with each other, and if R_3, R_4, and R_5 are selected correctly, the output voltages will have equal amplitudes. The output impedance from A to ground will be higher

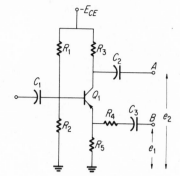

Fig. 13.27 Simple phase splitter circuit.

than the output impedance from the emitter to ground. Therefore, to ensure that the two transistors in push-pull see the same impedance, R_4 is placed in series with C_3 to increase the output resistance of the emitter circuit. This will in turn produce a voltage divider action on the output e_1, so R_5 is made larger than R_3 to compensate for this. Therefore R_4 is equal to the difference between the output resistances of the collector to ground and emitter to ground. Also, once the input resistances of the push-pull amplifiers are known, the voltage divider action of R_4 and R_{in} can be calculated, and R_5 can be raised in value accordingly.

13.10 Complementary Symmetry

Because transistors can be NPN or PNP types, this makes possible a push-pull circuit that has no vacuum tube counterpart. Consider the circuit shown in Fig. 13.28(a). Both Q_1 and Q_2 have their bases at zero bias and are therefore cut off. In the presence of a signal of positive polarity, for example, Q_2 will conduct while Q_1 will remain cut off. For a negative polarity signal, the reverse will hold true; Q_1 will conduct and Q_2 will remain below cutoff. This is just the push-pull action.

In this amplifier the configuration is that of the common collector. The load for the circuit is represented by R_E, and may be a loudspeaker voice coil or an output transformer. When Q_1 is conducting, the current flow through R_E is as shown in Fig. 13.28(a). For the transistor Q_2, it is in the opposite direction, and when no signal is present, the current through R_E is zero. Thus, in this circuit, the load R_E could very well be a loudspeaker voice coil, since the average dc current would be zero and there would be no static deflection of the speaker cone. This circuit suffers from crossover distortion because of the class B operation. Q_1 and Q_2 should each have a small forward-biasing current of the order of 1 to 5 ma.

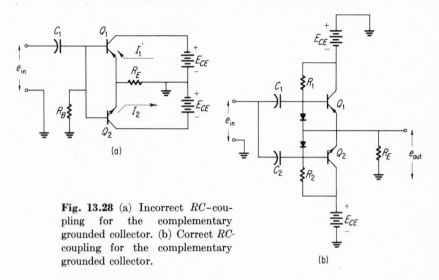

Fig. 13.28 (a) Incorrect RC-coupling for the complementary grounded collector. (b) Correct RC-coupling for the complementary grounded collector.

The circuit is modified as shown in Fig. 13.28(b) to produce the forward bias required for each transistor. The dc-biasing circuits are separated by coupling capacitors C_1 and C_2. Note that Q_1 requires a negative voltage on the base, and that Q_2 requires a positive voltage on the base. R_1 and R_2 are the biasing resistors. Diodes D_1 and D_2 are inserted to reduce the discharge time constants for C_1 and C_2 when Q_1 and Q_2 are cut off. The diodes are reverse-biased as far as the biasing circuits are concerned and do not affect them. R_1 and R_2 depend upon the idling base current desired and on the supply voltages. For example, if the bias current required is to be 2 ma and $E_{CE} = 10$ volts, then R_1 would be

$$R_1 = \frac{E_{CE}}{I_B} = \frac{10}{2 \times 10^{-3}} = 5 \text{ K}$$

D_1 and D_2 are chosen to have low reverse currents. As was mentioned before, since the output impedance of a grounded-collector power amplifier will be very low, this amplifier can be connected directly to the voice coil of a loudspeaker. One can then see that complementary symmetry has produced a push-pull amplifier that requires no phase splitter and no output transformer.

13.11 The Field-Effect Transistor (FET)

One of the deficiencies of a transistor is its relatively low input resistance. There is considerable loss in voltage when a phono-crystal is connected to a transistor amplifier because the internal resistance for such a device is usually much higher than the input resistance of the input transistor. This is compensated by adding another stage of amplification. For phonograph systems, this is an adequate solution, but in other instances, a high-input resistance is absolutely necessary. For example, a measuring device must have a very high input resistance so that it does not load down the circuit it is measuring. Also, some transducers have such high-output resistances that until now it was almost useless trying to match them to transistor amplifiers because of the large mismatch involved.

The field-effect transistor (FET) seems to solve this problem very nicely. Its input resistance can be as high as 10^8 to 10^{18} ohms. There is also negligible interaction between its input and output circuits. These two qualitities make it ideal for replacing vacuum tubes. Also, being a solid state device, it needs no filament voltage. Figure 13.29(a) shows diagrammatically the method of operation of a field-effect transistor. The particular one shown is called an n-channel FET. The major conducting block is made up of an n-type semiconductor. Attached to this semiconductor are two leads. The connections are ohmic, that is, current can pass equally well in both directions through the contact; there is no rectification. The n-block therefore acts simply as a resistance. With the battery V_D connected as shown, the left-hand lead is the source

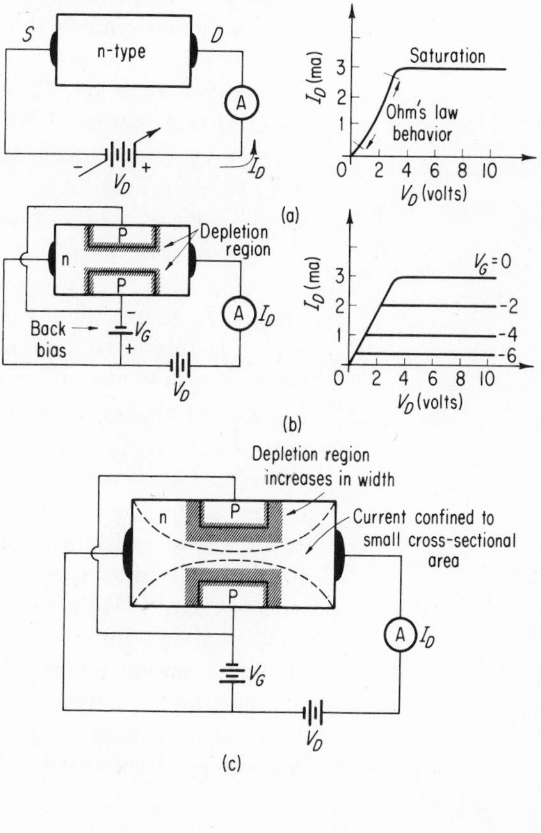

Fig. 13.29 (a) Characteristics for a simple semiconductor. (b) Introduction of gate electrodes to alter the characteristics. (c) Resistance increase with increase in the depletion region in a field-effect transistor.

lead and the right-hand lead is the drain lead. If V_D is varied and I_D measured for a small range of V_D, the material exhibits an Ohm's law behavior. The extrinsic carriers in the material are the predominant ones, and electrons flow from the source to the drain within the material. With further increase in V_D, the current quickly saturates because the available carriers in a doped semiconductor are limited. The saturation level reached indicates that as many carriers are being swept out by the drain as are injected by the source, and further increase in V_D produces very little change in I_D. Now if one places bars of p-type material in the n-block, as shown in Fig. 13.29(b), to form p–n junctions, and connects the two p-sections together, a field-effect transistor is formed.

The two p-type bars are termed the *gate*, and the voltage V_G, applied between the gate and source, is termed the *gate voltage*. Note that the gate voltage is a reverse bias for the p–n junctions formed. This means that there is no extrinsic conduction between gate and source, only intrinsic conduction. This intrinsic conduction is held to a very low value by using semiconductor materials having very few intrinsic conductors at room temperature. But the fact that there is intrinsic conduction means that FET are temperature-dependent. The value of this intrinsic conduction essentially determines the input resistance of an FET. In Fig. 13.29(b), if V_G is made more negative, this will increase the width of the depletion layer. This depletion layer extends into the n-type material and is formed because of the electric field configuration in the n-type material, between source and gate. The electric field is a retarding field for the electrons in the n-type material. Thus, as the reverse field is increased by increasing V_G, less and less carriers go from source to drain. If V_G is made large enough, the retarding field between source and drain can reduce the drain current I_D to practically zero. This is called pinchoff. Thus, the action is reminiscent of a vacuum tube triode, where the gate electrodes simulate the triode.

Figure 13.29(c) illustrates the constriction of the conduction path due to a high value of V_G. The depletion layer increases in width and the electrons in the n-type material are forced to the center of the channel. Since the cross-sectional area has been decreased, the resistance of the channel increases, and this is reflected in a smaller saturation value of I_D. There is a critical voltage in this setup. It is the potential difference between gate and drain. If this voltage difference is made too high, the intrinsic conductors that are swept across the depletion layer can obtain kinetic energies high enough to produce an avalanche breakdown. This breakdown voltage is of the order of 20 to 40 volts, depending upon the particular FET.

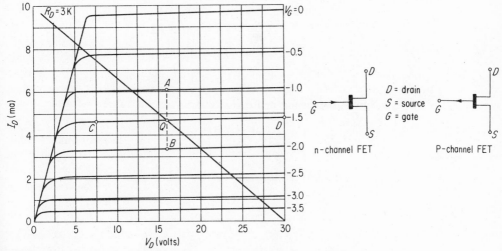

Fig. 13.30 Typical characteristics and schematic symbols for the field-effect transistor.

Figure 13.30 shows the symbol currently used for the FET and a typical set of output characteristics. The I_D and V_D characteristics are pentode-like, and the parameter that is changed to form the family of curves is the gate voltage V_G. A more detailed analysis would show that the ac equivalent circuit for the FET is similar to that of a triode. The quantities such as μ, g_m, and r_d are defined in the same way, graphically, where μ is the amplification factor, g_m the transconductance, and r_d the ac drain resistance. It would be of interest to see what the values of these parameters are for the characteristics shown in Fig. 13.30 at the point $V_D = 16$ volts, $V_G = -1.5$ volt, labeled Q:

$$r_d = \left(\frac{\Delta V_D}{\Delta I_D}\right)_{V_G} \tag{13.69}$$

$$g_m = \left(\frac{\Delta I_D}{\Delta V_G}\right)_{V_D} \tag{13.70}$$

$$\mu = \left(\frac{\Delta V_D}{\Delta V_G}\right)_{I_D} \tag{13.71}$$

1. We use points C and D to find a value for r_d.
   At point C:

   $$V_D = 7.5 \text{ volts}$$

   $$I_D = 4.62 \text{ ma}$$

   $$V_G = 1.5 \text{ volts}$$

At point D:

$$V_D = 30 \text{ volts}$$

$$I_D = 4.80 \text{ ma}$$

$$V_G = 1.5 \text{ volts}$$

Now substituting in Eq. (13.69),

$$r_d = \frac{(30 - 7.5) \text{ volts}}{(4.80 - 4.62) \times 10^{-3} \text{ amps}} = 125 \text{ K ohms}$$

2. We use points A and B to find a value for g_m.
   At point A:

$$V_D = 16 \text{ volts}$$

$$I_D = 6.08 \text{ ma}$$

$$V_\sim = 1.0 \text{ volt}$$

At point B:

$$V_D = 16 \text{ volts}$$

$$I_D = 3.39 \text{ ma}$$

$$V_G = 2.0 \text{ volts}$$

Now substituting into Eq. (13.70),

$$g_m = \frac{(6.08 - 3.39) \times 10^{-3} \text{ amp}}{[-1.0 - (-2.0)] \text{ volts}} = 2690 \text{ } \mu\text{mho}$$

Using the relationship $\mu = g_m r_d$ rather than Eq. (13.71) to evaluate μ, for accuracy,

$$\mu = 2.69 \times 10^{-3} \times 125 \times 10^3 = 336$$

This particular FET has output characteristics that resemble those of a pentode. The FET can be made with widely varying parameters, and amplification factors as high as 500 to 800 have been achieved, with correspondingly high r_d values.

A typical circuit utilizing an FET is shown in Fig. 13.31. A value of load resistance R_D equal to 3 K is shown. The load line is drawn in on the characteristics of Fig. 13.30. The transfer curve for this load line is shown in Fig. 13.31. This particular transfer curve is fairly linear between V_G values of -0.6 to -2.4 volts. Figure 13.31 shows that the gate voltage is obtained from a resistor-capacitor filter combination in the source lead, exactly similar to the cathode bias discussed for vacuum tubes. Using the transfer curve in Fig. 13.31 as an example, R_S can be calculated.

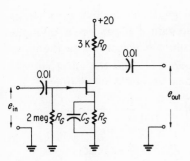

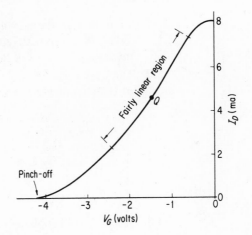

Fig. 13.31 Typical amplifier circuit and transfer curve for a field-effect transistor.

Selecting $V_G = -1.5$ volts as the gate voltage,

$$R_s = \frac{|V_G|}{I_D} = \frac{1.5 \text{ volts}}{4.62 \times 10^{-3} \text{ amp}} = 325 \text{ ohms}$$

The ac equivalent circuit calculations for the triode can be carried over to the FET. The midfrequency voltage gain for this circuit would be

$$A_v = \frac{\mu R_D}{r_d + R_D}$$

$$A_v = \frac{336 \times 3 \times 10^3}{(125 + 3) \times 10^3} = 7.9$$

A higher value of R_D will increase this low value of voltage gain significantly, since A_v depends upon the choice of R_D.

In the FET, the channel is used as a resistor; therefore the flow of charge depends upon the impressed electric field within the conductor. This means that the charges acquire a drift velocity. Regular transistors depend upon diffusion for charge motion. Since drift velocities are much greater than diffusion velocities, the transit times for FET are small and FET are capable of very high operating frequencies, up to several hundred megacycles.

There are several variations of the pattern that use special semiconductor materials to reduce reverse current in the gate circuit, thus producing extremely high input resistances, up to 10^{15} ohms. These are the MOSFET (Metal Oxide Semiconductor Field Effect Transistor) and the IGFET (Insulated Gate Field Effect Transistor). When these are used in cathode follower-like configurations, input resistances as high as 10^{18} ohms can be achieved.

Problems

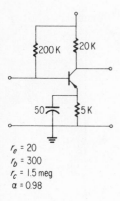

$r_e = 20$
$r_b = 300$
$r_c = 1.5$ meg
$\alpha = 0.98$

13.1

13.1 Find the available voltage gain for the following circuit. What is the overall voltage gain if a generator having an internal resistance of 2000 ohms is connected to the input?

13.2 Determine the overall voltage gain and the effective value of the output voltage for the accompanying circuit:

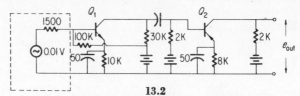

13.2

13.3 Recalculate Problem 13.2 if Q_2 has the following parameters: $r_e = 12$ ohms, $r_b = 800$ ohms, $r_c = 1.5$ meg, and $\alpha = 0.98$.

13.4 Compare the input resistances for the two transistors given in Problem 13.2.

13.5 Determine C_C for an f_1 value of 20 cps. If $C_{11} = 200$ pf and $f_{oob} = 1$ Mc, determine f_2 and the actual upper half-power point f_2'.

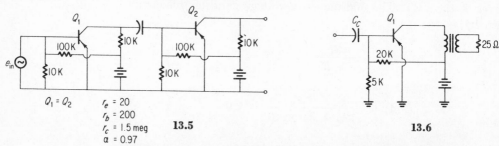

$Q_1 = Q_2$ $r_e = 20$
$r_b = 200$
$r_c = 1.5$ meg   **13.5**
$\alpha = 0.97$

13.6

13.6 For a typical power transistor, the T parameters are $r_e = 1$ ohm, $r_b = 4$ ohms, $r_c = 3280$ ohms, and $r_m = 3200$ ohms. Determine the input and output resistances of the transistor in a grounded emitter configuration, as shown.

13.7 Determine the input resistance for the circuit of Problem 13.6 if an unbypassed emitter swamping resistor of 3 ohms is inserted. Is the output resistance changed appreciably?

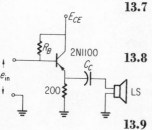

13.8

13.8 Using the 2N1100 transistor in the following circuit, determine (a) the resistors; (b) E_{CE} for a Q point of $E_C = -10$ volts, $I_B = -300$ ma; (c) the impedance of the loudspeaker for maximum power transfer.

13.9 Determine the proper base-current biasing so as to minimize crossover distortion if two 2N1068 transistors are to be operated in push-pull for an effective load of 50 ohms and $E_{CE} = 20$ volts.

14.1 The Oscilloscope

The cathode-ray oscilloscope is one of the most useful test and measurement instruments devised. Not only does it allow one to measure the magnitude of a test signal, but also it allows one to see how it varies with time. Oscilloscopes can easily measure time spans from seconds to 1×10^{-9} sec. Also, the quantities measured are limited only by the ingenuity of the one who desires the measurements. An oscilloscope is essentially a voltage-measuring device. It shows how voltages vary with time. But such parameters as current, light intensity, temperature, and sound pressure can easily be converted to voltages and measured by an oscilloscope. All one needs is a transducer that converts each of these parameters into a corresponding change in current in an electric circuit. One of the components of the electric circuit should be a resistor. Then the change in current produces a proportional voltage change across the resistor. This changing voltage is amplified and displayed by the oscilloscope.

The situation is shown in simplified form in Fig. 14.1. In this case, it is desired to display a sound wave on the oscilloscope. A transducer M is used to convert alternating pressure waves into a varying dc current. The circuit is a simple series circuit consisting of the transducer M, the bias voltage E, and the load resistor R. (The bias voltage is not always necessary. Some transducers are generators and produce their own emf.) The battery E produces a *dc* current in the simple series circuit and the transducer M modulates the current by varying its resistance in accordance with the impinging sound wave. The oscilloscope's measuring leads are placed across the load resistor R to detect the resultant voltage variation e_μ. This voltage variation is then amplified to a useful level that can be displayed on the oscilloscope screen.

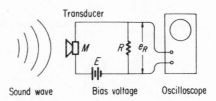

Fig. 14.1 A transducer is used here to convert sound pressure waves into a changing electric current.

Sound wave   Bias voltage   Oscilloscope

The oscilloscope is usually a high-impedance device and produces very little loading of any circuit it is measuring. The input impedance can vary from 1 meg and 25 picofarads in parallel to 100 meg and 4 picofarads in parallel. In essence, then, if a quantity can be changed into a proportional current, it can be made to produce a voltage drop across a load resistor, and this voltage can be displayed on the oscilloscope.

14.2 The Cathode-Ray Tube

The heart of the oscilloscope is the cathode-ray tube. It converts a varying voltage into a visible wave form. Usually, the varying voltage is plotted with respect to time. This type of display requires a two-coordinate system. The visible light is produced by the impact of electrons on a fluorescent screen. In order to define physically the displayed wave form, the electrons arrive at the fluorescent screen in a small diameter beam, producing a small visible spot where the beam strikes the screen. Figure 14.2(a) shows a simplified cathode-ray tube construction. The cathode-ray tube is a glass bulb, highly evacuated (so as not to slow down and spread out the electron beam), with an "electron gun" at one end and a fluorescent screen at the other end. The electrical connections are brought out of the glass bulb at the electron-gun end and terminate in a base with projecting pins so that it can be plugged into a mating socket. The electron gun produces a highly defined beam of electrons and directs them to the center of the

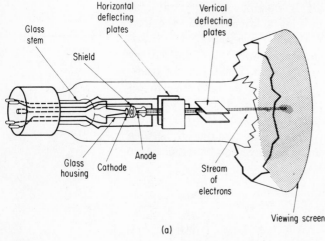

(a)

Fig. 14.2 Construction details for an electrostatic cathode-ray tube.

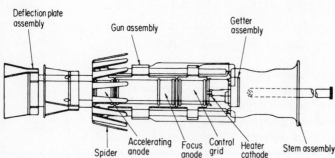

(b)

fluorescent screen. It is shown in more detail in Fig. 14.2(b). A heater-cathode assembly produces electrons and a control grid controls the intensity of the beam, and hence the intensity of the light given off by the fluorescent screen. An accelerating anode gives the electrons enough kinetic energy so that when they strike the fluorescent screen, they can initiate the fluorescent action. The spider, Fig. 14.2(b), is used to position and align the electron gun in the neck of the cathode-ray tube.

An electron beam can be deflected by means of an electric field or a magnetic field. Both methods are used in cathode-ray tubes. Most television tubes (which are cathode-ray tubes) use magnetic deflection, while most tubes used in oscilloscopes use electrostatic deflection. Electrostatic deflection is capable of higher frequency response and is therefore preferred for test equipment. The discussion will be confined to electrostatic deflection.

To secure the two-dimensional coordinate system required to display voltage wave forms, two sets of deflection plates are inserted between the electron gun and the screen. The deflection plates have a parallel-plate-capacitor geometry. When there is a difference of potential between these plates, an electric field deflects the electron beam. In Fig. 14.2, the vertical plates are the horizontal deflection plates. The electric field will be horizontal, and this deflects the beam. The horizontal plates are the vertical deflection plates.

14.3 Formation of Wave Forms

The display on a cathode-ray tube utilizes a right-handed coordinate system. One axis is vertical, the other horizontal. In most applications, displacement on the horizontal axis is proportional to time, and displacement on the vertical axis is proportional to the voltage being measured. If the horizontal displacement of the beam is directly proportional to time, then the electron beam will trace out a wave form on the screen. To secure a horizontal deflection that is proportional to time, the difference of potential across the horizontal deflection plates must change linearly with time. This is shown schematically in Fig. 14.3(a).

As the electric field is increased with time, the beam deflection increases, and the length of a line traced on the screen is proportional to time. The voltage that is applied to the horizontal deflection plates, and which is necessary to produce this type of deflection, is called a *sawtooth voltage*. It is shown in Fig. 14.3(b). As can be seen, during the time period T_1 the sawtooth voltage increases linearly with time from a to b. At b, the voltage drops suddenly to its starting value C. The time T_1 is called the *sweep*

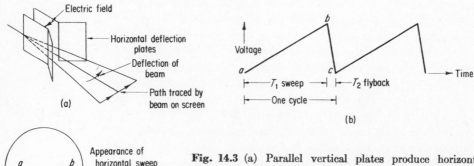

(a)

(b)

(c)

Fig. 14.3 (a) Parallel vertical plates produce horizontal deflection. (b) Sawtooth voltage applied to horizontal deflection plates. (c) Linear horizontal sweep produced by a sawtooth voltage.

time, since this is the time that the beam is tracing out a horizontal line. The time T_2 is small compared with T_1, and is called the *flyback time* because this is the time required to bring the beam back to its starting point. During this flyback time, the cathode-ray tube is cut off by applying either a negative pulse to the control grid of the electron gun or a positive pulse to the cathode of the electron gun. The cutoff pulse has the same duration as the flyback time, and is formed in wave-shaping circuits from the horizontal sweep voltage. Cutting off the cathode-ray tube in this manner keeps the return trace from appearing on the fluorescent screen.

The sawtooth voltage, when applied to an oscilloscope, is called a *sweep voltage,* and the horizontal trace on the screen, a *horizontal sweep.* If the sweep voltage is cyclic, then the beam will sweep from left to right across the face of the screen during time T_1, return suddenly during time T_2, and repeat. If the sweep is slow enough, several cycles per second, then the spot where the beam strikes the screen can be seen to move from left to right, suddenly return, and repeat. At higher sweep frequencies, greater than about 30 cps, the spot can no longer be distinguished because of the eye's persistence of vision, and one sees a line on the screen, as the cycles blend together, Fig. 14.3(c).

To show the desired voltage on the screen, the vertical deflection plates must have a difference in potential proportional to the desired voltage. This will cause the beam to deflect up and down. The combination of vertical and horizontal sweeps makes the beam move across the oscilloscope screen in such a way that the desired wave form is traced on the screen. Figure 14.4 will make this clear. A simple graphical construction allows one to predict the resultant motion of the spot on the screen of the cathode-ray tube. In Fig. 14.4(a), the horizontal sweep is shown to be a sawtooth wave form, and the desired signal is a sine wave. Using the same time intervals for each wave form, the corresponding points

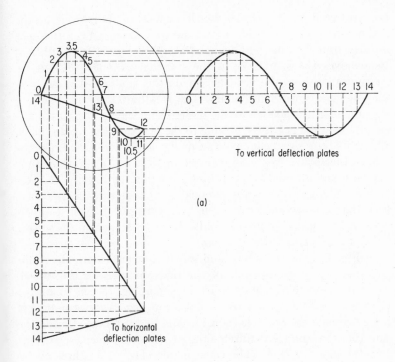

To vertical deflection plates

(a)

To horizontal
deflection plates

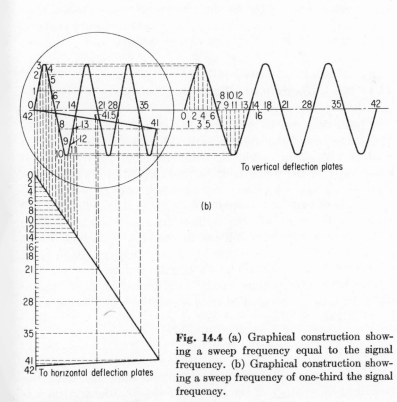

To vertical deflection plates

(b)

To horizontal deflection plates

Fig. 14.4 (a) Graphical construction showing a sweep frequency equal to the signal
frequency. (b) Graphical construction showing a sweep frequency of one-third the signal
frequency.

are projected to the circle, which represents the screen of the cathode-ray tube. The horizontal sweep voltage moves the spot linearly from left to right for points 1, 2, 3, $\cdots$, 10, 11, 12, and then moves the spot back to the starting point from 12 to 14. At the same time, the vertical signal (the sine wave) moves the spot up and down sinusoidally. The resultant trace is shown in the circle. It is a portion of the desired sine wave. The reason the entire sine wave is not seen is due to the fact that the retrace time, 12, 13, 14, is a significant portion of the sine wave period. On the screen, the spot moves from point 12 to 13 to 14. The retrace time has been accentuated here to show its action. Ordinarily, it takes only a few microseconds for the retrace. If both signals are cyclic, then the trace shown in the circle is repeated, each cycle is superimposed on the preceding one, and a sine wave is seen on the screen, owing to the persistance of vision.

In Fig. 14.4(a) the sweep and signal frequencies are the same, and therefore a single cycle of the signal is seen on the screen. To show more than one cycle of the signal, the sweep frequency must be a *submultiple* of the signal frequency. The formation of three cycles of the signal is shown in Fig. 14.4(b). Note that in the time that the signal goes through three cycles, the sweep (sawtooth) goes through one cycle. The construction shows that three cycles of the signal will appear on the screen before the process is repeated. So, in using an oscilloscope, the sweep frequency is always set equal to the signal frequency or a submultiple of it.

14.4 Block Diagram of a Basic Oscilloscope

The potentials applied to the horizontal and vertical deflection plates must be of the order of several hundred volts in order to produce full screen deflection. Ordinarily, the signal voltage is low. Therefore the signal must be amplified before it can be applied to the vertical plates. The sawtooth voltage used for the sweep is produced by a sawtooth oscillator in the oscilloscope and must also be amplified. The oscilloscope must therefore contain vertical and horizontal amplifiers. It would be of interest to describe the oscilloscope in terms of a block diagram. Each block will represent a circuit. The block diagram will be discussed first, so that the overall picture can be seen, then each block will be discussed in some detail. Figure 14.5 shows a typical oscilloscope in block diagram form.

ATTENUATOR

The input to an oscilloscope usually contains some means for controlling the amplitude of the signal connected to the vertical amplifier. This is necessary for two reasons: first, if the signal is too large, it may be clipped by the vertical amplifiers with the result that the display will be distorted, and second, the attenuator acts like a gain control, and helps determine the vertical size of the display on the screen.

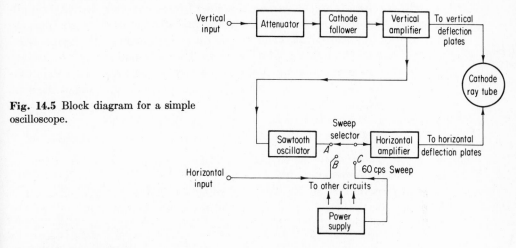

Fig. 14.5 Block diagram for a simple oscilloscope.

CATHODE FOLLOWER

To ensure that the input impedance of the oscilloscope will be high, the first stage is a cathode follower. This ensures that the input resistance will be several megohms. Consequently, the oscilloscope will produce little loading. Since complicated signals, containing high-order harmonics, will be studied, the cathode follower must present a constant impedance with respect to frequency. Since the input capacitance is low for a cathode follower, the impedance is fairly constant over a large range of frequencies (several megacycles).

VERTICAL AMPLIFIER

The vertical amplifier is used to amplify the signal to the point where it will produce full screen deflection when the vertical gain is turned up. Because the signals to be displayed are usually complex, the band width of the vertical amplifiers must be large, of the order of several megacycles. Ordinary *RC*-coupled amplifiers will not do the job. Special amplifiers called *video amplifiers* must be used. They will ensure that the display on the screen conforms closely to the input signal. The output of the vertical amplifier is connected to the vertical deflection plates.

SYNC CIRCUIT

If the horizontal oscillator is not synchronized, that is, made to have a frequency that is exactly equal to, or a submultiple of, the signal frequency, then the pattern on the cathode-ray tube will not be stationary. The horizontal sweep controls are varied to get the horizontal oscillator sweep frequency near the frequency desired, and the sync circuits will then "lock" the horizontal oscillator to the correct frequency. The input to the sync circuit is obtained from the vertical amplifier by RC coupling. In some simplified oscilloscopes, the vertical signal itself is used for the purpose of synchronization. And, in other oscilloscopes, the sync signal goes through an amplifier first. This is usually done to produce synchronization on either the positive-going cycle or on the negative-going cycle of the signal, depending on the setting of the sync control.

Sawtooth Oscillator

To produce the sawtooth voltage required for a linear sweep, an oscilloscope contains a sawtooth oscillator. The frequency of the oscillator can be controlled so that it can be made a submultiple of the signal frequency. The sawtooth oscillator output can be connected to the horizontal amplifier for further amplification.

Horizontal Amplifier

The horizontal amplifier is used to amplify the signal presented to the horizontal deflection plates. It also must have a wide band width, since its main purpose is to amplify the sweep signal without distortion. In some important applications a linear sweep is not used on the horizontal axis. For this reason there is provision to connect the horizontal amplifier to external leads so that an external horizontal signal of one's choosing can be inserted.

The selection of the appropriate input to the horizontal amplifier is made through a sweep selector switch. In position A, the input is the sawtooth voltage; in position B, the input is the external *Horizontal Input* lead; and in position C, the input is a 60-cps sine wave. The horizontal amplifier connects the amplified signal to the horizontal deflection plates.

60-Cycle Sweep

In some applications, one wants a 60-cycle horizontal sweep. This is produced by connecting a filament winding from the power transformer to the horizontal amplifier through the sweep selector switch.

Power Supply

The power supply is the source of the dc voltages and filament voltages for the remainder of the oscilloscope circuits. There are two ranges of dc voltage required: a 1500- to 20,000-volt high-voltage supply for the cathode-ray tube, and a 300- to 500-volt low-voltage supply to power the amplifier and sweep circuits. The input to the power supply is 115 volts at 60 cps.

14.5 Basic Circuits

ATTENUATOR AND CATHODE FOLLOWER

The attenuator, cathode follower, and vertical gain control are shown in Fig. 14.6. C_1 is the input-coupling capacitor. It serves to isolate the cathode-follower stage from any dc in the circuit

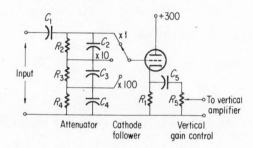

Fig. 14.6 An *RC* input attenuator.

being measured. The attenuator shown is a step attenuator. When R_2, R_3, and R_4 are selected correctly, it reduces the input voltage by a factor of 10 at every switch position except the first. The attenuator must be made frequency-insensitive so that it does not distort the input signal in any way. The capacitors C_2, C_3, and C_4 help accomplish this. To show how this is done, consider the two-step attenuator in Fig. 14.7. The output voltage is given by

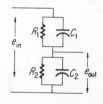

Fig. 14.7 A two-step input attenuator.

$$e_{out} = e_{in}\left(\frac{z_2}{z_1 + z_2}\right) \tag{14.1}$$

where

$$z_1 = \frac{-jR_1X_{C_1}}{R_1 - jX_{C_1}}$$

$$\tag{14.2}$$

$$z_2 = \frac{-jR_2X_{C_2}}{R_2 - jX_{C_2}}$$

Substituting those values into Eq. (14.1),

$$\frac{e_{out}}{e_{in}} = \frac{-jR_2X_{C_2}/(R_2 - jX_{C_2})}{[-jR_1X_{C_1}/(R_1 - jX_{C_1})] + [-jR_2X_{C_2}/(R_2 - jX_{C_2})]}$$

$$\tag{14.3}$$

Replacing X_{C_1} and X_{C_2} by

$$X_{C_1} = \frac{1}{4\pi f C_1}$$

$$X_{C_2} = \frac{1}{4\pi f C_2}$$

and factoring, Eq. (14.3) can be put in the form

$$\frac{e_{\text{out}}}{e_{\text{in}}} = \frac{R_2/(4\pi f R_2 C_2 - j)}{[R_1/(4\pi f R_1 C_1 - j)] + [R_2/(4\pi f R_2 C_2 - j)]} \tag{14.4}$$

A little reflection on will show that if

$$R_1 C_1 = R_2 C_2 \tag{14.5}$$

then Eq. (14.4) reduces to

$$\frac{e_{\text{out}}}{e_{\text{in}}} = \frac{R_2}{R_1 + R_2} \tag{14.6}$$

which is independent of frequency!

An extension of this discussion would lead to the condition

$$R_2 C_2 = R_3 C_3 = R_4 C_4 \tag{14.7}$$

for the attenuator in Fig. 14.6. The attenuator will present a constant impedance to any test circuit connected to it, irrespective of the position of the attenuator switch. This is so because the cathode-follower impedance is high and does not load the attenuator significantly. C_2, C_3, and C_4 are usually variable capacitors that are adjusted to compensate for the input capacitance of the cathode follower. In many circuits, C_2 *is* the interelectrode capacitance of the cathode follower, and the other capacitors are scaled from that by Eq. (14.7).

The cathode follower is standard with R_1 selected to give close to unity for the voltage gain. The output of the cathode follower is RC-coupled to a potentiometer. This potentiometer serves as the vertical gain control. Any fraction of the cathode-follower output can be fed into the vertical amplifiers. Thus the attenuator serves as a coarse vertical gain control, decreasing the gain in steps of one-tenth, and the vertical gain control serves as a vernier producing a smooth change in output.

Vertical Amplifier

The vertical amplifiers must have a wide band width so that complex voltages are not distorted when they are amplified. Compensated RC coupling is used to secure the wide band widths required. Depending upon the price and purpose of the oscilloscopes, band widths from 500 kc to several megacycles are used. Such an amplifier is called a *video amplifier*, a name chosen because such amplifiers were first needed in TV systems. Pentodes are used in video amplifiers because of their low interelectrode capacitances and high gain.

Two video amplifiers are shown in Fig. 14.8. In circuit (a), the changes are not obvious. The first change is the fact that R_L is made small. It was established in Example 7.6 that, for a given circuit, the gain band-width product will remain constant if C_{11} is not changed. Thus, lowering R_L will lower the gain per stage, but the band width goes up as a result. Values for R_L as low as 1800 ohms are used in circuit (a). This circuit also includes a low-frequency compensation circuit called a *bass boost circuit*. The addition of R_1 and C_1 comprises the bass boost. In the midfrequency range, and higher, capacitor C_1 keeps point A at ac ground, and the effective load is R_L. In the low-frequency range, the reactance of C_1 has increased, and the impedance of R_1 and C_1 in parallel is now in series with R_L, making the load greater. Thus, at low frequencies, the load impedance increases, raising the gain at low frequencies.

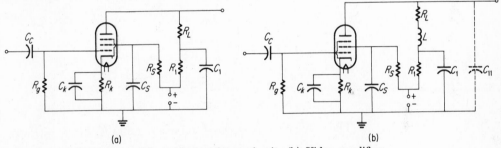

(a)   (b)

Fig. 14.8 (a) Pentode amplifier with bass-boost circuit. (b) Video amplifier utilizing bass boost and high-frequency compensation.

The circuit in Fig. 14.8(b) is a more efficient circuit. It has the bass boost, R_1 and C_1, and it has a new element, L. In the mid-frequency region, the inductor has negligible reactance and does not affect the gain. At the high frequencies, L is essentially in parallel with C_{11} and forms a low-Q, parallel resonant circuit with C_{11}. The result is that instead of the gain decreasing in the high-frequency range, the increased impedance of the parallel tuned circuit keeps the load impedance high and the upper half-power point is substantially increased. Coil L is called a *peaking coil*. Inclusion of the peaking coil allows one to use a larger value of R_L and still secure the desired wide band width. Thus circuit (b) would have more gain for a given band width than would circuit (a). Based on detailed calculations that will not be presented here, an optimum value for L is obtained for the condition

$$L = 0.5C_{||}R_L{}^2$$

and R_L is obtained by using the relation

$$R_L = \frac{1.8}{2\pi f_2 C_{||}}$$

where f_2 is the upper half-power frequency desired. R_L is used rather than $R_{||}$ in the preceding calculation because R_L is so small compared with r_p and R_g; therefore $R_{||} \cong R_L$.

Example 14.1 Design a video amplifier having an upper half-power frequency of 4 Mc and $f_1 = 1$ cps. Use a 6AK5 pentode and assume that $C_{||} = 30$ pf.

1. The characteristics for a 6AK5 are: plate, 180 volts; screen, 120 volts; grid, $R_K = 180$ ohms; plate, 7.7 ma; screen, 2.4 ma; r_p, 500 K; and g_m, 5100 μmho.

2. The circuit to be used is as shown below.

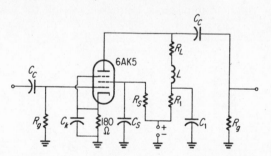

3. The value of R_g is selected to be $R_g = 500$ K. For $f_1 = 1$ cps,

$$C_C = \frac{1}{2\pi f_1 R_g}$$

$$= \frac{1}{2\pi \times 1 \times 500 \text{ K}} = 0.32\mu\text{f}$$

4. For the bass-boost circuit, the condition usually met is

$$R_1 C_1 \cong R_g C_C$$

where C_1 is found to be in the range 4 to 8 μf.

Let $C_1 = 4\mu$f; then

$$R_1 = R_g \left(\frac{C_C}{C_1} \right)$$

$$= 500 \text{ K} \left(\frac{0.32 \times 10^{-6}}{4 \times 10^{-6}} \right) = 40 \text{ K}$$

5. R_L is found by substitution in

$$R_L = \frac{1.8}{2\pi f_2 C_{11}}$$

$$= \frac{1.8}{2\pi \times 4 \times 10^6 \times 30 \times 10^{-12}} = 2380 \text{ ohms}$$

Choosing $R_L = 2.4$ K, L can be calculated from

$$L = 0.5 C_{11} R_L^2$$

$$= 0.5 \times 30 \times 10^{-12} \times (2.4 \times 10^3)^2$$

$$= 86 \ \mu\text{h}$$

6. R_S and C_S cannot be calculated until E_{bb} is selected. C_K would be inordinately large if calculated by

$$X_{C_K} \leq \tfrac{1}{10} R_K$$

Therefore, in video applications, one finds values from 100 μf to several thousand, depending upon the expense to which the manufacturer wishes to go. In each case there is still some degeneration at the very low frequencies. We shall choose $C_K = 1000$ μf at 5 volts, which will turn out to be a reasonable size, physically.

7. The completed circuit with all possible values calculated is shown below.

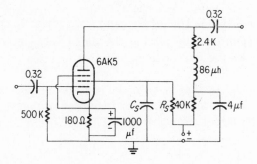

The vertical amplifier chain usually contains more than one amplifier, and the output is usually a push-pull connection to the vertical deflection plates. A more complete block diagram of a vertical amplifier chain is shown in Fig. 14.9. One or more video amplifiers may be used, depending upon the band width and overall gain required. The output circuit, shown in Fig. 14.10, is a cathode-coupled phase splitter. The theory of operation was covered in Chapter 8. This push-pull stage is also a video amplifier. It is direct-coupled to the vertical deflection plates.

Positioning of the beam, vertically, is accomplished by varying the bias on V_1 and V_2 by adjusting potentiometer R_3. When the

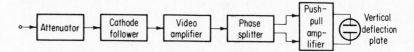

Fig. 14.9 Block diagram of vertical amplifiers in an oscilloscope.

movable contact of R_3 is toward V_1, the bias for V_1 will be due to the voltage drop across R_2, while the bias for V_2 will be due to the voltage drop across R_2 and R_3. This means that V_1 will conduct more than V_2, and consequently the plate voltage of V_1 will be lower than for V_2. Deflection plate D_1 will be at a lower potential than deflection plate D_2, and the beam will be moved away from D_1 and toward D_2. It takes only a small change in bias to change the plate voltage tens of volts; therefore R_3 need not be large. Capacitors C_4 and C_5 are used to bypass R_3 for ac signals, so that the feedback takes place across R_2 only. This keeps the gain for the two tubes from changing as R_3 is varied.

Sync Circuit

In order to keep the trace stationary on the cathode-ray tube screen, the signal frequency and sweep frequency must be synchronized. The sweep must be equal to, or a submultiple of, the signal frequency. In practice, a portion of the signal is RC-coupled from the output of the video amplifiers to the sync circuit and is used to synchronize the sawtooth oscillator. A thyratron sawtooth generator, described in Chapter 9, will be used. This type of circuit is easily synchronized by placing the signal voltage on the grid. This will make the ionization potential a function of the signal voltage.

To sync a signal, that is, to make it stationary on the screen, one adjusts the horizontal oscillator frequency so that it is slightly lower than the signal being observed. This will cause the thyratron

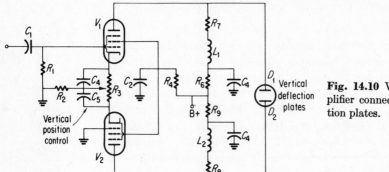

Fig. 14.10 Video push-pull amplifier connected to vertical deflection plates.

to fire during the positive excursion of each cycle of the signal frequency. In Fig. 14.11, the upper wave form is the signal to be observed, a sine wave. The bottom wave forms are the ionization potential (modified by the signal on the control grid of the thyratron) and the sawtooth output of the oscillator (superimposed). As can be seen, the oscillator is synchronized so as to produce a sawtooth having the same frequency as the signals. The dashed lines show the sawtooth operating at a frequency lower than the signal frequency, and the solid lines show the actual output due to the action of the sync signal.

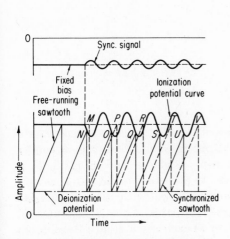

Fig. 14.11 Graphical construction showing synchronization of a thyratron sawtooth generator by means of a sinusoidal sync signal.

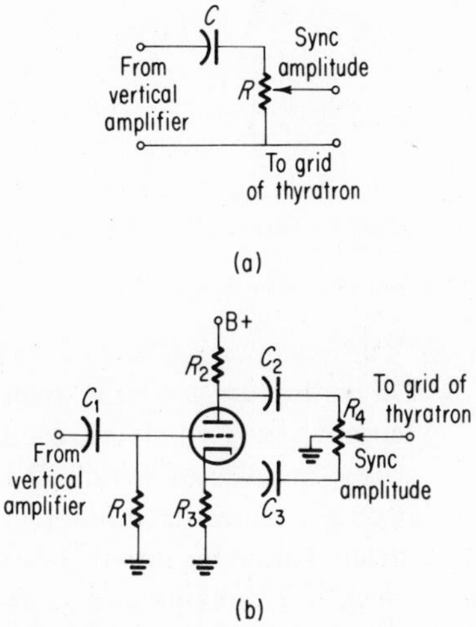

Fig. 14.12 Two methods of obtaining a variable-amplitude sync signal.

Figure 14.12 shows two methods of applying the sync signal to the grid of the sawtooth oscillator. The first method (a) is simply a potentiometer R, used to vary the amplitude of the sync signal applied to the grid of the sawtooth oscillator. Capacitor C is used to keep the dc voltages of the preceding circuit from appearing on the grid of the thyratron. In circuit (b), the vertical signal is applied to the grid of a paraphase amplifier; R_2 and R_3 have the same resistance. The output is applied to a potentiometer whose center tap is grounded. When the movable contact is at the center, the sync output is zero. With the contact moved up, the sync signal will increase in magnitude and have one polarity. With the contact moved down, the signal again increases in amplitude, but has the opposite polarity. This circuit, then, allows one to synchronize on either the positive- or negative-going portion of the signal, depending upon how the sync amplitude control is set.

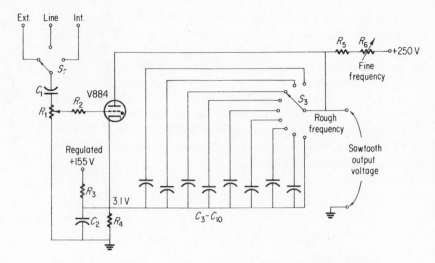

Fig. 14.13 Thyratron sawtooth sweep generator circuit.

SWEEP GENERATOR

The sweep generator, or sawtooth oscillator, is most generally a thyratron in a circuit such as described in Chapter 9. A similar circuit, as used in an oscilloscope, is shown in Fig. 14.13. S_2 is the sync selector switch. When it is connected to EXT, the connection goes to an external jack so that one can introduce a sync signal other than the signal for special purposes. When S_2 is connected to $LINE$, the input to C_1–R_1 is a 60-cycle signal from a transformer winding in the power supply. When S_2 is switched to INT, the sync signal is the signal voltage itself, derived from a vertical amplifier output. R_1 is the sync amplitude control. R_2 is used to keep the grid current at a safe level. The combination R_3, R_4, and C_2 is used to bias the cathode at 3.1 volts. This effectively places the control grid at −3.1 volts with respect to the cathode and determines the amplitude of the generated sawtooth. The capacitors C_3 to C_{10} and resistors R_5 and R_6 form the charging circuit that produces the sweep portion of the sawtooth. The sawtooth voltage is generated across the capacitors C_3 to C_{10}. The oscillator frequency is changed in large steps by switching in different values of C from C_3 to C_{10} by switch S_3.

HORIZONTAL AMPLIFIER

The main purpose of the horizontal amplifier is to amplify the sweep voltage. This is most usually a sawtooth voltage. Since the sweep voltage is usually a submultiple of the signal voltage, the frequency content is not so high, and therefore the horizontal amplifier need not have so wide a band width. Common band

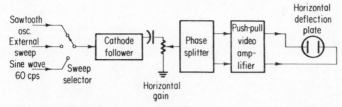

Fig. 14.14 Block diagram of horizontal amplifiers in an oscilloscope.

widths for horizontal amplifiers are from 250 to 500 kc. This still requires a video amplifier, but one whose load resistance can be higher than that used for a vertical amplifier. Therefore not so many stages of horizontal amplification are needed because each stage will have more gain.

The circuits used for horizontal amplifiers would be the same as for vertical amplifiers. Only the component values would be changed to secure the different band widths. A typical block diagram for a horizontal amplifier system is shown in Fig. 14.14. The sweep selector switch selects the desired sweep voltage. The choices usually are (1) a sawtooth, linear timebase; (2) a 60-cps sine wave; (3) an external connection so that other desired wave forms can be introduced.

POWER SUPPLY

The oscilloscope requires two separate dc-power supplies and a filament supply. The filament supply is furnished by stepdown windings on a power transformer and is used to heat the filaments of all the tubes used. The two dc supplies are (1) a low-voltage supply furnishing plate and screen voltages and bias voltages (both positive and negative) for the vacuum tube circuits, (2) a high-voltage supply ranging from 1.5 kv to about 20 kv, depending upon the oscilloscope. This high voltage is used on the focusing and accelerating anodes of the cathode-ray tube. The current required is very low, of the order of 50 to 100 μa.

14.6 Oscilloscope Applications

VOLTAGE MEASUREMENTS

For voltage measurements, the oscilloscope is used like a voltmeter that is, its vertical input leads are placed across the component that is being measured. For example, to measure the voltage across capacitor C in Fig. 14.15(a), a sawtooth oscillator circuit, the oscilloscope vertical input leads are placed across the capacitor, as shown. With the vertical gain, horizontal frequency, and sync

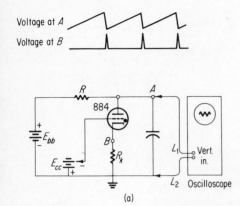

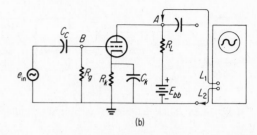

Fig. 14.15 (a) Use of an oscilloscope to determine voltage and current waveforms. (b) Use of an oscilloscope to determine the gain of an amplifier circuit.

controls set correctly, the screen will show several cycles of the sawtooth, which is the voltage across C as a function of time.

Similarly, to check the gain of an amplifier stage, Fig. 14.15(b), the oscilloscope leads would first be placed across the input (grid to ground) and then across the output (plate to ground). The ratio of the heights of these two signals on the screen will give A_v, the voltage gain, for the stage. Figure 14.14(b) shows the oscilloscope connected across the output, point A to ground. The voltage drop from point A to ground consists of E_{bb}, a dc voltage, and a varying dc voltage across R_L. Because the oscilloscope has an input capacitor, the dc component of this voltage will not be seen, but the coupling capacitor will allow the varying dc across R_L to appear at the grid of the first vertical amplifier. Thus, this oscilloscope will show only the ac component of a pulsating dc voltage.

To see the input voltage, the oscilloscope leads would be placed at point B and ground. Since the ground connection is common to both input and output observations, lead L_2 need not be moved. One simply moves lead L_1 to point B to observe the input, and then moves lead L_1 to point A to observe the output.

An oscilloscope can be calibrated to measure voltages directly. The calibration is effected by placing a sine wave or square wave of known amplitude on the vertical input and adjusting the vertical gain until the scale required is obtained. For convenience, oscilloscope screens usually have a clear plastic window with a rectangular grid printed upon it. The vertical gain is adjusted so that each grid spacing corresponds to a certain number of volts. The calibration would then be X number of volts per division. This is shown in Fig. 14.16.

Assume that one has access to a 1-volt peak-to-peak (1-volt pp) sine wave. Then, with this sine wave applied to the vertical input and the vertical gain adjusted so that the sine wave covers ten divisions, the oscilloscope will be calibrated to measure 0.1 volt

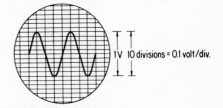

Fig. 14.16 Vertical gain adjusted so that a 1-volt peak-to-peak signal covers ten divisions.

1 V   10 divisions = 0.1 volt/div.

per division. The vertical gain control is not moved, once this setting has been achieved. The calibration voltage is now removed from the vertical input leads. One actually has a three-scale voltmeter. If the calibration is performed with the vertical input attenuator on the $X1$ position, then the calibration is 0.1 volt per division on that scale. If the attenuator is switched to the $X10$ position, the calibration will then be 1 volt per division, and on the $X100$ position, the calibration will be 10 volts per division.

Thus, although the vertical gain control should not be touched, the attenuator can be switched to change the calibration by factors of 10 or 1/10, depending upon which way one goes. If the calibration for 0.1 volt per division were made on the $X100$ position of the attenuator, then the $X10$ position calibration would be 0.01 volts per division and the $X1$ position would be 0.001 volt per division. As a result, the oscilloscope can be used as an ac voltmeter, but remember that the voltage is read peak to peak. Many oscilloscopes furnish a 1-volt pp calibrating-voltage connection on the face of the oscilloscope for just this purpose.

CURRENT MEASUREMENTS

One can use the property of a resistor to measure current. It is the most common, but not the only way, of using an oscilloscope to measure current. The resistor property used is the fact that $i = e/R$; that is, the current is proportional to the applied voltage. This equation also says that the wave forms for the two will be exactly the same. Therefore, if one obtains the voltage drop across a resistor as a function of time, the current has also been obtained as a function of time. It is given by

$$i = \frac{1}{R} e$$

Scaling the voltage down by a factor $1/R$ gives the current through the resistor. This application is shown in Fig. 14.15(a). Notice that a resistor R_K has been placed in the cathode lead of the 884 thyratron. A resistor is ordinarily not placed here, but R_K has been inserted into the circuit so that one can measure the current conducted by the thyratron. The resistor R_K is a very small value compared with the other resistances in the circuit, and therefore does not appreciably change the circuit. The oscilloscope leads are connected across R_K, L_1, to B and L_2 to ground. In this way the voltage across R_K will be displayed on the screen. If $R_K = 10$ ohms, the instantaneous current through R_K will be

$$i_{R_K} = \tfrac{1}{10}e_{R_K}$$

Since the thyratron is in series with R_K, this is also the current conducted by the thyratron.

In this particular circuit, the current will be seen to consist of sharp positive pulses, as shown in Fig. 14.15(a). This shows that the thyratron in this circuit conducts in pulses only. A knowledge of the circuit operation would determine that the thyratron conducts when it fires, discharging the capacitor C. This occurs when the sawtooth voltage has reached the ionization potential. This is why the two wave forms have been aligned to show the relative time relationships between the current through the thyratron and the voltage across the thyratron. Note that the voltage across the thyratron and the current through it certainly do not have the same wave shape, and therefore the thyratron does not satisfy the equation $i = e/R$. The thyratron cannot be thought of as a simple linear resistance. If the voltage calibration, when measuring e_{R_K}, shows that the peak voltage across R_K is 3 volts, then one can conclude that the peak current conducted by the thyratron is

$$i_{peak} = \tfrac{1}{10}e_{peak} = 0.3 \text{ amp}$$

Thus, placing a small resistance in parallel with the oscilloscope vertical input converts the oscilloscope to a low-resistance ammeter. Also note that, as in any ammeter application, the sensing resistor (in this example, R_K) goes in series with the component through which one wishes to measure the current.

LIGHT INTENSITY

With the aid of a photocell and associated circuitry, the oscilloscope can be used to measure the intensity of light as a function of time for rapidly changing intensities. The example shown in Fig. 14.17(a) represents an attempt to measure the light output from a flash bulb. This is a one-time event, that is, not cyclic, and so a special oscilloscope called a *synchroscope* must be used. The synchroscope has an added refinement in that it produces a single linear sweep when the operator wants it. Since in most applications the sweep is very fast, from milliseconds to as little as fractions of microseconds in duration, the eye cannot be used and the trace must be photographed. This allows one to examine and measure the result at leisure.

A simple photocell circuit consists of a series circuit, R, E, and the photocell. E is a dc-voltage source, either a battery or power

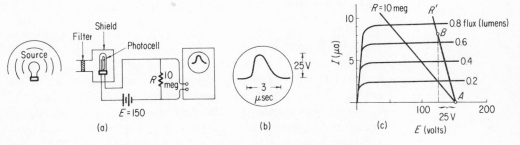

Fig. 14.17 (a) Test circuit for measuring the intensity of a flash bulb. (b) Calibrated scope trace. (c) Use of the photocell characteristics to compensate for the input impedance of the oscilloscope.

supply, R is the resistor across which one measures a voltage drop with the oscilloscope, and the photocell acts like a variable resistor whose resistance decreases as more intense light falls upon it. Note that a known filter has been placed between the photocell and light source so as to keep the intensity within range of the photocell characteristics; without this filter, distortion would result. The filter acts like a "gain control" so that the photocell circuit will not be overdriven and harmed. A typical trace as photographed at the screen of the oscilloscope is seen in Fig. 14.17(b). The flash-bulb intensity rises very rapidly and then decays slowly. The entire event lasts a few milliseconds.

In order to find the intensity of light falling on the photocell, its characteristics must be known. These are shown in Fig. 14.17(c). The ordinate is photocell current, the abscissa is the voltage applied across the photocell, and the third variable is the light flux, in lumens, falling on the photosensitive surface of the photocell. A load line for $R = 10$ meg is drawn for a supply voltage of $E = 150$ volts. With no light falling on the photocell, the operating point on the load line will be point A. As light falls on the photocell, the instantaneous operating point moves up the load line. This photocell is a small current device; the ordinate is in microamperes, and consequently R is large. Because R is so large, the oscilloscope will load down the circuit. The effective resistance that will be seen by the photocell generator, R' will be R and R_{in} of the oscilloscope in parallel:

$$R' = \frac{RR_{in}}{R + R_{in}}$$

The load line for R' is shown drawn in on the photocell characteristics and labeled R'. For rapidly changing voltages, as shown in Fig. 14.17(b), the operating point will move on the R' load line. In this manner, the peak intensity can be obtained by locating point B on the load line and reading off the maximum intensity. Correcting for the known filter, the actual intensity is found.

Frequency Measurements—Lissajous Figures

In many timing and calibrating applications it is necessary to know the frequency of a sine wave precisely. A simple but highly accurate method for comparing a known and unknown sine wave is the method of *Lissajaus figures*. Normally, the input to the horizontal amplifiers is the output of the sawtooth oscillator. In Lissajous figures, either the standard power-line frequency (60 cps sweep) or the output from a standard, precisely known sine wave oscillator is applied to the external horizontal input. The signal whose frequency is to be determined is applied to the vertical input. Usually, in this application, the gain controls are adjusted so that the two signals are of the same amplitude. If the signal frequency is unknown, then the standard should be variable in frequency.

Recognizable patterns are formed on the cathode-ray tube screen whenever a whole number ratio between known and unknown frequencies is presented. Some of these patterns are shown in Fig. 14.18. If the ratio between unknown and known frequencies is 1:1, the pattern will be anything from a straight line to an ellipse or circle, the differences being due to the relative phase between unknown and known. For the case where the ratio of unknown to known frequencies is 2:1, the pattern that results is shown to resemble a butterfly. Patterns for the ratios 5:1, 1:10, 3:5, 5:6, and 3:1 are also shown. If the frequencies are not exactly whole number ratios, but are very close, the pattern will appear to rotate slowly about a horizontal or vertical axis, depending upon which frequency is slightly higher. Two types of pattern for the 3:1 ratio are shown. To get one of these, imagine the other rotated about a vertical axis.

A trick can be used to determine the whole number ratio between unknown and known frequencies. In Fig. 14.18, imagine a horizontal line tangent to any of the patterns, and a vertical line tangent to the pattern. The number of times the pattern touches the horizontal line gives the whole number for the vertical signal, and the number of times the pattern touches the vertical line gives the whole number for the horizontal signal. This is illustrated for the 3:1 pattern. The pattern "touches" the horizontal line three times and the vertical line one time. Therefore the ratio of frequencies is 3:1. In the 3:1 pattern on the right, one must imagine that the pattern is superimposed on itself, which is what you would obtain if the 3:1 pattern on the left were rotated about a vertical axis. Check the other patterns to see if you can obtain the ratios shown. A ratio of 10:1 or 1:10 is about as high as one can go with this method. When the ratio is higher, it becomes difficult to count the number of times the pattern touches a vertical or horizontal tangent. The 60-cps frequency of the power line is kept at a precise

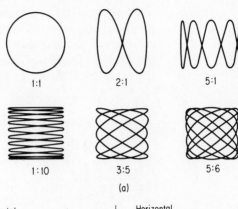

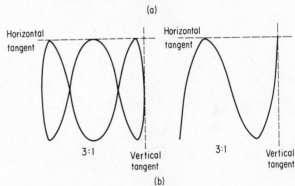

Fig. 14.18 Lissajous patterns for various frequency ratios.

value by the power company and can be used as a fairly accurate standard. With this, one can calibrate an audio oscillator, for example, at quite a number of frequencies. Some of these are tabulated below, with their whole number ratios for the range 10 to 60 cps. Only ratios where the numbers are less than or equal to 10 are shown.

Frequency of Audio Oscillator, cps	Ratio, Unknown:known
10	1:6
12	1:5
15	1:4
18	3:10
20	1:3
24	2:5
30	1:2
36	3:5
40	2:3
42	7:10
45	3:4
48	4:5
50	5:6
54	9:10
60	1:1

To show how the electron beam traces out such a pattern, the graphical construction in Fig. 14.4 is utilized. In this case, both signals will be sine waves. The construction for the case where the vertical frequency is lower than the horizontal frequency in the ratio 2:3 is shown in Fig. 14.19.

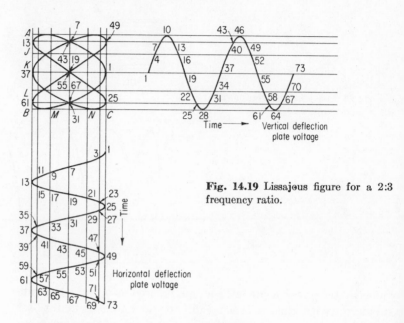

Fig. 14.19 Lissajous figure for a 2:3 frequency ratio.

In the vertical signal the sine wave completes two cycles while the horizontal signal is going through three cycles. Under the influence of these two signals, appearing as sinusoidal forces at right angles to each other at the deflection plates, the beam moves in the pattern shown.

Phase Measurement

The Lissajous figures can also be used to show the relative phase between two sine waves. For this use, the 1:1 ratio is used; that is, the vertical and horizontal signals are made the same frequency and the same amplitude. Although the amplitude is not critical, the accuracy of measurements is optimized when the amplitudes are the same.

A graphical example is worked out in Fig. 14.20 for the case where the relative phase angle is 90 deg. Whenever the phase angle is 90 deg, the pattern will be a circle. A study of the plotted points shows that the spot is moving clockwise about the circle. If the relative phase angle between the two signals were 270 deg, the result would again be a circle, but the spot would be moving counterclockwise. Thus, there would be ambiguous patterns, since

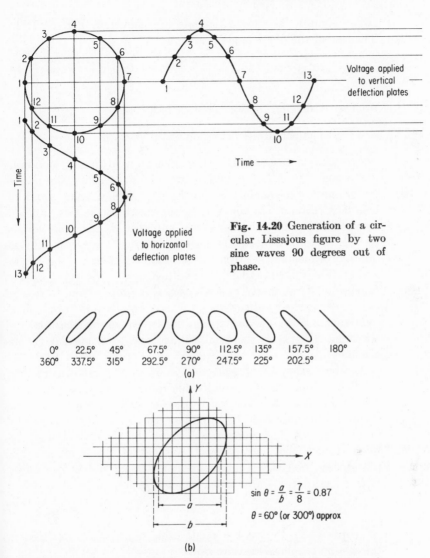

Fig. 14.20 Generation of a circular Lissajous figure by two sine waves 90 degrees out of phase.

$\sin \theta = \dfrac{a}{b} = \dfrac{7}{8} = 0.87$

$\theta = 60°$ (or 300°) approx

Fig. 14.21 (a) Resultant patterns when the phase difference between the unknown and standard signals varies in steps of 22.5 degrees. (b) Figure used in calculating phase difference.

a 270 deg lag is the same as a 90 deg lead. The system, then, cannot give the lag or lead unless the sense of the spot motion can be determined or other tests are made.

Figure 14.21(a) shows what the pattern looks like for several phase angles as the phase difference varies from 0 to 180 deg in 22.5 deg steps. This method is amenable to an analytical approach. Since the pattern is an ellipse rotated with respect to the axes, an exercise in analytical geometry shows that measurements on this pattern will allow one to calculate the phase difference between signals. The measurements are shown in Fig. 14.21(b). The sine of the phase angle is given by

$$\sin \theta = \frac{a}{b}$$

where a = distance between X-axis intercepts

b = maximum X extension of the pattern

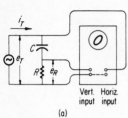

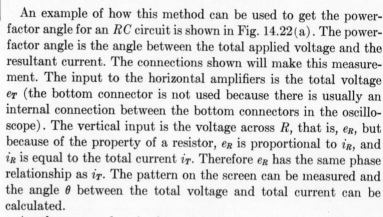

(a)

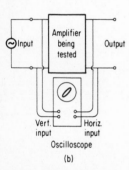

Vert. Horiz.
input input

Oscilloscope

(b)

Fig. 14.22 (a) Simple circuit used to illustrate phase shift. (b) Test circuit used to indicate phase shift in an amplifier.

An example of how this method can be used to get the power-factor angle for an RC circuit is shown in Fig. 14.22(a). The power-factor angle is the angle between the total applied voltage and the resultant current. The connections shown will make this measurement. The input to the horizontal amplifiers is the total voltage e_T (the bottom connector is not used because there is usually an internal connection between the bottom connectors in the oscilloscope). The vertical input is the voltage across R, that is, e_R, but because of the property of a resistor, e_R is proportional to i_R, and i_R is equal to the total current i_T. Therefore e_R has the same phase relationship as i_T. The pattern on the screen can be measured and the angle θ between the total voltage and total current can be calculated.

Another example of phase measurement is shown in Fig. 14.22(b), where the phase shift due to an amplifier is measured. The input signal is connected to the vertical input of the oscilloscope and the output of the amplifier is connected to the horizontal input. The resulting pattern will show the phase difference between input- and output-signal voltages of the amplifier. The pattern can be either a lag or lead, but this is not known unless other measurements are made. A plot of phase angle versus input frequency can be made in this manner for the amplifier under test.

Example 14.2 Derive the relationship $\sin \theta = a/b$ as used in the section on phase measurement.

1. In general, the input to the vertical and horizontal amplifiers will be two voltages having a phase angle between them. This can be written as

$$x = B \sin \frac{2\pi}{T} t$$

$$y = C \sin \left(\frac{2\pi}{T} t + \theta \right)$$

(1)

where B and C are the respective amplitudes. These are the parametric equations for an ellipse in the X–Y plane.

2. To show that these two equations will plot an ellipse, an example will be plotted for the case $\theta = \pi/4$.

$$x = B \sin \frac{2\pi}{T} t$$

$$(2)$$

$$y = C \sin \left(\frac{2\pi}{T} t + \frac{\pi}{4} \right)$$

These results are plotted as shown below, left.

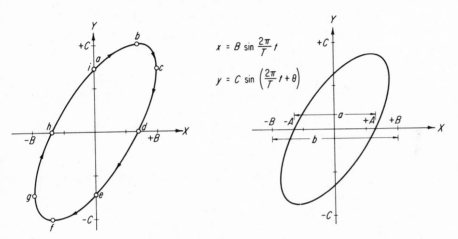

3. To obtain the equation $\sin \theta = a/b$, refer to the general curve given here defining the points in question. To find the intercepts $\pm A$, let $y = 0$. For point $x = +A$, let $t = t_1$.

$$A = B \sin \frac{2\pi}{T} t_1 \qquad (3)$$

$$0 = C \sin \left(\frac{2\pi}{T} t_1 + \theta \right) \qquad (4)$$

Expanding the second equation

$$0 = C \sin \frac{2\pi}{T} t_1 \cos \theta + C \cos \frac{2\pi}{T} t_1 \sin \theta \qquad (5)$$

Substituting Eq. (3) into Eq. (5) by use of the trigonometric construction,

$$\sin \frac{2\pi}{T} t_1 = \frac{A}{B}$$

$$(6)$$

$$\cos \frac{2\pi}{T} t_1 = \frac{\sqrt{B^2 - A^2}}{B}$$

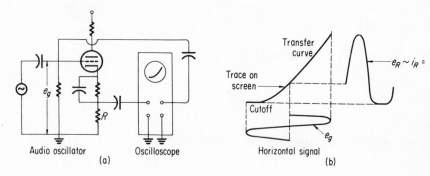

Fig. 14.23 A simple circuit used to obtain the transfer curve of an amplifie

$$0 = C\left(\frac{A}{B}\right)\cos\theta + C\frac{\sqrt{B^2 - A^2}}{B}\sin\theta \qquad (7$$

$$-A\cos\theta = \sqrt{B^2 - A^2}\sin\theta \qquad (8$$

Squaring both sides,

$$A^2\cos^2\theta = B^2\sin^2\theta - A^2\sin^2\theta \qquad (9$$

Grouping terms in A^2,

$$A^2(\sin^2\theta + \cos^2\theta) = B^2\sin^2\theta \qquad (10$$

Solving for $\sin\theta$,

$$\sin\theta = \frac{A}{B} = \frac{2A}{2B} = \frac{a}{b}$$

As can be seen, the result is independent of the amplitude in the y direction. However, the y amplitude is made other than zero so that the intercepts can be easily read off.

TUBE CHARACTERISTICS

As another example of an application where the linear time base is not used, the oscilloscope can be made to trace out tube characteristics dynamically. The simplest example is the formation of an I_B and E_C transfer curve with the oscilloscope.

The circuit connections used are shown in Fig. 14.23(a). The vertical input is the ac component of the voltage across R, that is, e_R. Since $e_R = i_R R$, this voltage is directly proportional to the current in the plate circuit. The horizontal input is the ac signal on the grid (this measurement is usually made in the midfrequency range). The graphical construction showing how a transfer curve is formed on the screen is shown in Fig. 14.23(b). In this case, e_g

vas made large enough so that it ranged from $E_C = 0$ to below
:utoff. The oscilloscope can be calibrated along the two axes to
ead grid voltage and plate current. Plate characteristics can also
)e traced out, but the setup is more complicated and will not be
:overed here.

14.7 Electronically Voltage-Regulated Power Supply

In many applications involving oscillators and sweep circuits,
the supply voltage must be very precise, with no drift; otherwise,
the timing in the circuit will be affected. A simple VR tube regulator will keep a supply voltage constant only to within ±0.25
volt at best. This represents excellent regulation, but is not good
enough for some applications. For this reason, electronic regulation involving dc amplifiers has been devised to produce a control
that keeps the supply voltage constant within millivolts, or in
some applications within microvolts.

A standard three-tube circuit is shown in Fig. 14.24. The voltage
E_{in} is the output of the power supply. The voltage regulator consists of tubes V_1, V_2, V_3, and their associated circuitry. The load
R_L represents the effective resistance of the electronic circuitry
operated by E_{out}. Tube V_3 is a VR tube. Its function is to furnish
a reference voltage, E_{VR}. Any change in E_1, with respect to E_{VR}
will be amplified by V_1 and V_2, and compensated for, in microseconds. A circuit like this gives excellent regulation because of
the amplification involved, and the response of the system is extremely fast. Since it is dc-coupled throughout, the response is
excellent to dc or slowly varying transients as well as to high-
frequency transients.

Qualitatively, the operation of the circuit is as follows: The
cathode of V_1 is held at a constant voltage by VR tube V_3. The
resistor R_3 is used to start the VR tube V_3. E_1 is adjusted, with the
proper selection of R_A and R_B, so that the control grid of V_1 is
negative with respect to its cathode. Therefore $E_1 < E_{VR}$. Tube

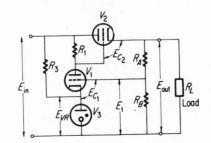

Fig. 14.24 Simple series voltage
regulator.

V_1 is in a conducting state, producing a voltage drop across R_1
This determines E_{c_2} and the operating point of V_2. If E_{out} tends
to go down, the voltage E_1 is reduced proportionately. This in
creases the bias on V_1. The current through V_1 is reduced and its
plate voltage rises. This, in turn, makes E_{c_2} less, and the resistance
of V_2 decreases, thereby raising E_{out}. The feedback from E_1 to
E_{c_2}, then, is such as to cancel out any output voltage variations.
A detailed step-by-step procedure will show that any input voltage
variation will also be compensated for by this circuit. In this case,
the resistance of V_2 will increase so as to keep the output voltage
constant. The overall effect is that V_2 acts like a variable series
resistance. Its resistance is changed by changing E_{c_2} in such a
way that E_{out} will remain constant for either an input or output
voltage variation.

14.8 Voltage Regulator—DC Calculations

Example 14.3 Design an electronic voltage regulator under the
following conditions: The input voltage is 350 volts and one wishes
to regulate this at 250 volts at 50 ma. Use a 2A3 power tube for
the series tube (V_2), a 6SJ7 as the dc amplifier (V_1), and a VR75
as reference tube (V_3).

1. The block diagram, then, is as given here.

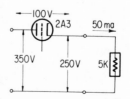

The effective load resistance is given by

$$R_L = \frac{E_{out}}{I_L} = \frac{250}{50 \times 10^{-3}} = 5 \text{ K}$$

2. Starting with the 2A3, the voltages are as shown below and
on the facing page.

The plate voltage for V_2 is $E_{b_2} = 100$ volts. On the plate characteristics shown below, a load line for $R_L = 5$ K has been drawn

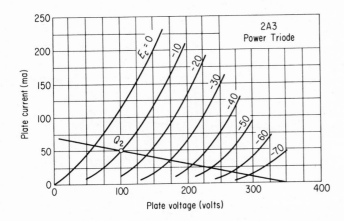

n. Since $E_{b_2} = 100$ volts, the point where $E_{b_2} = 100$ volts intersects the load line gives the operating point for V_2. This has been labeled Q_2. It turns out to be a good choice, at $E_{C_2} = -10$ volts. This allows a leeway of at most 10 volts above this point (to $E_{C_2} = 0$), which will suffice, since transients this big are usually not encountered. This construction has established E_{C_2} and therefore the voltage at the plate of V_1, since the difference between the voltage at the plate of V_1 and E_{out} established E_{C_2}. To make $E_{C_2} = -10$ volts, the voltage at the plate of $V_1 = 240$ volts.

3. Thus the voltages obtained so far are as shown in the diagram.

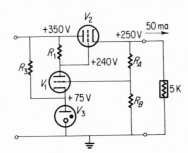

From this, one can obtain the plate voltage for V_1:

$$E_{b_1} = 240 - 75 = 165 \text{ volts}$$

The effective supply voltage for V_1 is given by

$$E'_{bb} = 350 - 75 = 275 \text{ volts}$$

This is the voltage from the cathode of V_1 to the top of R_1.

4. The operating point of V_1 can now be decided upon. We go to the plate characteristics of V_1 as shown on page 506.

The value $E'_{bb} = 275$ volts is put in, and labeled point A. Now a Q point must be chosen. Through these two points, the load line

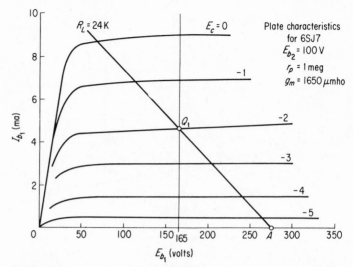

can be drawn, and R_1 deduced. Remember that the plate voltage $E_{b_1} = 165$ volts, as deduced in step 3. The choice of E_{C_1} is left to the designer. A good choice, roughly in the middle of the transfer curve, is $E_{C_1} = -2$ volts. This point is labeled Q_1. The resultant plate current is $I_{b_1} = 4.65$ ma.

5. The load resistor R_1 can now be calculated, since the current through R_1 and the voltage across R_1 are known.

$$R_1 = \frac{E_{R_1}}{I_{R_1}} = \frac{350 - 240}{4.65 \times 10^{-3}} = 23.7 \text{ K}$$

A value of $R_1 = 24$ K will be used.

6. Observe that the plate characteristics for the 6SJ7, V_1 are obtained for $E_{S_1} = 100$ volts. To make the screen grid positive with respect to the cathode by 100 volts means that $E_{S_1} = 175$ volts with respect to ground. A 6SJ7 screen grid conducts about 1 ma at 100 volts (tube manual), so R_S can be calculated because the current through it and voltage across it are known:

$$R_S = \frac{350 - 175}{1 \times 10^{-3}} = 175 \text{ K}$$

7. To calculate R_3, the VR tube will be operated at 20 ma, its design center current. This current is due to the pentode current and the current through R_3. The current through the pentode is the sum of plate and screen currents:

$$I_{V_1} = I_{b_1} + I_{S_1} = 5.65 \text{ ma}$$

This means that the current through R_3 is

$$I_3 = I_{\text{VR}} - I_{V_1} = 14.35 \text{ ma}$$

$_3$ can now be calculated:

$$R_3 = \frac{E_{R_3}}{I_3} = \frac{350 - 75}{14.35 \times 10^{-3}} = 19.2 \text{ K}$$

value $R_3 = 20$ K will be used. This reduces I_{VR} slightly.

8. For calculating R_A and R_B, voltages required from the voltage divider R_A and R_B are shown.

The choice of current through the voltage divider is up to the designer. The rule of thumb is 1 ma. This value is low enough so that it is much less than the load current and will not necessitate going back and recalculating new Q points for V_1 and V_2. Also, it is large enough so that any filter capacitors across E_{out} will be discharged in seconds. Thus,

$$R_A + R_B = \frac{E_{out}}{I_{A+B}} = \frac{250}{1 \times 10^{-3}} = 250 \text{ K}$$

From the voltage divider,

$$73 \text{ volts} = 250 \text{ volts} \left(\frac{R_B}{R_A + R_B} \right)$$

Solving for R_B (remember $R_A + R_B = 250$ K), $R_B = 73$ K and $R_A = 177$ K. These values must be precise. For this reason, instead of putting in a 73-K resistor and a 177-K resistor, one puts in a variable resistor as shown at right, so that these values can be obtained by adjusting the resistor.

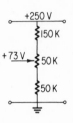

This almost makes the calculation of R_A and R_B academic. It helps one choose a value for the variable resistor so correct operation is approximately in the middle of its resistance range.

9. The completed circuit becomes as shown below.

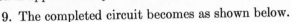

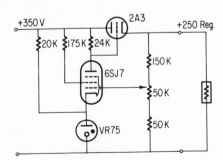

This step-by-step graphical procedure is the most direct and accurate procedure that can be used in the design of a particular voltage regulator. There are many linear approximation methods, but they are second-order approximations at best. Because the devices are nonlinear, the graphical method is the preferred one.

14.9 Voltage Regulator—AC Calculations

Of prime interest in the electronic voltage regulator is its effec
on ripple, that is, its filter factor. This factor relates a change i
input voltage to a change in output voltage:

$$\Delta E_{\text{in}} = F \, \Delta E_{\text{out}}$$

where F = filter factor for the voltage regulator. This is essentiall
an ac calculation, and the problem will be solved by calculation
on the ac equivalent circuit for Fig. 14.24. The ac equivalent circui

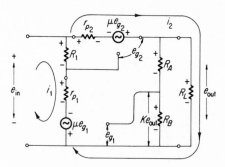

Fig. 14.25 AC equivalen
circuit of the voltage regula
tor in Fig. 14.24.

is shown in Fig. 14.25. Using positive branch currents, the follow
ing equations can be written down:

$$e_{g_1} = K e_{\text{out}} \qquad (14.8)$$

where $0 < K < 1$, and is the fraction of e_{out} that appears across R_B

$$K = \frac{R_B}{R_A + R_B} \qquad (14.9)$$

The remaining equations are

$$e_{\text{in}} - i_1(R_1 + r_{p_1}) + \mu_1 e_{g_1} = 0 \qquad (14.10)$$

$$e_{\text{in}} - i_2(r_{p_2} + R_L) + \mu_2 e_{g_2} = 0 \qquad (14.11)$$

$$e_{g_2} = -e_{\text{out}} - \mu_1 e_{g_1} + i_1 r_{p_1} \qquad (14.12)$$

$$e_{\text{out}} = i_2 R_L \qquad (14.13)$$

Because $(R_A + R_B) \gg R_L$, the current through the $(R_A + R_B)$
branch is considered to be small and, essentially, i_2 flows through
R_L.

Eliminating e_{g_1} and e_{g_2} from Eqs. (14.10) and (14.11) by using
Eqs. (14.8) and (14.12), and rewriting these equations and Eq.
(14.13) in determinantal form:

$$e_{\text{in}} = i_1(R_1 + r_{p_1}) \qquad + \mu_1 K e_{\text{out}} \qquad (14.14)$$

$$e_{\text{in}} = -i_1\mu_2 r_{p_1} + i_2 r_{p_2} + (1 + \mu_2 + \mu_1\mu_2 K)e_{\text{out}} \qquad (14.15)$$

$$0 = i_2 R_L - e_{\text{out}} \qquad (14.16)$$

Solving for e_{out}:

$$e_{\text{out}} = \frac{\begin{vmatrix} R_1 + r_{p_1} & 0 & e_{\text{in}} \\[6pt] -\mu_2 r_{p_1} & r_{p_2} & e_{\text{in}} \\[6pt] 0 & R_L & 0 \end{vmatrix}}{\begin{vmatrix} R_1 + r_{p_1} & 0 & \mu_1 K \\[6pt] -\mu_2 r_{p_1} & r_{p_2} & 1 + \mu_2 + \mu_1\mu_2 K \\[6pt] 0 & R_L & -1 \end{vmatrix}}$$

$$= \frac{e_{\text{in}}[R_L(R_1 + r_{p_1}) + R_L\mu_2 r_{p_2}]}{(R_1 + r_{p_1})[r_{p_2} + R_L(1 + \mu_2 + \mu_1\mu_2 K)] + \mu_1\mu_2 K R_L r_{p_1}}$$

$$(14.17)$$

The filter factor F is given by

$$F = \frac{e_{\text{in}}}{e_{\text{out}}} \qquad (14.18)$$

or

$$F = \frac{(R_1 + r_{p_1})[r_{p_2} + R_L(1 + \mu_2 + \mu_1\mu_2 K)] + \mu_1\mu_2 K R_L r_{p_1}}{R_L[R_1 + (1 + \mu_2)r_{p_1}]}$$

$$(14.19)$$

For the circuit obtained in Example 14.2, one can get an order of magnitude for F. Using the following values,

$$K = \frac{R_B}{R_A + R_B} = 0.282 \qquad r_{p_1} = 1 \text{ meg}$$

$$R_L = 5 \text{ K} \qquad\qquad \mu_1 = 1650$$

$$R_1 = 24 \text{ K} \qquad\qquad r_{p_2} = 1 \text{ K}$$

$$\mu_2 = 4.5$$

put these values in Eq. (14.19):

$$F = 1500$$

This means that

$$e_{\text{out}} = \frac{1}{1500} e_{\text{in}}$$

which is a significant reduction in ripple content. A circuit of this type will have a ripple output in the range of 1 mv (rms).

For greater values of F, a second stage of dc amplification can be added. This will allow one to get values for F as high as 20,000 to 30,000. For greater current-carrying capacity, V_2 can consist of two or more similar tubes in parallel.

TRANSISTORIZED VERSION

There is practically a one-to-one relationship between components for a transistorized version of a regulated power supply and the vacuum tube type just discussed. The only difference is that transistors are low-voltage devices, so that this type of regulator would be used to regulate voltage outputs of the order of 50 volts and lower. Figure 14.26 shows a typical series regulator. The reference voltage is furnished by D_1, a Zener diode referencing at about 1 to 2 volts. R_1 serves as a collector load for Q_1. Since Q_1 is in series with R_1, this chain serves to operate D_1. Q_2 is the series power transistor. R_A and R_B form the voltage divider needed to bias Q_1 properly. The regulating action is the same as that for the vacuum tube type.

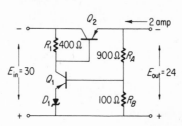

Fig. 14.26 Transistorized voltage regulator.

Problems

14.1 In a typical attenuator circuit, the total resistance desired is 4 meg and $C_{in} = 30$ pf. Determine R_1, R_2, R_3, C_1, C_2, and C_3 so that each step reduces the output by a factor of one-tenth.

14.2 Using the circuit in Fig. 14.8(b), determine R_L and L for an upper half-power frequency of 10 Mc. Assume that the tube is a 6AK5 and that $C_{||} = 30$ pf.

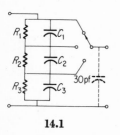

14.1

14.3 A test is conducted on an oscilloscope as follows: Using an input voltage of 1 volt pp, the vertical gain is adjusted so that the signal fills ten spaces vertically. Then switch SW-1 is opened and the signal is reduced on the screen to eight divisions. Determine the input resistance of the oscilloscope.

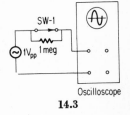

Oscilloscope

14.3

14.4 Using the graphical method shown in Fig. 14.19, determine the Lissajous figure when (a) the vertical signal is given by

$$e_{\text{vert}} = \sin \omega t$$

and the horizontal signal is given by

$$e_{\text{horiz}} = \sin \left(\omega t + \frac{\pi}{4} \right)$$

and (b) the vertical signal is given by $e_{\text{vert}} = \sin 2\omega t$ while the horizontal signal is given by $e_{\text{horiz}} = \sin 4\omega t$.

14.5 Redesign the voltage regulator of Example 14.2 for an output voltage of 200 volts at 100 ma.

APPENDIX

APPENDIX I

Natural Trigonometric Functions

Degrees	Sine	Cosine	Tangent	Degrees	Sine	Cosine	Tangent
0	0.000	1.000	0.000	46	0.719	0.695	1.03
1	0.017	1.000	0.017	47	0.731	0.682	1.07
2	0.035	0.999	0.035	48	0.743	0.669	1.11
3	0.052	0.999	0.052	49	0.755	0.656	1.15
4	0.070	0.998	0.070	50	0.766	0.643	1.19
5	0.087	0.996	0.087	51	0.777	0.629	1.23
6	0.105	0.995	0.105	52	0.788	0.616	1.28
7	0.122	0.993	0.123	53	0.799	0.602	1.33
8	0.139	0.990	0.141	54	0.809	0.588	1.38
9	0.156	0.988	0.158	55	0.819	0.574	1.43
10	0.174	0.985	0.176	56	0.829	0.559	1.48
11	0.191	0.982	0.194	57	0.839	0.545	1.54
12	0.208	0.978	0.213	58	0.848	0.530	1.60
13	0.225	0.974	0.231	59	0.857	0.515	1.66
14	0.242	0.970	0.249	60	0.866	0.500	1.73
15	0.259	0.966	0.268	61	0.875	0.485	1.80
16	0.276	0.961	0.287	62	0.883	0.469	1.88
17	0.292	0.956	0.306	63	0.891	0.454	1.96
18	0.309	0.951	0.325	64	0.899	0.438	2.05
19	0.326	0.946	0.344	65	0.906	0.423	2.14
20	0.342	0.940	0.364	66	0.914	0.407	2.25
21	0.358	0.934	0.384	67	0.920	0.391	2.36
22	0.375	0.927	0.404	68	0.927	0.375	2.48
23	0.391	0.920	0.424	69	0.934	0.358	2.61
24	0.407	0.914	0.445	70	0.940	0.342	2.75
25	0.423	0.906	0.466	71	0.946	0.326	2.90
26	0.438	0.899	0.488	72	0.951	0.309	3.08
27	0.454	0.891	0.510	73	0.956	0.292	3.27
28	0.469	0.883	0.532	74	0.961	0.276	3.49
29	0.485	0.875	0.554	75	0.966	0.259	3.73
30	0.500	0.866	0.577	76	0.970	0.242	4.01
31	0.515	0.857	0.601	77	0.974	0.225	4.33
32	0.530	0.848	0.625	78	0.978	0.208	4.70
33	0.545	0.839	0.649	79	0.982	0.191	5.14
34	0.559	0.829	0.675	80	0.985	0.174	5.67
35	0.574	0.819	0.700	81	0.988	0.156	6.31
36	0.588	0.809	0.727	82	0.990	0.139	7.12
37	0.602	0.799	0.754	83	0.993	0.122	8.14
38	0.616	0.788	0.781	84	0.995	0.105	9.51
39	0.629	0.777	0.810	85	0.996	0.087	11.4
40	0.643	0.766	0.839	86	0.998	0.070	14.3
41	0.656	0.755	0.869	87	0.999	0.052	19.1
42	0.669	0.743	0.900	88	0.999	0.035	28.6
43	0.682	0.731	0.933	89	1.000	0.017	57.3
44	0.695	0.719	0.966	90	1.000	0.000	∞
45	0.707	0.707	1.000				

$$e^{-x}$$

x		.0	.1	.2	.3	.4	.5	.6	.7	.8	.9
0		1.000	.9048	.8187	.7408	.6703	.6065	.5488	.4966	.4493	.4066
1		.3679	.3329	.3012	.2725	.2466	.2231	.2019	.1827	.1653	.1496
2		.1353	.1225	.1108	.1003	*9072	*8208	*7427	*6721	*6081	*5502
3	0.0	4979	4505	4076	3688	3337	3020	2732	2472	2237	2024
4	0.0	1832	1657	1500	1357	1228	1111	1005	*9095	*8230	*7447
5	0.00	6738	6097	5517	4992	4517	4087	3698	3346	3028	2739
6	0.00	2479	2243	2029	1836	1662	1503	1360	1231	1114	1008
7	0.000	9119	8251	7466	6755	6112	5531	5004	4528	4097	3707
8	0.000	3355	3035	2747	2485	2249	2035	1841	1666	1507	1364
9	0.000	1234	1117	1010	*9142	*8272	*7485	*6773	*6128	*5545	*5017
10	0.0000	4540	4108	3717	3363	3043	2754	2492	2254	2040	1846

$$e^{x}$$

x	.0	.1	.2	.3	.4	.5	.6	.7	.8	.9
0	1.000	1.105	1.221	1.350	1.492	1.649	1.822	2.014	2.226	2.460
1	2.718	3.004	3.320	3.669	4.055	4.482	4.953	5.474	6.050	6.686
2	7.389	8.166	9.025	9.974	11.02	12.18	13.46	14.88	16.44	18.17
3	20.09	22.20	24.53	27.11	29.96	33.12	36.60	40.45	44.70	49.40
4	54.60	60.34	66.69	73.70	81.45	90.02	99.48	109.9	121.5	134.3
5	148.4	164.0	181.3	200.3	221.4	244.7	270.4	298.9	330.3	365.0
6	403.4	445.9	492.7	544.6	601.8	665.1	735.1	812.4	897.8	992.3
7	1097	1212	1339	1480	1636	1808	1998	2208	2441	2697
8	2981	3295	3641	4024	4447	4915	5432	6003	6634	7332
9	8103	8955	9897	10938	12088	13360	14765	16318	18034	19930

Four Terminal Network Transformation Equations

$h_{11} = r_{11} - (r_{12}r_{21}/r_{22})$ $r_{11} = h_{11} - (h_{12}h_{21}/h_{22})$

$h_{12} = r_{12}/r_{22}$ $r_{12} = h_{12}/h_{22}$

$h_{21} = -r_{21}/r_{22} = -\alpha$ $r_{21} = -h_{21}/h_{22}$

$h_{22} = 1/r_{22}$ $r_{22} = 1/h_{22}$

$$g_{11} = \frac{r_{22}}{r_{11}r_{22} - r_{12}r_{21}} \qquad r_{11} = \frac{g_{22}}{g_{11}g_{22} - g_{12}g_{21}}$$

$$g_{12} = \frac{-r_{12}}{r_{11}r_{22} - r_{12}r_{21}} \qquad r_{12} = \frac{-g_{12}}{g_{11}g_{22} - g_{12}g_{21}}$$

$$g_{21} = \frac{-r_{21}}{r_{11}r_{22} - r_{12}r_{21}} \qquad r_{21} = \frac{-g_{21}}{g_{11}g_{22} - g_{12}g_{21}}$$

$$g_{22} = \frac{r_{11}}{r_{11}r_{22} - r_{12}r_{21}} \qquad r_{22} = \frac{g_{11}}{g_{11}g_{22} - g_{12}g_{21}}$$

$h_{11} = 1/g_{11}$ $g_{11} = 1/h_{11}$

$h_{12} = -g_{12}/g_{11}$ $g_{12} = -h_{12}/h_{11}$

$h_{21} = g_{21}/g_{11} = -\alpha$ $g_{21} = h_{21}/h_{11}$

$h_{22} = g_{11} - (g_{12}g_{21}/g_{11})$ $g_{22} = h_{22} - (h_{12}h_{21}/h_{11})$

BASING DIAGRAM

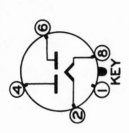

KEY

TERMINAL CONNECTIONS

Pin 1—No connection

Pin 2—Filament

Pin 4—Plate Number 2

Pin 6—Plate Number 1

Pin 8—Filament

5U4-GA

Full-wave high-vacuum rectifier

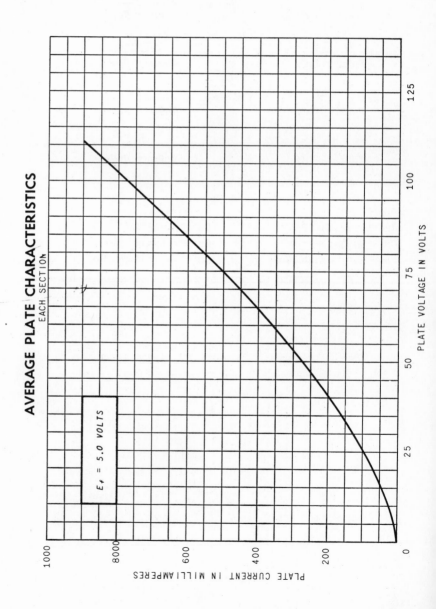

AVERAGE PLATE CHARACTERISTICS

EACH SECTION

$E_f = 5.0\ VOLTS$

PLATE CURRENT IN MILLIAMPERES

PLATE VOLTAGE IN VOLTS

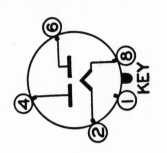

TERMINAL CONNECTIONS

Pin 1—No connection

Pin 2—Filament

Pin 4—Plate Number 2

Pin 6—Plate Number 1

Pin 8—Filament

5Y3-GT

Full-wave rectifier

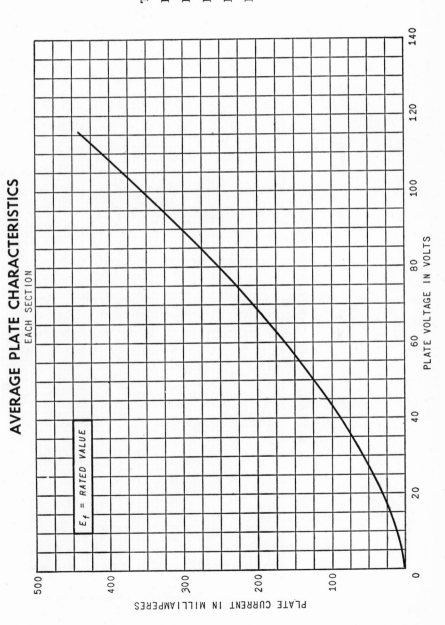

AVERAGE PLATE CHARACTERISTICS
EACH SECTION

E_f = RATED VALUE

PLATE CURRENT IN MILLIAMPERES

PLATE VOLTAGE IN VOLTS

BASING DIAGRAM

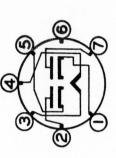

TERMINAL CONNECTIONS

Pin 1—Cathode (Section 1)

Pin 2—Plate (Section 2)

Pin 3—Heater

Pin 4—Heater

Pin 5—Cathode (Section 2)

Pin 6—Internal shield

Pin 7—Plate (Section 1)

6AL5

3AL5

12AL5

Twin diode

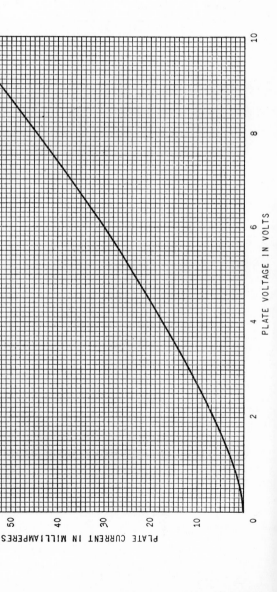

AVERAGE PLATE CHARACTERISTICS
EACH SECTION

E_f = RATED VALUE

PLATE VOLTAGE IN VOLTS

PLATE CURRENT IN MILLIAMPERES

6AQ5

5AQ5

Beam power pentode

BASING DIAGRAM

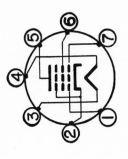

TERMINAL CONNECTIONS

Pin 1—Grid Number 1

Pin 2—Cathode and beam plates

Pin 3—Heater

Pin 4—Heater

Pin 5—Plate

Pin 6—Grid Number 2 (screen)

Pin 7—Grid Number 1

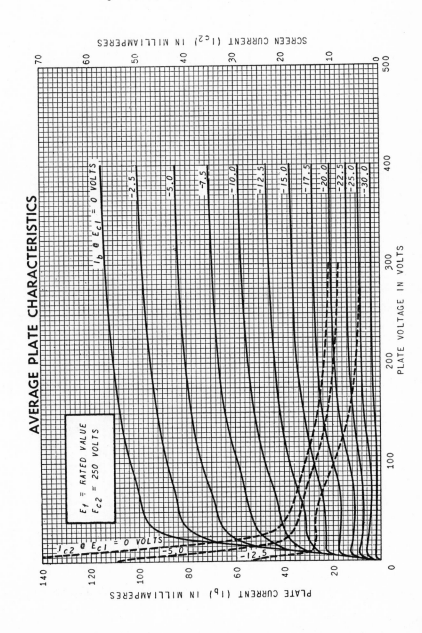

AVERAGE PLATE CHARACTERISTICS

BASING DIAGRAM

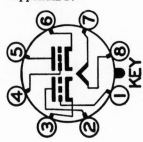

TERMINAL CONNECTIONS

Pin 1—Grid (Section 2)

Pin 2—Plate (Section 2)

Pin 3—Cathode (Section 2)

Pin 4—Grid (Section 1)

Pin 5—Plate (Section 1)

Pin 6—Cathode (Section 1)

Pin 7—Heater

Pin 8—Heater

6AS7-GA

Low-mu twin triode

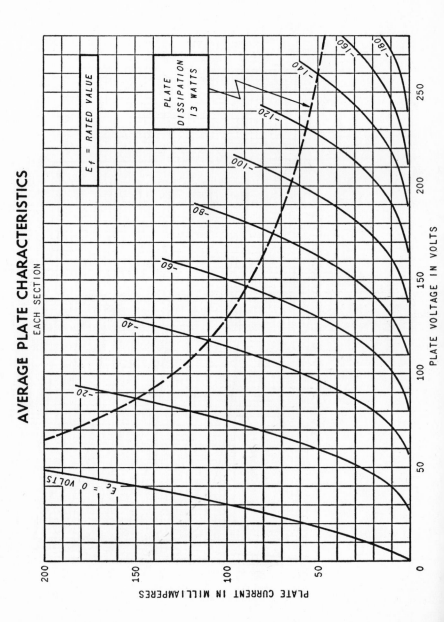

AVERAGE PLATE CHARACTERISTICS

EACH SECTION

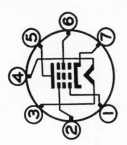

TERMINAL CONNECTIONS

Pin 1—Grid Number 1

Pin 2—Internal shield and grid
Number 3 (suppressor)

Pin 3—Heater

Pin 4—Heater

Pin 5—Plate

Pin 6—Grid Number 2 (screen)

Pin 7—Cathode

6AU6 — 3AU6 — 12AU6

Miniature sharp-cutoff pentode

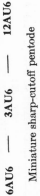

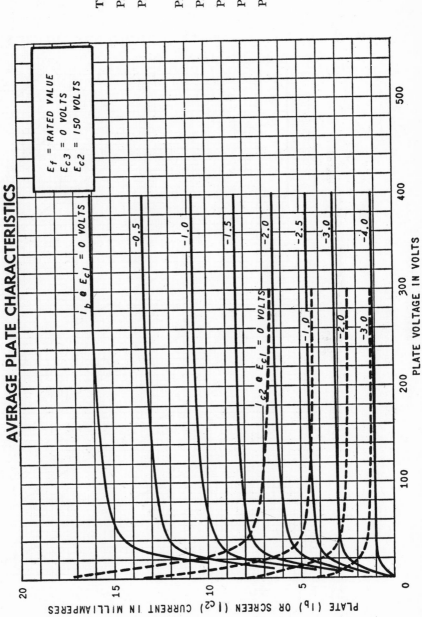

E_f = RATED VALUE
E_{c3} = 0 VOLTS
E_{c2} = 150 VOLTS

AVERAGE PLATE CHARACTERISTICS

PLATE VOLTAGE IN VOLTS

PLATE (I_b) OR SCREEN (I_{c2}) CURRENT IN MILLIAMPERES

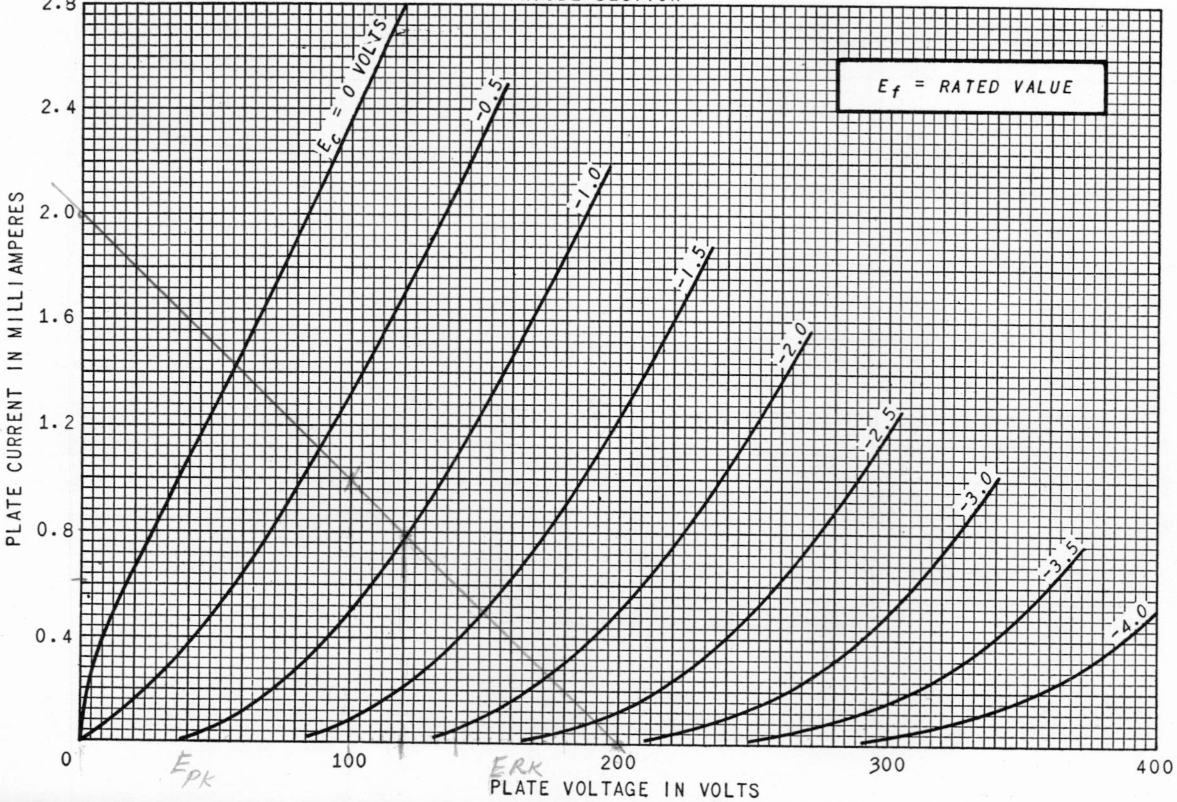

6AV6 —— 3AV6 —— 12AV6

Miniature duplex-diode, high-mu triode

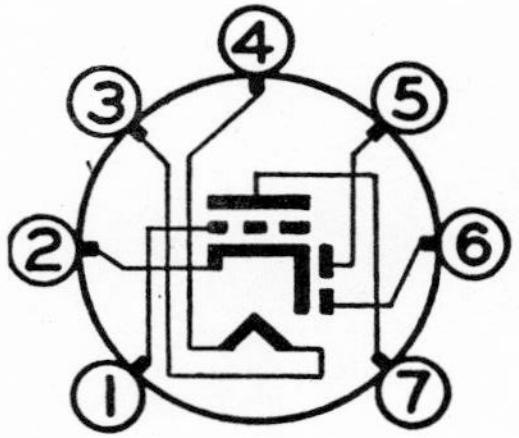

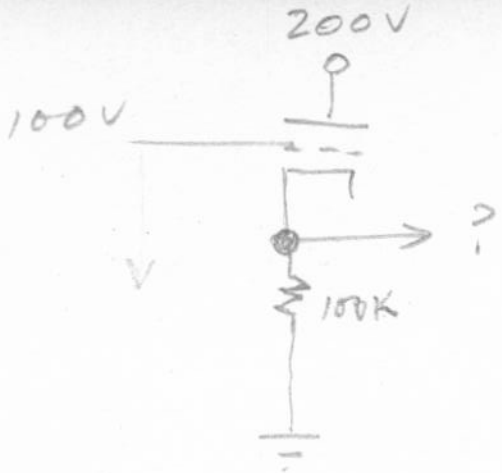

AVERAGE PLATE CHARACTERISTICS

TRIODE SECTION

Ef = RATED VALUE
Ec = 0 VOLTS
-0.5
-1.0
-1.5
-2.0
-2.5
-3.0
-3.5
-4.0
PLATE CURRENT IN MILLIAMPERES
2.8
2.4
2.0
1.6
1.2
0.8
0.4
0
100
200
300
400
PLATE VOLTAGE IN VOLTS

BASING DIAGRAM

1
2
3
4
5
6
7

TERMINAL CONNECTIONS

Pin 1—Triode grid

Pin 2—Cathode

Pin 3—Heater

Pin 3—Heater

Pin 4—Heater

Pin 5—Diode Number 2 plate

Pin 6—Diode Number 1 plate

Pin 7—Triode plate

BASING DIAGRAM

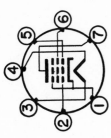

TERMINAL CONNECTIONS

Pin 1—Cathode and beam plates

Pin 2—Grid Number 1

Pin 3—Heater

Pin 4—Heater

Pin 5—Grid Number 1

Pin 6—Grid Number 2 (screen)

Pin 7—Plate

6CA5 — 12CA5 — 25CA5

AVERAGE PLATE CHARACTERISTICS

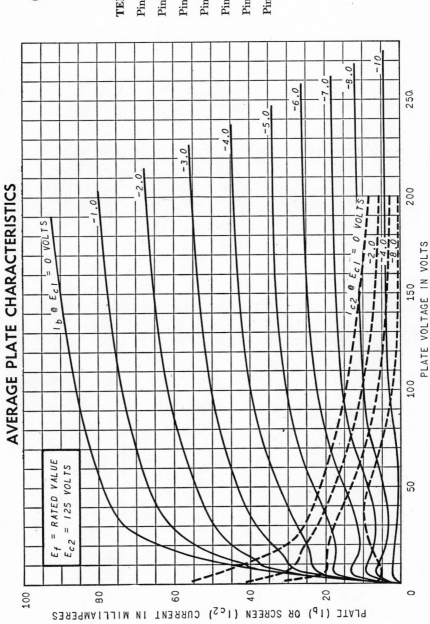

BASING DIAGRAM

TERMINAL CONNECTIONS

Pin 1—Internal connection—do not use

Pin 2—Cathode

Pin 3—Grid

Pin 4—Heater

Pin 5—Heater

Pin 6—Grid

Pin 7—Internal connection—do not use

Pin 8—Internal connection—do not use

Pin 9—Plate

6S4 Power-amplifier triode

AVERAGE PLATE CHARACTERISTICS

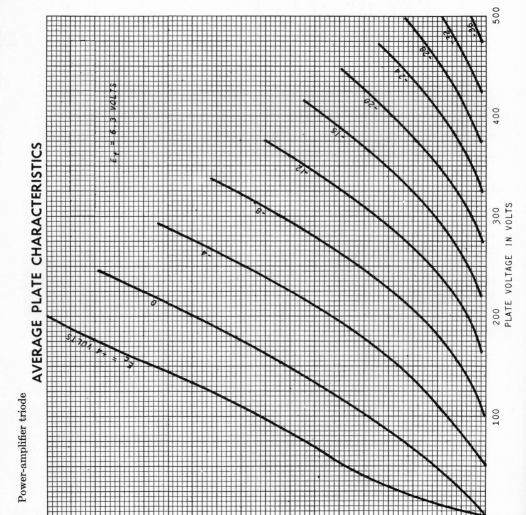

$E_f = 6.3$ volts

PLATE CURRENT IN MILLIAMPERES

PLATE VOLTAGE IN VOLTS

BASING DIAGRAM

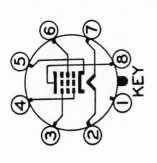

TERMINAL CONNECTIONS

Pin 1—Shell and internal shield

Pin 2—Heater

Pin 3—Grid Number 3 (suppressor)

Pin 4—Grid Number 1

Pin 5—Cathode

Pin 6—Grid Number 2 (screen)

Pin 7—Heater

Pin 8—Plate

6SJ7-12SJ7

Pentode

For audiofrequency and radiofrequency amplifier applications

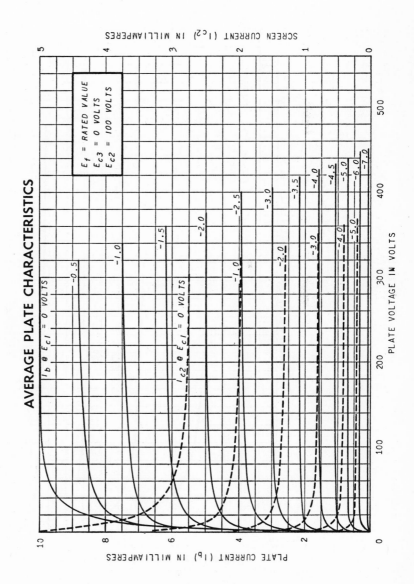

AVERAGE PLATE CHARACTERISTICS

BASING DIAGRAM

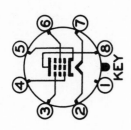

TERMINAL CONNECTIONS

Pin 1—Shell and internal shield

Pin 2—Heater

Pin 3—Grid Number 3 (suppressor)

Pin 4—Grid Number 1

Pin 5—Cathode

Pin 6—Grid Number 2 (screen)

Pin 7—Heater

Pin 8—Plate

6SK7-12SK7

Remote cutoff pentode

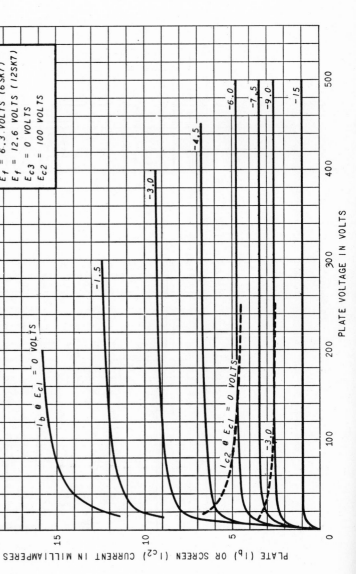

AVERAGE PLATE CHARACTERISTICS

$E_f = 6.3 \cdot VOLTS \; (6SK7)$
$E_f = 12.6 \; VOLTS \; (12SK7)$
$E_{c3} = 0 \; VOLTS$
$E_{c2} = 100 \; VOLTS$

BASING DIAGRAM

TERMINAL CONNECTIONS

Pin 1—Grid (Section 2)

Pin 2—Plate (Section 2)

Pin 3—Cathode (Section 2)

Pin 4—Grid (Section 1)

Pin 5—Plate (Section 1)

Pin 6—Cathode (Section 1)

Pin 7—Heater

Pin 8—Heater

6SL7-GT-12SL7-GT

High-mu twin triode

AVERAGE PLATE CHARACTERISTICS

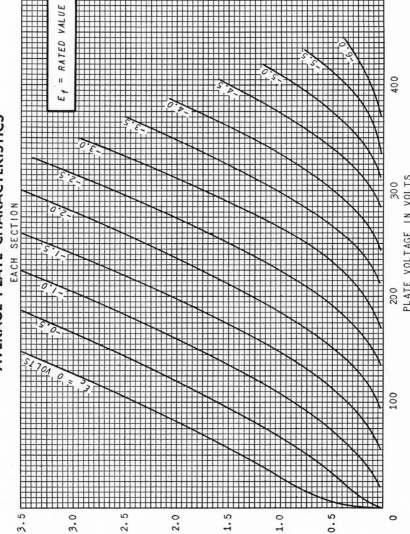

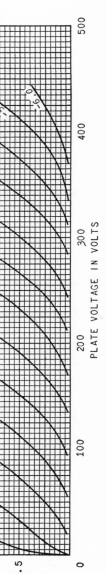

BASING DIAGRAM

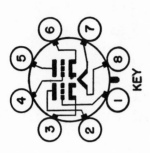

KEY

TERMINAL CONNECTIONS

Pin 1—Grid (Section 2)

Pin 2—Plate (Section 2)

Pin 3—Cathode (Section 2)

Pin 4—Grid (Section 1)

Pin 5—Plate (Section 1)

Pin 6—Cathode (Section 1)

Pin 7—Heater

Pin 8—Heater

Medium-mu twin triode

6SN7-GTB-6SN7-12SN7-GTA

AVERAGE PLATE CHARACTERISTICS

EACH SECTION

E_f = RATED VALUE

PLATE VOLTAGE IN VOLTS

PLATE CURRENT IN MILLIAMPERES

BASING DIAGRAM

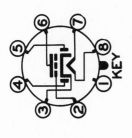

TERMINAL CONNECTIONS

Pin 1—Shell

Pin 2—Triode grid

Pin 3—Cathode

Pin 4—Diode Number 2 Plate

Pin 5—Diode Number 1 Plate

Pin 6—Triode plate

Pin 7—Heater

Pin 8—Heater

6SQ7-12SQ7

Duplex-diode triode

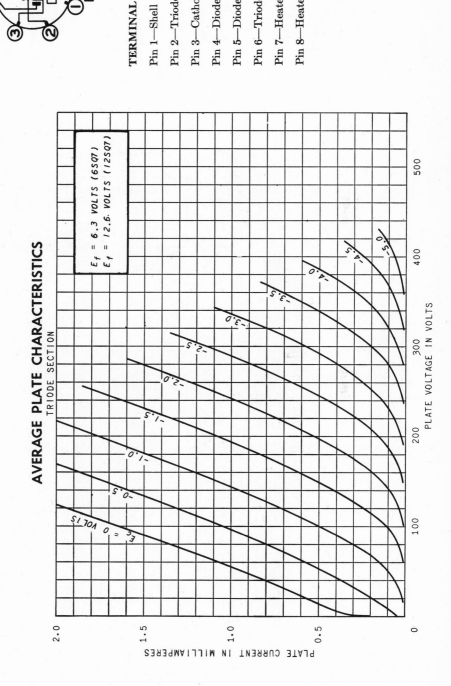

AVERAGE PLATE CHARACTERISTICS

TRIODE SECTION

$E_f = 6.3$ VOLTS (6SQ7)
$E_f = 12.6$ VOLTS (12SQ7)

$E_c = 0$ VOLTS

−0.5
−1.0
−1.5
−2.0
−2.5
−3.0
−3.5
−4.0
−4.5
−5.0

PLATE VOLTAGE IN VOLTS

PLATE CURRENT IN MILLIAMPERES

BASING DIAGRAM

TERMINAL CONNECTIONS

Pin 1—Plate (Section 2)

Pin 2—Grid (Section 2)

Pin 3—Cathode (Section 2)

Pin 4—Heater

Pin 5—Heater

Pin 6—Plate (Section 1)

Pin 7—Grid (Section 1)

Pin 8—Cathode (Section 1)

Pin 9—Heater center-tap

12AU7-A-12AU7-7AU7

Miniature medium-mu twin triode

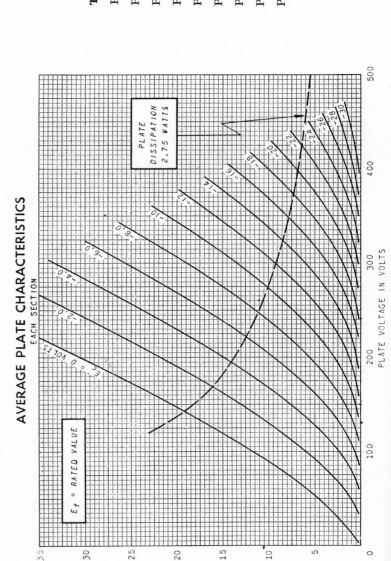

AVERAGE PLATE CHARACTERISTICS

BASING DIAGRAM

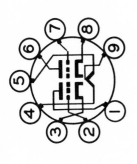

TERMINAL CONNECTIONS

Pin 1—Plate (Section 2)

Pin 2—Grid (Section 2)

Pin 3—Cathode (Section 2)

Pin 4—Heater

Pin 5—Heater

Pin 6—Plate (Section 1)

Pin 7—Grid (Section 1)

Pin 8—Cathode (Section 1)

Pin 9—Heater center-tap

12AX7

Miniature high-mu twin triode

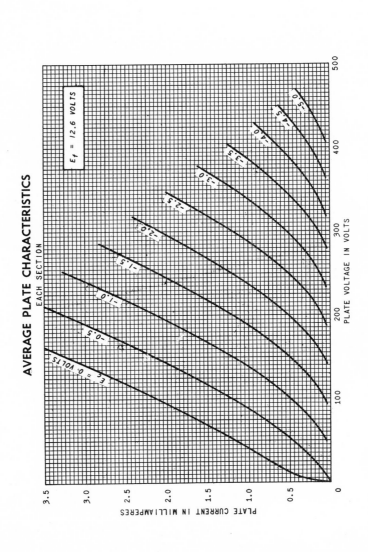

AVERAGE PLATE CHARACTERISTICS

PNP JUNCTION TRANSISTOR

The General Electric type 2N76 germanium fused junction transistor triode is a P-N-P unit particularly recommended for intermediate-gain, low-power applications. A hermetic enclosure is provided by use of glass-to-metal seals and resistance-welded seams. This transistor is capable of dissipating 50 mw in 25°C free air.

Specifications

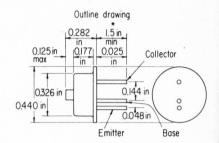

Outline drawing

Absolute Maximum Ratings

Collector voltage (referred to base), V_c	−20 volts
Collector current, I_c	−10 ma
Emitter current, I_e	10 ma
*Junction temperature, T_j	60° C

*Junction temperature may be determined by the method outlined in curve number 6. As an alternative method, a small thermocouple may be attached to the transistor shell (allowing 0.2° C/mw temperature drop between junction and shell). Rating may not be exceeded when soldering into circuit or during operation.

Average Characteristics

(Common Base, $T_j = 30$°C, $f = 270$ cps)	Design center	Typical production spread		
		max	min	
Collector voltage	−5.0			volts
Emitter current	1.0			ma
Output admittance (input open circuit). h_{22}	1.0	2.0	0.5	μmhos
Current amplification (output short circuit), h_{21}	−0.95	−0.99	−0.90	
Input impedance (output short circuit), h_{11}	32	40	25	ohms
Voltage feedback ratio (input open circuit), h_{12}	3×10^{-4}	5×10^{-4}	1×10^{-4}	
Collector cutoff current, 1_{co}	5	10	1	μa
Output capacitance, C_c	40	50	30	mmf
Noise figure (V_c, − 1.5 V; I_e, 0.5 ma; f, 1 kc; BW, 1 ∼), NF	18	30	10	db
Maximum power gain (common Emitter)	38	42	34	db
†Frequency cutoff, f_{co}	1.0	2.5	0.5	mc
Temp. rise/unit collector dissipation (in free air)	0.5			°C/mw
‡Temp. Rise/Unit collector dissipation (infinite heat sink)	0.2			°C/mw

†Frequency at which the magnitude of h_{21} is 3 db down from its 270 cps value.
‡Temperature rise with transistor clamped to metallic heat sink.

Typical Operation. (Small Signal Amplifier)

($T_j = 30$° C, $f = 1$ KC)	Common base	Common emitter	Common collector	
Collector voltage	−5	−5	−5	volts
Emitter current	1.0	1.0	1.0	ma
Input impedance	55	700	15,000	ohms
Source impedance	100	600	15,000	ohms
Load impedance	50,000	30,000	600	ohms
Power gain (PG)	28	38	12	db

2N406

TRANSISTOR

Germanium *p-n-p* type used in class *A* audiofrequency driver-amplifier applications in battery-operated portable radio receivers.

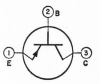

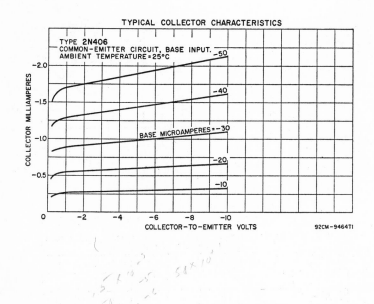

2N647

TRANSISTOR

Germanium *n-p-n* type used in large-signal audiofrequency amplier applications. It is designed especially for use with its *p-n-p* counterpart, RCA-2N217
in class *B* complementary symmetry power-output stages of compact, trans-
formerless, battery-operated portable radio receivers, phonographs, and
audioamplifiers operating at battery-supply voltages up to 9 volts. This type
can also be used in conventional class *B* push-pull and class *A* audioamplifier
circuits.

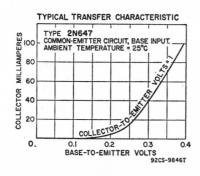

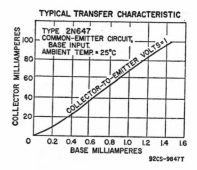

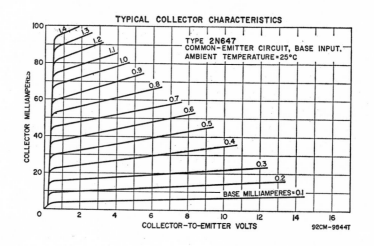

2N1493

TRANSISTOR

Silicon *n-p-n* type used in a wide variety of high-frequency and very high frequency applications in industrial and military equipment. It is used in large-signal power-amplifier, video-amplifier, oscillator, and mixer circuits over a wide temperature range. This type can also be used in switching service in circuits requiring transistors having high voltage, current, and dissipation values.

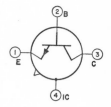

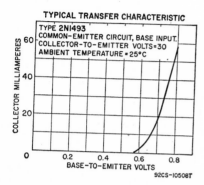

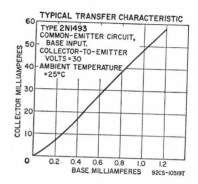

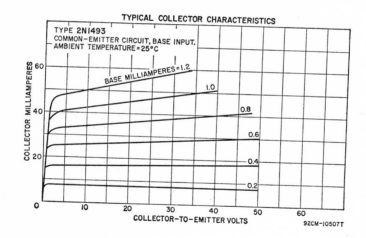

Answers to Odd-Numbered Problems

Chapter 1

.1 $R_1 = 2$
  $R_2 = 4$
  $R_3 = 4$
.3 $I_3 = 20/3$, $I_5 = 15$, $I_6 = 10/3$
.5 (a) 2.5Ω (b) 5Ω
.7 $I_8 = 10/26$ down, $I_4 = 14/26$ up, $I_6 = 9/26$
  down
.9 $E_{AB} = -155/41$
.11 260/3
.13 100/13 volts
.15 1 amp
.17 $E_0 = 6$ volts, $R_0 = 16/3\Omega$
.19 20,000 ohms/volt
.21 Shunt $= 0.001\Omega$
.23 (a) 49.8 volts   (b) 49.5 volts

Chapter 2

.1 628, 942, 1256 ohms
.3 (a) $5 + j62.8$ (d) $100 + j29.2$
  (b) $100 - j159$ (e) $100 - j361$
  (c) $-j255.2$
.5 (a) $8 + j10$ (g) $0.87 - 0.4j$
  (c) $2 - j2$ (i) 41
  (e) $-9 + j42$ (k) $0.2 + 0.93j$
.7 $i = 7.07 \sin(\omega t + \tan^{-1} 4/3)$
.9 $I = 0.71$, $E = 61.2 + j35.3$
.11 $E = -96 + j108$, $|E| = 203$, 0.371
.13 $-j962$, $-j118$, $+j460$
.15 16 watts, $I = 0.145 - 0.545j$
.17 $I = 2$, $Z = 100 - j401$,
  $E = 200 - j802$, $P = 400 - j1604$
.19 1.15 amp
.21 $R = 25$, $L = 0.08h$, $C = 66.5\mu f$
.23 $X = 47.1\Omega$, $L = 0.125h$
.25 0.447, 0.632
.27 12.5 watts

Chapter 3

3.1 31.8kc, low pass
3.5 6.28Ω
3.11 (a) $f_R = 1/2\pi\sqrt{L'C}$, $L' = L_1L_2/(L_1 + L_2)$
  (b) $f_R = 1/2\pi\sqrt{LC - R^2C^2}$
  (c) $f_R = 1/2\pi\sqrt{L(C_1 + C_2)}$
3.13 8.94 K, 3.461 kc, 3.639 kc
3.15 $|e_{out}/e_{in}| = 1/\sqrt{1 + R^2\omega^2C^2}$

Chapter 4

4.1 0.045 amp
4.3 44, 23, 16.3 ohms
4.5 0.145 amp, 5.38 volts

Chapter 5

5.1 $\tau = RC_1C_2/(C_1 + C_2)$
5.5 (a) 183.8 volts,   (b) 0.058 volt
5.7 $R = 1.6\ K$, $C = 24\ \mu f$
5.9 100 k, 1.91 k
5.13 (a) 424 volts   (b) positive
  (c) 212 volts

Chapter 6

6.1 100, 1910 μmho, 48.3 k
6.3 24, 8830 μmho, 2.5 k
6.5 4000 μmho
6.7 2730, 1500, 1090, 508 μmho, -18 volts
6.9 770Ω
6.11 127 ohms, 42.9 k, 3.5 ma
6.13 0.09 ma, 296 volts
6.15 $E'_{bb} = 900$, $R_k = 286$, $A_v = 18.7$

Chapter 7

7.1 13.3, 40 volts
7.3 3.3 k
7.5 26.1 Kc
7.7 404 pf
7.9 (a) 1660 pf,   (b) 6120 pf,
  (c) 33.3 k,   (d) 50
7.11 25 megohm, 383 ohm
7.13 $\mu(\mu + 1)\ R_L/[R_L + (\mu + 2)\ r_p]$
7.15 -12.5, 38.2 kc
7.17 $e_0 = (e_1 + e_2)\ r_pR_k/[1 + 2g_m)(r_p + 2R_k)]$

Chapter 8

8.1 900 volts, 0.62 watts, 5.4 percent distortion,
  $N_p/N_s = 23/1$
8.3 (a) -6 volts,
  (b) -2.5 volts, $R_L = 4.45$ k, 16.6:1
8.7 $R_{AG} = R\ [r_p + R_k\ (\mu + 1)]/[r_p + R + R_k\ (\mu + 1)]$
8.9 $R_k = 164\Omega$, $A_V = 0.93$

Chapter 9

9.10 2.17 kc to 13 kc, $A_V \pm 29$
9.3 $L = 79.6\ \mu\text{h}, C = 318\ \text{pf}$
9.5 990 cps, 60 volts
9.7 (a) 109 cps (b) 295 volts

Chapter 10

10.3 (a) 155 vpp (b) square wave
(c) 1.62 μsec

Chapter 11

11.3 (a) 45 at 10 ma to 80 at 90 ma
11.7 52
11.9 0.997
11.11 (a) 338 k (b) 13.5 volts, (c) 40
11.13 $1/(1 - \alpha)$

Chapter 12

12.1 $g_{11} = 8.34 \times 10^{-2}, g_{12} = -4.66 \times 10^{-5}$,
$g_{21} = -8.09 \times 10^{-2}, g_{22} = 4.75 \times 10^{-5}$
12.3 $r_{11} = 115, r_{12} = 77.5, r_{21} = 2 \times 10^6$,
$r_{22} = 2.04 \times 10^6$

12.5 $r_e = 17, r_b = 300, r_c = 1 \times 10^6$,
$r_m = 9.5 \times 10^5$
12.7 (a) $A_i = (r_{\text{in}} + R_B)/R_B\ (1 - \alpha)$
(b) $A_i = r_m\ (r_{\text{in}} + R_B)/R_B\ (R_c + r_c - r_m)$
12.9 (a) 19 (b) -204 (c) 746

Chapter 13

13.1 $A_V = -648, A_{OV} = -202$
13.3 3540, 16.4 volts
13.5 0.9 μf, $f_2 = 1.1 \times 10^6$ cps, $f'_2 = 3 \times 10^4$
cps
13.7 153Ω, yes
13.9 13 ma

Chapter 14

14.1 $C_1 = 30$ pf, $R_1 = 3.6 \times 10^6$; $C_2 = 300$ pf,
$R_2 = 3.6 \times 10^5$; $C_3 = 2700$ pf, $R_3 = 4 \times 10^4$
14.3 4 megohm
14.5 $E_{\text{in}} = 350$ volts, $R_1 = 35$ k, $R_3 = 19.2$ k,
$R_s = 175$ k, $R_A = 127$ k, $R_B = 73$ k

Index

A

Addition of sine waves, 53, 57
Alpha cutoff, 448–449
Alternating current circuits, 42–73
Ammeter, 14–17
Amplification factor
  defined 180
  pentode, 197–200, 249
  triode, 180, 182–185, 209, 249
Amplifier
  beam power, 248–250
  cathode follower, 234–241, 481, 483
  circuits, 192, 201, 206, 207, 208, 367, 376, 380,
    406, 431, 433, 436, 439, 463, 468
  class A, 196–197
  class B, 196–197, 272
  class C, 197
  current gain, 405, 407, 415, 420
  frequency characteristics, 221–227
  grounded base, 367, 373, 413–419
  grounded cathode, 206
  grounded collector, 375–379, 419–423
  grounded emitter, 353, 356–358, 406–413
  high-frequency characteristics, 214, 219–221,
    446–447
  high-frequency compensation, 485
  intermediate frequency characteristics, 214,
    218–219
  linear, 206–208
  low-frequency characteristics, 213–218,
    445–446
  low-frequency compensation, 485
  mid-frequency characteristics, 218–219, 222,
    229, 231, 233, 445
  multistage audio
    transistor, 435–442
    vacuum tube, 194, 212–213, 280, 282–284
  overdriven, 312
  pentode, 201
  phase characteristics, 221
  power transistor, 451–466
  power vacuum tube, 247
  push-pull transistor, 463–468
  push-pull vacuum tube, 271–277
  RC coupled, 192–194, 388, 458, 466, 468
  transient response, 313–315
  transformer coupled, 247, 278, 388, 458
  triode, 186, 206

video, 485–488
voltage gain, 187, 189–192, 216, 219, 220, 234,
  353–358, 373, 378, 405, 407, 415, 419, 432

B

Band width
  amplification and, 221–227, 449
  of resonant circuit, *see* Resonance
Barkhausen condition for oscillation, 288, 293
Base characteristics, 367–371
Beam power amplifiers, *see* Amplifier
Beam power tube, 248–250
Bias
  transistor, 342, 355, 358–366, 373–375,
    380–383, 467
  vacuum-tube, 194–196, 274, 276
Blocking oscillator, *see* Oscillator
Bridge rectifier, *see* Rectifier circuits
Bypass capacitor
  cathode, 206–208
  emitter, 387, 450
  screen, 201

C

Capacitance, 51
  coupling, 192–194, 216, 217, 450
  input, 211
  interelectrode, 210, 442–443
  Miller, 210–212
  shunt, 213–214, 219, 241, 313, 443
  wiring, 212
Capacitive reactance, *see* Reactance
Capacitor charge-discharge curves, 146, 148–151,
  299, 303, 305, 313–315, 317–321, 324
Capacitor input filter, *see* Power supply
Carrier
  majority, 130–132
  minority, 132, 133
Cathode follower, *see* Amplifier
Cathode-ray oscilloscope, 475–504
  as ammeter, 493–494
  block diagram, 480–483
  as detector of oscillations, 294
  electrostatic sweep, 477–478
  as frequency meter, 496–498
  phase measurements, 498–502
  sweep circuit, 482, 490
  tube characteristics displayed, 502–503
  as voltmeter, 491–493
  waveform analysis by, 135, 477–480